Water Policy Science and Politics

Water Policy Science and Politics

An Indian Perspective

M. Dinesh Kumar
Institute for Resource Analysis and Policy,
Hyderabad, India

Elsevier
Radarweg 29, PO Box 211, 1000 AE Amsterdam, Netherlands
The Boulevard, Langford Lane, Kidlington, Oxford OX5 1GB, United Kingdom
50 Hampshire Street, 5th Floor, Cambridge, MA 02139, United States

Library of Congress Cataloging-in-Publication Data
A catalog record for this book is available from the Library of Congress

British Library Cataloguing-in-Publication Data
A catalogue record for this book is available from the British Library

ISBN: 978-0-12-814903-4

For information on all Elsevier publications
visit our website at https://www.elsevier.com/books-and-journals

Publisher: Candice Janco
Acquisition Editor: Louisa Hutchins
Editorial Project Manager: Jennifer Pierce
Production Project Manager: Denny Mansingh
Cover Designer: Matthew Limbert

Typeset by SPi Global, India

Dedication

Dedicated to my loving nephew, Arjun.

Contents

List of Figures

List of Tables

List of Abbreviations

ADB	Asian Development Bank
AMUL	Anand Milk Union Limited
AP	Andhra Pradesh
APRLP	Andhra Pradesh Rural Livelihoods Project
ATREE	Ashoka Trust for Research in Ecological Economics
AUD	Australian Dollar
AWM	Agricultural Water Management
BCM	Billion Cubic Metre
BJP	Bharatiya Janata Party
CAG	Comptroller and Auditor General of India
CEA	Central Electricity Authority
CGWB	Central Ground Water Board
CPCB	Central Pollution Control Board
CU	Consumptive Use
CWC	Central Water Commission
CWR	Centre for Water Resources
DFID	Department for International Development
DTW	Deep Tube Well
EPW	Economic and Political Weekly
ET	Evapo-Transpiration
FAO	Food and Agricultural Organization
GCA	Gross Cropped Area
GCMMF	Gujarat Cooperative Milk Marketing Federation
GDP	Gross Domestic Product
GHI	Global Hunger Index
GOI	Government of India

Continued

GOM	Maharashtra Groundwater (Development and Management) Act, 2009
GRACE	Gravity Recovery and Climate Experiment
GSDA	Groundwater Surveys and Development Agency
GW	Ground Water
GWIA	Ground Water Irrigated Area
HDI	Human Development Index
HDR	Human Development Report
HPI	Human Poverty Index
ICAR	Indian Council of Agricultural Research
ICID	International Commission on Irrigation and Drainage
ICRISAT	The International Crops Research Institute for the Semi-Arid Tropics
IDF	Intensity Duration Frequency
IFPRI	International Food Policy Research Institute
IGCC	Integrated coal-gasification combined cycle
IIT	Indian Institute of Technology
ILR	Interlinking of Rivers
IMT	Irrigation Management Transfer
INR	Indian Rupee
IRAP	Institute for Resource Analysis and Policy
IRENA	International Renewable Energy Agency
IRRI	International Rice Research Institute
IWMI	International Water Management Institute
IWMP	Integrated Watershed Management Project
IWP	Irrigation Water Productivity
IWRM	Integrated Water Resources Management
KAWAD	Karnataka Watershed Development
LCOE	Levelized cost of energy
MCM	Million Cubic Metre
MDG	Millennium Development Goals
MGNREGS	Mahatma Gandhi Rural Employment Guarantee Scheme
MLD	Million Litres per day
MNREGA	Mahatma Gandhi National Rural Employment Guarantee Act

MNREGS	National Rural Employment Guarantee Scheme * 1
MOWR	Ministry of Water Resources
MP	Madhya Pradesh
MPR	Monopoly Price Ratio
NAC	National Advisory Council
NGO	Non Governmental Organization
NIA	National Irrigation Administration
NRLP	National River Linking Project
NRRA	National Rainfed Area Authority
NWA	National Water Academy
NWC	National Water Commission
NWDB	National Water Development Board
NWSDPRA	National Watershed Development Programme for Rain fed Areas
OECD	Organization for Economic Co-operation and Development
OW	Open Well
PET	Potential Evapo-Transpiration
PIM	Participatory Irrigation Management
PPP	Public Private Partnership
PV	Photo Voltaic
PWC	Private Sector Participation
RBI	Reserve Bank of India
RD & GR	River Development & Ganga Rejuvenation
RH-AM	Relative Humidity (Morning)
RH-PM	Relative Humidity (PM)
RRA	Revitalizing Rainfed Agriculture
RWH	Rain Water Harvesting
RWHS	Rain water harvesting system
SOLAW	State of the World's Land and Water Resources for Food and Agriculture
SSA	Sub-Saharan Africa
SSP	Sardar Sarovar Project
STW	Shallow Tube Well
SWH	Small Water Harvesting

Continued

SWI	Soil Water Index
SWID	State Water Investigation Directorate
SWUI	Sustainable Water Use Index
TDS	Total Dissolved Solids
TMC	One thousand million cubic feet
TN	Tamil Nadu
UAE	United Arab Emirates
UN	United Nations
UNDP	United Nations Development Programme
UNEP	United Nations Environmental Programme
UNESCO	United Nations Educational, Scientific and Cultural Organization
UP	Uttar Pradesh
WB	West Bengal
WDP	Watershed Development Project
WH	Water Harvesting
WHO	World Health Organization
WHS	Water Harvesting Systems
WP	Water Productivity
WPI	Water Poverty Index
WPI	Wholesale Price Index
WRD	Water Resources Department
WRIS	Water Resources Information System
WSD	Water Shed Development
WSI	Water Security Index
WUA	Water Users' Association
WUE	Water Use Efficiency

Preface

The idea of this book came toward the beginning of 2017 when a member of the board of Institute for Resource Analysis and Policy (IRAP), Hyderabad, Prof *R. Maria* Saleth, also one of my mentors, encouraged me to put together a book based on the several short articles I posted on LinkedIn dealing with the contemporary issues in water, energy, agriculture, and environment, during one of the board meetings. Prof Saleth, one of the internationally renowned experts on water economics, was so fascinated by some of the articles that he found merit in converting them into professional papers for the benefit of the larger water community around the world.

But the very idea of posting articles on LinkedIn had its genesis in a workshop held in the Groundwater Surveys and Development Agency (GSDA) of Maharashtra (Pune) to discuss about the latest developments in the water management sector in September 2015. While GSDA put together a good program and organized the event very well, I was very disappointed by the triviality of the issues raised by some of the speakers, when compared with the sheer magnitude of the problems being faced by a large country like India. I found some of the ideas quite absurd. I felt the need to provide a rebuttal, in addition to responding to the speakers concerned during the event itself, which I did. The brainstorming during the workshop was on how participatory aquifer mapping would promote sustainable groundwater management in India, how understanding the spring shed is important for the protection of springs and water supply systems in hilly areas, and how improved understanding of aquifers through mapping would help boost ecological sanitation. I found that most of the ideas discussed there were either so dated or completely disconnected from the real problems on the ground and therefore inconsequential to India's fast-changing water management landscape.

In fact during the previous 8–10 years or so, the water management discourse in India was almost hijacked by those who were champions of these ideas. India's water policy makers were listening only to them. In fact, to a great extent, even the policy-making body was dominated by these civil society representatives. There was a dire need to bring the discourse back on track to look at some of the biggest water management challenges facing the country, be it water scarcity, energy security and role of clean energy, groundwater depletion, growing environmental water stress, agricultural stagnation, and institutional capacity building challenges in the water sector. Soon after coming back from

Pune, I wrote my first article "Dancing with the invisible in the theatre of the absurd" and posted on "LinkedIn" on the very next day, i.e., on September 22, 2015, and this echoed some of my concerns on the course that some professionals in the sector of India wanted the water management debate to take.

Posting such articles on contemporary issues, which most of the time was a response to either an article published in the popular media or a new policy proclamation or the launching of a new government scheme, became a regular event. By February 2017, I had already written 22 articles on LinkedIn. By the time of writing the first article, the atmosphere in the water circles in India had become much vitiated by persuasive and motivated writings from propagandists of certain water management paradigms and models of agricultural growth driven by what they consider as "successful models of water management." A weekly magazine that deals with politics and economics in India was flooded with articles from these writers, and the publisher hardly took any special interest in publishing articles from those belonged to a different worldview. To get a rejoinder published, it took not less than 10–15 weeks. The same was the case with mass media. So much was the clout enjoyed by the civil society groups on the media that getting a 1000-word article, a critique of the Mihir Shah committee report on restructuring of Central Water Commission and Central Ground Water Board, published in a national daily in English turned out to be an impossible task.

Those who were concerned with water sector in India knew only one version of the story about what is happening with the sector, particularly "what is working and what is not working," and that eventually became the truth. I was often reminded of what the great German philosopher *Friedrich Nietzsche* once said: "All things are subject to interpretation. Whichever interpretation prevails at a given time is a function of power and not truth." The speakers at many conferences that discussed water issues of India within the country and abroad during the past many years parroted the ideas propagated by the champions of the new "water management paradigm." Under these circumstances, many of the network colleagues found my LinkedIn articles thought-provoking. If not, in their opinion, these articles surely marked a significant departure from the "run-of-the mill" pieces that were appearing in the magazines and popular media that claimed to be "silver bullets" for all water- and energy-related problems confronting India. Their comments encouraged to continue writing such short articles on LinkedIn.

Most of these LinkedIn articles were small thought pieces or excerpts from longer research articles written in the recent times, touching upon different themes in the water sector, such as canal irrigation, microirrigation, tank management, dairy farming, irrigation sector reform, groundwater management, land-use hydrology, urban water management, climate variability and water management, and water-energy nexus. But they all are on a central theme, i.e., science and politics of policy making in India's water sector. Their key messages were "how politics becomes central to policy making in the country's water sector" and "how scientific data and information relating water are collected, analyzed, interpreted, and put in the public domain by various interest groups in a manner that it suits certain political interests." They exposed the serious problems with the data and assumptions that formed the basis for certain key reforms, policy changes, and projects in the interrelated sectors of water, agriculture, energy, and environment.

When I first approached my esteemed colleague from *Elsevier*, Ms. Louisa Hutchins, who is their commissioning editor, she expressed great interest in publishing a synthesis of these articles into a book but on the condition that I would convert each one of them into professional book chapters with proper data and references. I agreed to these conditions instantly. It was a daunting task to convert these short pieces into full-fledged articles, and some of the articles had to be clubbed to make a story interesting enough for the professionals and academics. At the end, there were 17 articles, including one on the international implications of the political process of water policy making in India.

I sincerely hope this book *Water Policy, Science, and Politics: An Indian Perspective* would raise the level of international debate on improving the rigor of policy-making process in the water sector in developing countries so as to frame sound water management policies.

M. Dinesh Kumar

Acknowledgment

First of all, I am extremely thankful to the five reviewers, who had gone through the book proposal, and the editorial office of Elsevier Science, who had gone through the sample chapter, for offering invaluable comments and suggestions, which had immensely helped in framing the "problem" that the volume addresses, sharpening the focus of the contents and enhancing the value of the final manuscript. All the reviewers were quite meticulous in their observations, and I was very glad to accept most of them.

I am also extremely thankful to my great colleagues, Meera Sahasranaman, K Sivarama Kishan, and Nitin Bassi, for offering valuable comments on the earlier (LinkedIn) versions of many of the chapters included in this book, which immensely helped me draft them.

Thanks are also due to many of my esteemed colleagues in the water sector who kept encouraging me to write such articles on LinkedIn for consumption by general readers, which ultimately led to the drafting of the book manuscript. A few names that come to my mind are Dr. S.A. Prathapar, Shri Chetan M. Pandit, Shri K.V. Dinesh, Shri M. Gopalakrishnan, Mr. Chandrakant Kumbhani, Dr. V. Ratna Reddy, and Prof. A. Narayanamoorthy.

I owe a lot to my beloved family, especially my wife Shyma and my daughter Archana and my beloved mother who stood by me during the difficult times, provided the moral strength, and became constant source of motivation. Lastly, it is a matter of great bliss for me to dedicate this book to my loving nephew, Arjun, who is very close to my heart.

M. Dinesh Kumar

Chapter 1

Introduction

M. Dinesh Kumar

1.1. CONTEXT

This book is about the interface between the three domains, that is, science, politics, and policy making in India's water sector. Its key message is about how politics becomes central to policy making in the country's water sector, and scientific data and information relating water resource status, development, use, and management are collected, analyzed, interpreted, and put in the public domain by various interest groups in the academics, civil society, and water administration in a manner that suits certain political interests. It exposes many problems with the types, analysis, and interpretation of the data that formed the basis for certain key reforms, policy changes, and projects in the interrelated sectors of water, agriculture, energy, and environment.

Issues concerning water have always been a part of the political discourse in independent India, as it serves as basic input for the survival of the people and their socioeconomic development and an environmental good (Iyer, 2005, 2011; Kumar and Pandit, 2016). It is quite imperative that access to and control over water resources is politically contested (Roth et al., 2005). This political discourse had included themes as varied as the technologies for harnessing water (large dams vs. small dams), agricultural development (irrigation vs. rainfed), investments in water resources development (public vs. private), water pricing (actual cost of production vs. subsidized price), basic paradigms for water resources development and management—centralized versus decentralized government and management of water supplies especially for irrigation and drinking water, role of communities in water management—(Kumar and Pandit, 2016), and institutional reforms in the water sector. This political discourse is not only between political parties and movements but also between state and union governments, between government and Civil Society Organizations (CSOs) and groups, and between government and academia and even between academic groups.

In water policy making, which is an outcome of such political processes, science has often been a casualty (Molle, 2008). The growing influence of civil society groups with strong ideological positions and the increasing presence of academics and researchers, who peddle falsehood, in the policy forums, and the

Water Policy Science and Politics. https://doi.org/10.1016/B978-0-12-814903-4.00001-4

"academic-bureaucrat-politician nexus" have become the defining feature of this process. There is little attention being paid to hydrologic sciences, climate science, economics, and other social science disciplines that are critical to designing sustainable systems for water management. This applies to overall water resources development strategy, strategies for irrigation development, agricultural development and poverty reduction strategies for backward regions, pricing of water and subsidies, energy pricing in agricultural sector, institutions for managing water supplies, and long-term strategies for dealing with floods and droughts.

The book would discuss the nature of the political discourse on water management in India, what characterizes this discourse, and how this discourse had influenced the process of framing water-related policies that produce certain outcomes in terms of policies, strategies, paradigms, and activities. It provides a framework to explain how certain paradigms and ideas are pursued in the water resources development and management sector internationally, using several of the concepts evolved in the recent past (Molle, 2008; Molle et al., 2008; Whitty and Dercon, 2013). While these paradigms and ideas also find resonance in India's water policy debates, certain unique characteristics are encountered in the policy-making process in India. They are explained by four distinct phenomena: (i) the "academic-bureaucrat-politician" nexus; (ii) the increasing presence of academic who spreads falsehood in policy forums; (iii) the growing influence of the civil society groups on policy makers; and (iv) the poor application of social sciences in policy formulation, producing certain policy outcomes that are not supported by sufficient scientific evidence. As a result, the policy outcomes and paradigms in the water and related sectors are also unique. The volume presents several cases from the interrelated sectors of water resources, water supplies, energy, and climate, including some of the most controversial ones to illustrate these phenomena, and discusses how the science has become a casualty in the design of various government programs, projects, schemes, and strategies for these sectors. The book also suggests technical, institutional, legal, and policy alternatives for addressing the problems in the water management sector and specific areas such as irrigation, watershed management, rural and urban water supply, climate mitigation and adaptation, and flood control.

In this chapter, we discuss the dominant views that characterize the water policy debate in India, analyze the reasons for the ineffectiveness of the national and state water policies, and present a conceptual framework to explain the increasing choice of certain paradigms and ideas in the water management sector internationally. The framework includes certain additional aspects that characterize the political processes that are unique to water policy making in India. Finally, the contents of the volume are discussed chapterwise.

1.2. INDIA' WATER POLICY: GROWING DEBATE

India poses a unique challenge to water resource managers and policy makers with great spatial heterogeneities in hydrologic regimes, geohydrologic environments, climates, physiographic conditions, and socioecological and cultural environments. The complex socioecology of water in the country is evident from the following facts: The mean annual rainfall in the country varies from as low as 100mm in Jaisalmer of western Rajasthan to 11,700mm in Cherrapunji of Meghalaya; the country's river systems include one of the most complex and mightiest river systems of the world, that is, Brahmaputra, to several hundreds of marginalized ephemeral streams that see flows only for a few hours in the whole year; its groundwater resources include the richest Gangetic alluvium to some of the lowest yielding hard-rock aquifers in the plateau; its human habitations extend from mountainous sub-Himalayan region to the coastal plains; its farming systems include one of the most productive farming systems of the world (in Punjab) to one which is least productive (eastern India); its agricultural withdrawal of groundwater is as high as 1280 m^3 per capita per annum in water-scarce Punjab to as low as 130 m^3 per capita per annum in water-rich Bihar; and its urban areas include large metros with 30,000 plus persons per square kilometer, to towns that are sparsely populated with less than 1000 persons per square kilometer. The country has provinces that have human development index as high as some of the eastern European countries (such as Kerala) to those that have very low indicators of human development and comparable with some of the East African countries. All these make water management decisions extremely complex not only at the national level but also at the regional and local level.

Given the scenario, a water policy for the country, which aims at sustainable, efficient, equitable, and harmonious use of water, should not try to prescribe what should be the water management action for a given situation. Instead, it should guide the water resource and service managers to take appropriate policy decisions to fit the socioecological context of their geographic areas of operation. The guiding principle for policy formulation should be that the policy document should provide the framework for decision-making about water management, rather than rigid rules and dogmatic norms. Therefore, the policy objectives and criteria for decision-making are important and not the action.[1] However, India's water policy debate is largely driven by often diametrically opposite views on policy actions and is often characterized by ideological positions on several fundamental issues (see Iyer, 2011).

The following are some of the dominant view that characterizes the debate: (1) who should have ownership of and control over natural resources, especially water resources (the national government or the state governments or the local communities); (2) who can govern and manage water resources (i.e., whether the professionally managed centralized agencies of the state or institutions of local

1. "Water use" can be taken as an example to illustrate this point. Given the wide variation in water supply-demand situations across the country, the universal policy objective can be an economically efficient use of water, with overall water availability situation and or the cost of production of water to be an important criteria for deciding on what is efficient use of water for a region or locality and not water pricing per se! Here, pricing will be an instrument.

self-governance or local communities); (3) the ideal scale at which water can be managed (i.e., whether at the local level in the village or watershed or at the river basin level or at the subnational or national level); (4) which technology should be used to augment water supplies, whether large water storage/diversion systems or small water harvesting structures; (5) whether water should be treated as a "social good" or "economic good"; (6) whether water should be treated as a "public good" or a "private good"; and (7) how to regulate water use, whether through top-down state regulations or through social regulation by local communities.

For instance, the fiercest opposition to privatization of water supply services in India comes from those who believe that water should be treated both as a "social good" and a "public good." They believe that privatization of water services would eventually lead to commodification of water with private utilities increasing their profit margins through monopolistic prices, seriously jeopardizing the goal of "equitable access" and depriving the poor communities of the water for their basic survival. The suggestions for larger institutional reforms such as "instituting water rights" are also met with similar opposition by these groups, which claim that it would lead to complete privatization of water. They instead advocate community management of water resources and water services, without ever prescribing what should be the basis or norm for the allocation of water across users within the community and where the legal sanctity for such allocation mechanisms could come from (Iyer, 2011; Shah, 2013).

On the other hand, the proponents of water supply privatization also have too little to show that it is the only alternative model available to achieve greater affordability of water and improved efficiency of the utilities (Biswas and Tortajada, 2003).

The main opposition to building of large dams comes from certain civil society groups that view democracy as one where the "communities" are the ultimate owners of natural resources, and all the powers for taking decisions regarding their development and use should be vested with them or by village-level local self-governments. They perceive the state as an aggressor (Kumar and Pandit, 2016), which leads to the untenable position that water development technologies, which require the involvement of arms of the state for execution, be it large dams with canals or mega water supply projects, are considered detrimental because of the human displacement and ecological destruction, which ultimately lead to social conflicts. They opine that a major transformation of our thinking about water is needed; as according to them, the legal, institutional, and procedural changes would not work to resolve such conflicts beyond a point without such a transformation (see Iyer, 2005).[2]

2. That transformation would include an awareness and understanding of water as a scarce and precious resource to be conserved, protected, and used with extreme economy; an integral part of nature; a sacred resource; a common pool resource to be managed by the community or held as a public trust by the state; primarily a life-support substance and only secondarily anything else (economic good, social good, etc.); a fundamental human and animal right; and a bounty of nature to be gratefully and reverentially received and shared with fellow humans (within the state or province or country or beyond the borders of the country), future generations of humans, and other forms of life (Iyer, 2005).

Some believe that decisions to build large irrigation/multipurpose projects are part of "electoral politics" to win voters and are often outcomes of the politician-bureaucrat-technocrat-consultant-contractor nexus (see Bosshard, 2004; Iyer, 2005), which breed corruption. The other points of contention are the absence of mechanisms to compensate for those who are displaced by reservoirs and canals, the unscientific criteria followed for evaluating the cost of large water projects (Shah and Kumar, 2008), and the shoddy rehabilitation carried out in large schemes in the past (Kumar and Pandit, 2016). As noted by Kumar and Pandit (2016), the latter are valid concerns, though the recent experience in the case of the multipurpose Sardar Sarovar Project (SSP) in addressing the rehabilitation issues has been positive (Jagadeesan and Kumar, 2015).

Corruption is a major concern in all public works in India and needs to be tackled through concerted efforts. But as noted by Kumar and Pandit (2016), the "corruption" argument is not used against decentralized water conservation projects implemented by village Panchayats under MGNREGA, which are often characterized by rampant corruption, nepotism, and favoritism, leading to bad planning decisions, poor designs, poor siting of structures, and inferior quality of construction (World Bank, 2011). They ignore the fact that achieving transparency and accountability in such activities in rural areas is next to impossible, as one cannot even properly estimate the amount of earthwork involved (see, e.g., Ambasta et al., 2008). Instead, the operational guidelines for the implementation of MGNREGA were revised, and the new guidelines were supposed to have come out with mechanisms for eliminating the scope of corruption (see Shah, 2012), though nothing much has changed on the ground. The belief is that the Panchayats would always act in the interests of the communities and are relatively corruption-free, and even if it is not so, enough checks and balance could be created at the local level to protect the community interests.

On the other hand, the proponents of these models of water resources development, which involve building of large reservoirs in water-rich catchments and canal systems for transfer of water to water-scarce regions, still do not employ criteria for evaluating such projects, which are comprehensive enough to capture all costs and benefits (Shah and Kumar, 2008), while resettlement and rehabilitation remain as contentious issues for many large projects.

1.3. INEFFECTIVE NATIONAL AND STATE WATER POLICIES

India had its first National Water Policy in 1987, the second one in 2002 and the third one in 2012. All these policy documents highlighted the importance of provision of water for basic survival needs and had given top most priority to drinking water (water for human needs) in water allocation decisions. However, neither did these policies prescribe nor these policies were followed by creation of rules and norms and institutional mechanisms for ensuring water allocation for ensuring drinking water security. The water policies of the provincial governments also give top most priority to drinking water. However, when it comes to norms and institutional mechanisms for water allocation, nothing really exists on the ground. Once "drought" is declared in a district, the district collectors have the power to freeze surface reservoirs for making the water available for drinking purpose. However, the existing laws are inadequate to deal with groundwater, as the de facto rights in groundwater are still attached to land ownership rights. No state government has so far been able to effectively regulate groundwater mining for irrigation during droughts and earmark wells for drinking water supply.

An important prescription in the water policy documents of 2002 (Government of India, 2002) and 2012 (Government of India, 2012) is the need for taking the river basin as the unit for planning of water resources, in order to promote sustainable water resources development. However, so far, no attempts were made to implement this idea, and various agencies concerned with water resources development and management in India continue to act independently in a sectoral and segmented manner.

The resource assessment and planning of groundwater and surface-water resources are carried out separately (by the respective agencies concerned both at the central level and state level), without caring for their interconnectedness. The hydraulic interconnectedness of surface water and groundwater is not considered in the water allocation decisions of interstate water dispute tribunals (Ranade, 2005). Studies have shown that intensive use of groundwater in the upper catchments of river basins results in reduced streamflows, as excessive groundwater draft reduces the lean season flows (base flow) (Kumar, 2010).

Similarly, the wing of Water Resources Department of various state governments, which carry out catchment-wise assessment of surface-water potential and plan large and medium irrigation/multipurpose water projects, does not take cognizance of the myriad of minor irrigation and watershed development projects that are being planned within these catchments by the Minor Irrigation Department and Watershed Development Agency, respectively, and vice versa (Kumar, 2010). Such uncoordinated planning leads to overappropriation of the resource at the basin level. Intensive watershed development and desilting of tanks, etc. causes reduce inflows into reservoir downstream (Batchelor et al., 2003). In summary, the policies have not resulted in any organizational restructuring within India's water administration for coordinated basin-wide planning of water resources.

The water policies of 2002 and 2012 did recognize the need to treat water as an economic good and that the pricing of water to reflect its scarcity value. Yet, no state government is willing to charge for water supplied from public irrigation schemes on volumetric basis. The water cess charged on the basis of crop area and crop type is heavily subsidized, not even covering the full operation and maintenance costs of the schemes (Kumar, 2017). In very few cities and towns in India, the municipalities/corporations or the autonomous water utilities charge for water supplies on volumetric basis and that too not all domestic water connections are metered (Asian Development Bank, 2007; Kumar, 2014b).

This is in spite of the growing evidence of the price elasticity of domestic water demand from other parts of the world (Hoffman et al., 2006) and the importance of water pricing in managing demand. Water prices that are not based on actual volumetric consumption leave no incentive among the consumers to use water efficiently and reduce wastage.

In the farm sector, electricity supplied for groundwater pumping is not charged on prorata basis. While some state power utilities charge for electricity on the basis of connected load, many states are offering free electricity to farmers (Kumar, 2017).[3] What emerges from these examples is that the water policy is never taken seriously by the governments.

1.4. CONCEPTUAL FRAMEWORK FOR ANALYSING POLICY MAKING IN THE WATER SECTOR

Government decision-making and policy formulation increasingly call for gathering scientific evidence (Graffy, 2008; OECD, 2015). Yet internationally, water resources development and management sector is influenced by paradigms and ideas that manifest themselves through ubiquitous and proliferating "success stories," "best practices," "bright spots," or "promising technologies," promoted as universal and transferable to other contexts (Molle, 2008). Molle (2008) discussed three different types of concepts that shape policy and decision-making in the contemporary water sector. The first is the *nirvana concepts*, supporting overarching frameworks that promote or strengthen certain *narratives* or *storylines*—that is, simple, causal, and explanatory beliefs—and legitimize specific blueprints or *models* of both policies and development interventions.

After Molle (2008), nirvana concepts, narratives, and models of development are all ideational and ideological objects that emerge at some point in time to characterize a certain view, approach, or "solution." Over time, such conceptual objects tend to acquire a life of their own. The eventual fate of these concepts depend on many factors, not least their usefulness for particular actors and constituencies who may reappropriate, repackage, and integrate them into their discourses and strategies. One such paradigm that has gained widespread acceptance in India is participatory watershed development model.

As noted by some scholars, influential concepts in policy making emanate from complex webs of interests, ideologies, and power. These interests, ideologies, and power also shape the ways things are framed; options are favored, disregarded, or ignored (Keeley and Scoones, 1999; Kumar, 2014a; Kumar and Pandit, 2016; Molle, 2008), and particular groups are empowered or sidelined (Keeley and Scoones, 1999; Molle, 2008). In the case of India, the ideologies of people who became part of both the extra constitutional body of the National

3. In fact, only West Bengal had successfully introduced metering of electricity in the farm sector for groundwater pumping (way back in 2006) with 100% coverage and charge electricity tariffs that are comparable with the cost of production and supply (Kumar, 2017).

Advisory Council (NAC) and the Planning Commission of the Govt. of India, who are ardent supporters of local self-governance and community management of natural resources, influenced many of the key decisions (during the rule of United Progressive Alliance from 2004 to 2014) that had significant implications for water sector such as the National Rural Employment Guarantee Scheme (NREGS). It was also responsible for the government running cold feet on the National River Linking Project in spite of the supreme court's directive in 2012 to move fast on it.

Molle (2008) reviewed how the three types of concepts populate the water sector, investigated how they spread, and then examined the implications of this analysis for applied research on policy making and practice. The three well-known models of policies and developmental interventions that are outcomes of applications of these concepts are Mexican model of irrigation management transfer (IMT), Chilean model of water markets, and the concept of river basin organizations (Molle, 2008). However, it did not discuss the process of policy making and the role of different actors, especially for countries like India that offers a complex ecosystem with a mix of epistemic community, political parties, local and international donors, civil society groups, and academic institutions.

Further, not all policy decisions in the water sector can be explained by the above theory though. There are large areas of international development where decision-makers are forced to make decisions purely on gut feeling and ideology not because they wish to, but because of the lack of "proper evidence." Evidence empowers the decision-maker to be able to make better choices. As noted by Whitty and Dercon (2013), in every discipline, in every country, where rigorous testing of expert solutions has started, many ideas and strategies promoted by serious and intelligent people with years of experience have been shown not to work because of the lack of sufficient evidence. Over and above, the communities we seek to assist are more vulnerable, denying the luxury of taking time to generate evidence.

One reason, according to Whitty and Dercon (2013), is the reductionist and misinformed view of evidence as purely "technical" or as being only about "what works." At least in the case of decentralized water harvesting interventions, this seems to be the case. Many state governments that embraced this model of water management as part of their water management solutions did so as they found it "working," though it only meant that several thousands of structures were built with some involvement of local communities and NGOs. The policy makers and senior officials often negated the arguments about negative social, ecological, and economic impacts of large-scale water harvesting on the ground that evidence to the effect was lacking. Even when evidence was provided, they were refuted as insufficient and localized.

Whitty and Dercon (2013) point out that the process of policy making should also include generating evidence and learning about why certain interventions and approaches may or may not work, for which we collectively have the capacity, but attempt is often not made to this affect. It is often argued that it is not worth trying to provide the best and most rigorous evidence to those who need to make difficult decisions because they will have other influences as well. They suggest that where evidence is clear-cut, we should be making that plain to decision-makers and, where it is not, we should say that as well; be honest about what is there and try to get better evidence for the future. If the academic community is serious about trying to assist those working in the field to make the most informed possible decisions available for their own development, we should be putting our greatest efforts into supporting decision-makers to use the best evidence and finding better methodologies in areas where we currently have very weak evidence. There are many, and this should be tackled as a matter of urgency.

The defining features of water policy making for many countries, especially in the developing world, are the dramatic changes in management paradigms, policy priorities, and worldviews about water governance since the early 1970s. We will elaborate them in the following paragraphs.

The dramatic expansion of irrigation was followed by the recognition that institutional reforms were needed to ensure better management and physical and financial sustainability of irrigation systems (Meinzen-Dick, 1997).

When competition and conflicts over the use of water between different sectors, especially irrigation and drinking water, grew, a competing set of solutions to problems related to water-use efficiency and allocation emerged in the late 1980s. Principles such as the user-pay and polluter-pay principles and economic tools including stricter cost-benefit analyses, pricing, polluter taxes, tradable rights or entitlements, and markets were proposed as means to regulate demand. Although seen in the 1990s as a new opportunity for making riches by the transnational water industries, the privatization of urban water supply largely backfired (Molle et al., 2008).

As water was at the heart of many environmental debates and controversies, engineers had to accommodate more ecocentric paradigms. Responding to this change, environmentalists promoted a more holistic view of ecosystems, introduced the concepts of environmental flows and environmental services, and often found common ground with resource economists in their promotion of pricing and taxes as means of regulating water diversion and pollution (Disco, 2002).

Environmental or livelihood struggles contributed to revealing the contested and political nature of water. The plurality of worldviews, ideologies, interests, and discourses related to water is manifested in countless conflicts and negotiation processes aimed at solving competing claims for the resource at the level of households, distribution systems, watersheds, and river basins (Molle, 2008). The need for negotiation and conflict resolution inspired theoretical and action research on ideas of coconstruction of knowledge or comanagement, mediation of conflicts through various participatory techniques, and multistakeholder platforms or dialogues, propelled by concepts of deliberative democracy and communicative rationality (Warner, 2008).

Modes of water governance at various scales are constantly challenged and changed as both the society and the environment undergo transformations. Water governance regimes result from, or often are associate parcels of, competing models that alternatively promote management by the state, regulation by the market, and empowerment of users, communities, and civil society at large (Merrey et al., 2007; Mollinga and Bolding, 2004). Although, ideally, the deciding on a particular mode of water governance should be left to the self-determination of site-specific institutions, this process is frequently distorted by lopsided power relations and negotiated by frontal oppositions of viewpoints and ideologies that are at times uncompromising (Molle et al., 2008).

All the combined dysfunctions of the water sector have given rise to the concept of integrated water resources management (IWRM) that emphasizes integrated management of land and water, surface water and groundwater, upstream and downstream uses, different sectoral uses, economic production and environmental sustainability, and coordinated actions of the state and nonstate actors (Biswas, 2004). There is massive promotion of IWRM by mainstream institutions including international donors and multilateral agencies (Warner, 2008; Molle et al., 2008).

The result is that many of these concepts and ideas (equity, economic efficiency, environmental sustainability, community participation in water management, stakeholder consultation and involvement, participatory management, decentralized governance of water, etc.) started echoing in the working documents of the water administration of several countries, especially those in Asia, Latin America, and Africa. IWRM has been adopted as a principle and practice in the National Water Policy or National Water Law or Water Resources Management strategies or all the three 42 countries (source, United Nations Educational, Scientific and Cultural Organization, 2009).

Having provided some theoretical explanations on how certain ideas in water governance and management find appeal in policy making internationally, we will discuss certain aspects that characterize the political process of policy making that are unique to India.

1.5. UNDERSTANDING POLICY MAKING PROCESS IN WATER AND RELATED FIELDS IN INDIA

1.5.1 Civil Society Views on Water

Inconsistent views and incoherent arguments dominate the civil society debate on water policy in India. This is mainly due to limited conceptual clarity on many water-related issues. For instance, most of those who consider water resources as a "common good" and stand strongly against privatization of water services promote decentralized water management at the village level under the pretext that it promotes community participation and thereby equity in distribution of the newly created water source. This is evident from the National Framework Law on Water, and Groundwater Model Bill. But, what it actually does is privatization of surface water, which is a common property resource, and greater inequity in access to the scarce resource, through "elite capture" by allowing private initiatives to collect runoff from common land and put it underground to recharge the irrigation wells or store in small farm reservoirs located in private farms. It is hardly realized that decentralized management of the resource at the local level and equity in resource distribution are often oxymoron.

Further, those who are ardent promoters of access equity in resource use blindly support heavy public subsidies in water supply services in both domestic and irrigation sector and energy subsidies in agriculture, under the pretext that such subsidies would help poor communities to access water at low costs (Mukherji et al., 2012; Shah, 2016b; Shah et al., 2016a,c). But, it is a well-established fact that a large share of the subsidy benefits in irrigation and electricity supply in the farm sector is actually appropriated by the large resource-rich farmers (Vashishtha, 2006a,b), whereas the major chunk of the subsidies in water supply is cornered by rich households that consume large amount of water, while the poor who do not even enjoy water connections hardly gets a share of the subsidy benefits (Kumar and Pandit, 2016).

Intriguingly, those who are against water sector privatization support intensive use of groundwater through well, which are almost entirely privately owned in India, and individual/community initiatives for recharging groundwater using runoff that is in a common property resource. More importantly, they are vociferous in their criticism of public investments in irrigation (Shah, 2016a; Shah, 2011). Interestingly, some others who are against public investment in irrigation openly advocate private sector involvement in provision of irrigation services to farmers (see Shah, 2011). The deferred investment in public irrigation is resulting in the government gradually losing control over the use of water.

The debate on whether water is a "social good" or "economic good" is often used to argue for or against pricing of water. The fact that water is an economic good, when used in production functions such as irrigated agriculture, dairy farming, and industry, is not suggestive of the need to price the resource. Instead, the marginal returns from the use of water for economic production functions can be a basis for fixing the prices, if there is a significant cost involved in its production and supply for these economic activities, as the "marginal returns" give an indication of the price users would be able to pay. In other words, when the resource becomes scarce, with high cost of production and supply, only those uses of water, which yield marginal returns higher than the marginal costs, can be sustained (Aw-Hassan et al., 2014). Conversely, if water is available in plenty in the natural system, pricing of water may not be a big concern, even when allocated for productive purposes, if the cost of production and supply of water is insignificant.

On the other hand, water becomes a "social good" when it is allocated for human consumption that helps improve community health and results in overall social advancement. However, this recognition doesn't take away the right of the water utilities to charge a price for it, if the aim is to recover the cost of production and supply.

However, affordability should be a major concern in fixing the unit price of water. This is because due to information imperfections (relating to the health benefits of consuming good quality water and not being able to effortlessly observe quality aspects of the water consumed), consumers may respond to higher prices for piped water by consuming too much low-quality water from contaminated alternative sources (Noll et al., 2000). This can be at the cost of progress in the much needed social development, and at times, it can even threaten the very survival of the community itself. This essentially means that mechanisms should be devised to offer targeted subsidies for the poor, when prices become high so as to protect the "right to life" and not free water for all.

1.5.2 Increasing Influence of Civil Society

Civil society groups often attach many virtues to decentralized management of rural water supply services; community management of water resources as against centralized water management institutions; import of agricultural commodities (virtual water) from water-rich regions to water-scarce regions, as against physical transfer of water; small water harvesting and recharge initiatives against large dams; and rainfed farming against irrigated agriculture (source, based on Iyer, 2008; Kulkarni and Vijayshankar, 2009; Shah, 2009). According to them, these virtues include efficiency, equity, and sustainability of decentralized management of drinking water services; greater basin-/aquifer-wide equity in access to water in the case of community management of resource, which can also address community interests and concerns through their participation (Kulkarni and Vijayshankar, 2009); ecological sustainability of virtual water trade, which is also capable of averting huge submergence of land and social displacement that large water transfer projects would involve (Iyer, 2008); cost-effectiveness of small water harvesting and recharge systems and their ability to produce high incremental returns per unit of water generated (Shah, 2009); and ecological sustainability of rainfed farming (International Institute for Economic Development (IIED), 2015; Parthasarathy Committee, 2006).

Such propaganda about alternatives in water management continues in spite of the fact that there isn't much evidence to support the claims, especially on matters concerning community management of water supplies and decentralized water management. On the contrary, there is enough evidence to prove some of these fundamentalist views are flawed—be it the argument about the potential of virtual water trade (Kumar and Singh, 2005), rainfed farming to feed the country, and rainwater harvesting to augment water supplies for expanding irrigation in arid and semiarid regions (Glendenning and Vervoort, 2011; Kumar et al., 2008a). However, such advocacy groups have been highly influential in shaping water and energy-related policies in India, as is evident from many policies, programs, and projects of the government of India during the past couple of decades (they would be discussed in the subsequent chapters). Part of the reason is the acceptance of some of these ideas by the international development agencies and donor bureaucracies and the legitimacy they enjoy in the international arena (see Barron, 2009; Shah, 2014). Another reason is the enormous clout these advocacy groups enjoy in government and policy circles.

Similarly, some members of the civil society had used every opportunity to push their own agenda for the water sector while being part of government committees for suggesting reforms in the water and agricultural sectors. The latest example is the Mihir Shah Committee report on restructuring of Central Water Commission (CWC) and Central Ground Water Board (CGWB). Instead of suggesting changes based on a proper diagnosis of the problems facing these sectors and the agencies, facts were often misrepresented. For wider acceptability of their prescriptions that are based on outdated concepts, the ideas were sold as "new paradigm in water management" (see Shah, 2016a). The media was heavily used to publicize their views and to put pressure on the governments to buy into their ideas (see EPW, 2016, pp. 19–62).

1.5.3 The Academic-Bureaucrat-Politician Nexus

Water security is important for human development and economic growth (Grey and Sadoff, 2007; Kumar et al., 2008b). Water security drives economic growth through the human development route, by impacting on food security and nutrition and education (Kumar et al., 2008a). Energy security is also crucial for economic growth, and the energy sector's impact on the economy is greater than the sum of its parts (World Economic Forum, 2012). India's ability to maintain good economic growth rates and move people out of poverty requires solving several complex problems relating to demand and supply of water, energy, and food. This more often requires hard decisions on the part of the state bureaucracies dealing with water, energy, and agriculture, supported by strong political leadership (Kumar, 2014a).

A growing proportion of the Indian academia believe that for a democratically elected government to survive, the decisions concerning the management and use of resources, which directly affect the livelihoods of the millions of ordinary people, should be driven by popular concerns rather than the macroeconomic interests and that these decisions have to be "politically correct." The latter suites the political class better than the former, as they produce immediate electoral gains. A related assumption is that the ordinary people are only interested in measures that produce immediate gains and are not concerned with long-term economic management.[4]

---

4. As Kumar (2014a) notes, to a great extent, this is true for India, where a large proportion of the people living in villages and cities are concerned with their immediate survival needs. However, it is also true that there is a large chunk of the population that is quite concerned with the long-term economic growth needs, and their numbers are growing as the economy is on the upward swing. Good politics also implies that this segment of the population should not simply be ignored for the sake of the rest. Conversely, it is also true that continuing with such populist measures may lead to crippling of the economy, forcing the governments to resort to harsher measures and leading to "political backlash."

In good statecraft, it is also the responsibility of the technocrats, bureaucrats, and the partners in policy making, backed by research-based knowledge, to educate the rulers of the need to have pragmatic policies that strike a balance between "populism" and long-term economic management and to inform them that such an approach would be in the interest of the ruling political class. The political class in turn can educate the electorate. This can help obtain mass support for policy reforms (Kumar, 2014a).

It is the duty and responsibility of the state to frame policies, rules, administrative structures, and laws from time to time to make sure that the subjects have more or less equal opportunities to access common property resources like water for beneficial uses (Ford Foundation, 2010). At the same time, it is the duty of the academicians and scholars to help governments design the right kind of policies, legal frameworks, institutions, pricing mechanisms and taxes on development, allocation and use of these resources based on considerations of productivity, equity, sustainability, and social justice. But the state should make sure that the right kind of institutional regime exists to bring checks and balance for ensuring quality in academic research (Kumar, 2014a).

But, as Kumar (2014a) notes, the academics advise the politicians and bureaucrats that providing free power, fuel subsidies, free power connections, and free water would help eradicate poverty and bring more votes and that market-based solutions would harm the interests of the poor. This "politics-bureaucracy-academics" nexus is costing the economy hugely. As further noted by Kumar (2014a), while the argument of inefficient and unsustainable resource use is being used by a few scholars and development thinkers to counter the government policy of the pervasive subsidy (see Gulati, 2011; Saleth, 1997), the counterargument is that attempts to do away with subsidies in water and electricity sectors would be "political suicide" for ruling parties (see, e.g., Shah et al., 2004).

1.5.4 Spreading of Falsehood?

In the past, some Indian researchers were successful in influencing policy making in water, agriculture, and energy sectors at the national and state level through their persuasive writings in the mass media. The normal course chosen by them is to overplay the "positive impacts" of some of the schemes in these sectors either supported or implemented by the government, through articles in the popular media, without subjecting their work to any peer review for scientific accuracy and analytic rigor. Some of the schemes and interventions are the *Jyotigram* scheme for separation of feeder line for domestic electricity supply from that for agriculture attempted in Gujarat (Shah and Verma, 2008), which is being replicated in many other states (Khanna, 2013); decentralized water harvesting and water conservation through small structures (launched in Gujarat and later on picked up by many western, central, and southern states) (Kale, 2017; Kumar et al., 2008a); promotion of microirrigation systems through capital subsidies (all over the country) (Malik et al., 2016); solar pumps for irrigation (Gujarat, Rajasthan, and Chhattisgarh) (Shah et al., 2016a, c); and "irrigation sector reforms" in Madhya Pradesh (Shah et al., 2016b). However, there are no well-documented studies showing any positive impacts of these interventions to date.

It was argued that limited hours of quality power supply to agriculture under *Jyotigram* Yojna lunched in 2003 significantly reduced groundwater pumping in Gujarat (Shah and Verma, 2008), decentralized water harvesting through check dams improved groundwater recharge mainly in Saurashtra while pushing agricultural growth in the state (Jain, 2012; Shah et al., 2009), solar PV systems would increase incentive for groundwater and electricity conservation (Kishore et al., 2014; Shah et al., 2016a), and irrigation sector reforms led to remarkable increase in canal-irrigated area in Madhya Pradesh (Shah et al., 2016b).

On the contrary, available evidence suggests that after the implementation of *Jyotigram*, the power consumption in agriculture actually went up in Gujarat, and power subsidies increased.[5] Decentralized water harvesting had no impact in terms of mitigating water scarcity situation in Saurashtra region, where massive work of construction of check dams was taken up by the government (Kumar et al., 2010). During the drought of 2012, the region was badly hit with acute drinking water shortage, and water from Sardar Sarovar project was the only source of drinking water supply. In the case of Madhya Pradesh, as we will see in Chapter 8, the irrigation expansion in the recent past was found to be because of the completion of large reservoir projects (meant for irrigation) and not because of any reforms per se.

1.5.5 Role of Social Sciences in Water Management

While the conventional engineering solutions to water management are increasingly becoming inadequate to solve the growing water problems, the institutions that were built on the foundations of civil engineering and infrastructure building, with the purpose of water supply augmentation, are becoming less effective (Kumar, 2010). It is increasingly being acknowledged that the solutions to the complex problems of water management cannot be found within the disciplines of science/engineering (hydrology, geohydrology, and water engineering/technology) alone and that there is a need to integrate knowledge from social science disciplines, namely, economics, sociology, public administration, and law, with that from the traditional disciplines.

5. The Reserve Bank of India identified Gujarat as one of the states where power sector reforms failed to create the "desired impact" on the financial position of the state power utilities (SPUs). The report says "subsidies to SPUs/State Electricity Boards (SEBs) have been rising over the years for Gujarat and Karnataka, which had unbundled utilities." Particularly blaming "huge cash losses" of the SPUs on "nonrevision of tariffs over extended period of time," the RBI report adds that such losses, coupled with simultaneous "nonrealization of subsidies from the state government," are leading to a situation where deterioration in financial performance of SPUs has "significant implications for state finances" (Reserve Bank of India, 2012).

As water allocation and water resources management take precedence over water development under growing scarcity problems (Saleth and Dinar, 1999), this interdisciplinary knowledge is required in the field of water resource economics, environmental economics, water law, organizational behavior, etc. for valuation of water and design of economic instruments for water demand management, institutions, laws/regulations, and policies (Kumar et al., 2012). The reason is the latter combined with the technical system make a good water management system. Such a need is increasingly felt with the acceptance of concept of IWRM at the highest levels in the water resources bureaucracy and government policy-making bodies, which call for interdisciplinary approach in water management.

However, the Indian experts from social and management sciences have limited understanding of water management institutions born mainly out of their experience working with small and microlevel community organizations (such as watershed management committees and water user associations in canal commands with limited operational domain). They lack experience in crafting modern, higher order institutions that are capable of bringing about behavioral changes of water users such as water rights systems, water tax, water resource fee, pollution taxes, and water resource regulatory agencies. They are also unable to impress upon the traditional water bureaucracies that are dominated by civil engineers, due to the lack of interdisciplinary perspective in their work. One reason for this is their insufficient knowledge of the fundamental and technical issues in water management.

1.6. THE CONTENTS OF THIS VOLUME

The book is a collection of cases, which would discuss how the strong "academic-bureaucrat-politician nexus," ever-increasing presence of civil society groups in policy space and peddling of falsehood by influential academicians and researchers who engage in persuasive writing in media, had resulted in government policies in many domains of water management (technology for water harnessing, technology for water saving in irrigation, water pricing, investments for water resources development, and community participation in water management) and projects in water and other sectors (forestry, energy, and climate adaptation) that have significant implications for water management. Other theories that explain how certain paradigms and views tend to dominate the water sector internationally (Molle, 2008; Molle et al., 2008) and that also found some appeal among the water community in India and how policy decisions are taken by the epistemic community (Whitty and Dercon, 2013) are also sparingly used. The cases also illustrate how basic scientific principles and theories in hydrology, water engineering, economics, energy, and climate governing the effectiveness of water management systems were ignored in the design of these "solutions."

These cases include the recent water sector reform initiative undertaken by the Ministry of Water Resources, government of India (Shah, 2016a), involving restructuring of two central water agencies, namely, CWC and CGWB; recommendations of Parthasarathy Committee (2006), dealing with long-term strategies for agricultural development in rainfed areas; strategies for reforming public irrigation sector (Shah, 2016a); strategies for agricultural growth in backward region of eastern India (Mukherji et al., 2012; Shah, 2016b; Shah et al., 2016c); and strategies promoted by the civil society and academic for improving climate resilience in water and energy sectors (Shah et al., 2016a, c) and for improving groundwater governance (Shah and Verma, 2008; Kulkarni et al., 2011; Kulkarni and Vijayshankar, 2009); the government initiated the $2 billion project of *Mission Kakatiya* to rehabilitate tanks in Telangana and government-sponsored projects to improve water-use efficiency and save water. Each chapter also illustrates how an integrated understanding of the problems from a physical and social science perspectives can help frame meaningful policies and at the same time belie the long-held views.

There are attempts from some quarters to misinterpret the concept of water sector reforms in order to move away from infrastructure development and serious institutional change processes and policy measures. They merely parade vacuous slogans such as participatory management, "community management," and "making water everybody's business" and label them as "paradigm shift" in water management (see Iyer, 2005, 2011; Shah, 2013, 2016a). Such attempts are driven by a pathological hatred toward good water science, professional engineers, and hydrologists (Kumar and Pandit, 2016). Politicians and policy makers get carried away by such catchy slogans and even use them to masquerade their lack of ability to fix the problems in the under-performing sector.

Chapter 2 of this book discusses the findings of earlier analyses using global data sets showing how human development, economic growth, and poverty alleviation were linked to water security, particularly how building large water systems helps achieve human development and economic growth in arid and semiarid tropics by improving water security. It also presents the recent micro-level empirical evidence pertaining to various positive impacts of large water storages that correspond to variable that can affect human development, economic growth, and poverty reduction, from intensive primary research carried out in India in order to revalidate the findings from global data analysis.

India, with one-sixth of the world population still has low water security and stands high in ranking in terms of global hunger index. Chapter 3 takes a hard look at India's future water scenario with regard to potential supplies to meet the likely demand for water from various sectors and the pattern of demand growth with respect to regions and sectors. It also builds four key arguments in favor of interbasin transfer of water for achieving future water, food, and energy security in India, which is characterized by unequal distribution of water and land resources.

Chapter 4 argues that the problems of droughts experienced in many parts of India in the recent past are a result of poor statecraft, rather than a natural calamity, and could be avoided if India builds adequate infrastructure for transfer of water from water-rich to water-scarce regions. The growing influence of CSOs in government policy making, the "academics-politician-bureaucrat nexus," and a weak institutional regime that controls the quality of research used for public policy are the defining features of this statecraft.

While many parts of India experience perennial water scarcity and periodic droughts and floods, the working of some of the oldest federal agencies in the water sector has come under severe criticism. Chapter 5 critiques the report on reforms of water institutions in India, prepared by a committee on restructuring CWC and CGWB in India. The chapter examines the validity of the findings of the committee vis-à-vis the problems facing India's water sector and the potential impact of the "new paradigm in water management" proposed by the committee on the sector performance, particularly the new institutional structure proposed for the technical agencies concerned with water resources development and management at the central level.

Chapter 6 provides illustrative examples to highlight the prevailing "misconceptions" in the water management sector relating to forest-water interactions, basin-wide impacts of water harvesting, and causes of streamflow reduction in rivers and how researchers and CSOs succeed in influencing policies relating to water management and climate adaptation and mitigation. It discusses some of the projects and schemes that were recently launched by the provincial governments in response to lobbying by researchers and CSOs, without the backing of hard evidence to the effect. The chapter also argues that there is a need for strengthening the institutional regime that scrutinizes research for their quality, before their recommendations are taken onboard by governments for policy changes. The chapter also suggests a few ways of achieving it.

The Mission Kakatiya of the government of Telangana envisages enhancing the agriculture-based income of small and marginal farmers by accelerating the development of minor irrigation infrastructure and strengthening community-based management of tank systems. The government has prioritized the restoration of minor irrigation tanks to enhance their effective storage capacity, so as to fully utilize Telangana's allocation of 255 TMC of water from Godavari and Krishna. In Chapter 7, we examine the constraints to achieving the intended hydrologic and socioeconomic benefits from this ambitious scheme and what needs to be done to improve the effectiveness of the tank rehabilitation program in the state.

In the recent past, there were attempts by researchers to argue on the basis of crude data on canal-irrigated area that increasing investments in public irrigation was not leading to improved performance of that sector (see Shah, 2009) and this "poor performance of public irrigation" (mainly surface irrigation) was attributed to poor and outdated management models adopted by the state irrigation bureaucracy. The factors such as changes in land use in the catchment, larger hydrologic changes, upstream water diversion, and changes in water allocation from public reservoirs, which can impact severely on the amount of water that can be released for irrigation (Kumar et al., 2012), were never considered in such analysis. Chapter 8 examines whether doing a diagnosis of India's irrigation sector problem using the simple criterion of comparing the "irrigation potential utilized" against the "potential created" is correct or not. It then takes a serious look at the miracle performance shown by MP's public irrigation sector and the factors that are driving this impressive performance.

"Rainwater harvesting" has become a buzzword in India's water management debate during the past three decades or so, due to constant campaigning by civil society groups, who are against large water systems, in favor of the former. But this is based on very limited understanding of India's rainfall characteristics and their implications for the hydrologic feasibility and economic viability of various small-scale rainwater harvesting systems. In Chapter 9, we analyze how this obsession of civil society activists toward "rain water harvesting" and their increasing influence on the civil society at large weaken the institutional capability of water sector agencies to evolve long-term strategies to deal with floods and droughts and achieve water security, using the case of Chennai floods.

As a knee-jerk reaction to the growing global concern about climate change and its impact on water, energy, and food systems, many economic decisions are being taken by national and state governments in water, agriculture, and energy sectors. This is being done without having comprehensive and integrated understanding of the likely impact of climate change on these sectors, with the sole purpose of convincing the international development community. Many of them are seriously flawed, affecting the power sector economy, financial institutions, and India's manufacturing sector adversely (Iyer, 2005, 2017a,b). Chapter 10 discusses some of the fundamental flaws in climate research in the Indian context and analyzes the physical efficacy and economic viability of some of the climate mitigation and adaptation strategies that are based on such research. It also looks at some of the alternative ways of reducing carbon emissions that would help the economy and save precious natural resources.

In Chapter 11, we argue how doing comparison between private well irrigation and large-scale public surface irrigation using simple statistics of irrigated area expansion offers a distorted view of Indian irrigation and highlights some of the unique benefits of large surface irrigation, which are largely ignored by researchers who promote private well irrigation as an alternative to public irrigation. It also highlights the equity concerns posed by intensive groundwater irrigation, which is supported by heavy public subsidies for electricity, and also how groundwater markets become highly monopolistic under the current institutional and policy regimes governing the use of groundwater, characterized by flat rate pricing of electricity and the lack of well-defined water rights that favor large well owners. The chapter also discusses the institutional reforms needed to address the governance challenges tried successfully in some of the developed countries.

The problem of agricultural stagnation in eastern India, particularly Bihar, West Bengal, and Eastern UP, has been a topic for research for many development economists for the past two to three decades. However, their analyses, which merely focused on public policies for promoting equity in access to groundwater for irrigation, had failed to capture the key factors responsible for agrarian crisis in that region and also the potential policy impacts of public policies in irrigation on the functioning of groundwater markets. Chapter 12 brings out the flaws in their analyses of the agrarian problems in the region, assesses the likely outcomes of the policies vis-à-vis access equity in groundwater, and comes out with concrete suggestions on what need to be done to break the agricultural stagnation in the region.

In Chapter 13, we question the criterion used by government of India to classify agricultural areas into "rainfed" and "irrigated." This criterion fails to consider the agroclimate and hydrometeorology of the area that decide whether crops can be grown under rainfed conditions or require irrigation. We argue that watershed development interventions, which are usually prescribed for agricultural development of rainfed areas, should be based on agroecology and hydrometeorology. We also propose the types of water management interventions for different types of watersheds based on agroecology and hydrometeorology.

In Chapter 14, we discuss the complex concept of "water saving" through microirrigation systems based on the different notions of "water saving" in agriculture, review the related concept of "rebound effect" relevant to microirrigation system adoption in agriculture to analyze its impacts at the system and basin level, examine the limited global evidence on the water-saving impacts of drip systems to identify the determinants of change in actual water-use post adoption, and analyze how the "rebound effect" gets played out in different situations in India.

Dairy production has already emerged as an important subsector of India's agricultural economy and is emerging as a major source of livelihood for the small and marginal farmers, especially in the semiarid and arid regions. Yet, the impact of milk production on the water environment in India has been least studied. The general belief is that consuming vegetarian diet will create lower water footprint as compared with having a meat-based diet. This is true only for regions with temperate climate, for which most of the available data on water intensity of milk production are available. In Chapter 15, we look at the drivers of milk demand in India, the water intensity of dairy production in different climatic conditions, the key determinants of dairy production, and their overall implications for the future of dairy production and water use in that sector. The chapter also discusses the ways to promote improved water-use efficiency in milk production.

Most countries of sub-Saharan Africa score very low on water security, affecting food production and water use for domestic and livestock sector adversely with an overall impact on the economy. The region is riddled by conflicts, political instability, poor governance, corruption, politics of exclusion, and high rural poverty and weak human resource capacities and is not comparable with India when it comes to government policy making and public sector investment in water resources development. But because of the heterogeneous conditions prevailing across India vis-à-vis agroclimate, socioeconomic status, and overall water security, the country's experience provides important learnings for sub-Saharan Africa. In Chapter 16, we would draw some important lessons on water management strategies for that region. Finally, some policy directions are given for the fledgling economies of Sub-Saharan Africa, and parts of South and Southeast Asia manage the political process of policy making well.

In the concluding chapter (Chapter 17), we would summarize the key findings emerging from each chapter and offer some clues on how we maneuver the difficult task of managing the political process of policy making in the water sector, in a manner that is participatory and at the same time doesn't get hijacked by influential vested interests.

REFERENCES

Ambasta, P., Vijayshankar, P.S., Shah, M., 2008. Two years of NREGA: the road ahead. Econ. Polit. Wkly. Insight XLIII (8), 41–50.

Asian Development Bank, 2007. 2007 Benchmarking and Data Book of Water Utilities in India, a Partnership Between the Ministry of Urban Development. Government of India and the Asian Development Bank. November.

Aw-Hassan, A., Rida, F., Telleria, R., Bruggeman, A., 2014. The impact of food and agricultural policies on groundwater use in Syria. J. Hydrol. 513 (2014), 204–215.

Barron, J. (Ed.), 2009. Rainwater Harvesting: A Lifeline for Human Wellbeing, A Report Prepared for UNEP by Stockholm Environment Institute. United Nations Environment Programme.

Batchelor, C., Rama Mohan Rao, M.S., Manohar Rao, S., 2003. Watershed development: a solution to water shortages in semi-arid India or part of the problem? Land Use Water Resour. Res. 3, 1–10.

Biswas, A.K., 2004. Integrated water resources management: a reassessment. Water Int. 29 (2), 248–256.

Biswas, A.K., Tortajada, C., 2003. Colombo's Water Supply: A Paradigm for the Future? Special Feature. Asian Water.

Bosshard, P., 2004. The World Bank at 60: A Case of Institutional Amnesia? A CRITICAL LOOK at the Implementation of the Bank's Infrastructure Action Plan. International Rivers Network. April 2004.

Disco, C., 2002. Remaking "nature": the ecological turn in Dutch water management. Sci. Technol. Hum. Values 27 (2), 206–235.

Economic and Political Weekly, 2016. Water governance. Econ. Polit. Wkly. 51 (52), 19–62 (Special Issue).

Ford Foundation, 2010. Expanding Community Rights over Natural Resources: Initiative Overview. Ford Foundation, New York, USA.

Glendenning, C.J., Vervoort, W., 2011. Hydrological Impacts of Rainwater Harvesting (RWH) in a Case Study Catchment: The Arvari River, Rajasthan, India. Part 2. Catchment-Scale Impacts. Agricultural Water Management, pp. 715–730.

Government of India, 2002. National Water Policy. Ministry of Water Resources, Government of India, New Delhi.

Government of India, 2012. National Water Policy, 2012. Ministry of Water Resources, Government of India, New Delhi.

Graffy, E., 2008. Meeting the Challenges of Policy-Relevant Science: Bridging Theory and Practice, Thinking about Public Administration in New Ways. US Geological Survey.

Grey, D., Sadoff, C., 2007. Sink or swim: water security for growth and development. Water Policy 9 (6), 541–570.

Gulati, M., 2011. 50% Savings, 100% Free Power. The Times of India. Retrieved from: http://epaper.timesofindia.com.

Hoffman, M., Worthington, A.C., Higgs, H., 2006. Urban Water Demand with Fixed Volumetric Charging in a Large Municipality: The Case of Brisbane, Australia. Faculty of Commerce Papers, University of Wollongong, Brisbane, Australia.

International Institute for Economic Development (IIED), 2015. Reviving Knowledge: India's Rainfed Farming, Variability and Diversity, Briefing. Food and Agriculture, Policy and Planning, IIED.

Iyer, R.R., 2005. The Politicisation of Water. Info Change News & Features.

Iyer, R.R., 2008. Water: a critique of three basic concepts. Econ. Polit. Wkly. 43 (1), 15–17.

Iyer, R.R., 2011. National water policy: an alternative draft for consideration. Econ. Polit. Wkly. XLVI (26–27).

Iyer, S.A., 2017a. Dark Side of Solar Success: It May Kill Thermal Power, Banks. Swaminomics. Times of India.

Iyer, S.A., 2017b. 100,000MW of Costly Solar Power Can Sink "Make in India." Swaminomics, Times of India.

Jagadeesan, S., Kumar, M.D., 2015. The Sardar Sarovar Project: Assessing Economic and Social Impacts. Sage Publications, New Delhi.

Jain, R.C., 2012. Role of Decentralized Rainwater Harvesting and Artificial Recharge in Reversal of Groundwater Depletion in the Arid and Semi-Arid Regions of Gujarat, India. IWMI-Tata Water Policy Research Highlight # 49, IWMI Field Office, Anand, Gujarat, India.

Kale, E., 2017. Problematic uses and practices of farm ponds in Maharashtra. Econ. Polit. Wkly. LII (3), 20–22.

Keeley, J., Scoones, I., 1999. Understanding Environmental Policy Processes: A Review. IDS Working Paper 89. Environment Group, Institute of Development Studies, University of Sussex, Brighton, UK.

Khanna, A., 2013. Separate Power Feeders can Greatly Improve Rural Electrification. Down to Earth.

Kishore, A., Shah, T., Tewari, N., 2014. Solar PV systems would increase incentive for groundwater and electricity conservation. Econ. Polit. Wkly. 49 (10), 55–62.

Kulkarni, H., Vijayshankar, P.S., 2009. Groundwater: towards an aquifer management framework. Econ. Polit. Wkly. XLIV (6).

Kulkarni, H., Vijayshankar, P.S., Sunderrajan, K., 2011. India's groundwater challenge and the way forward. Econ. Polit. Wkly. XLVI (2).

Kumar, M.D., 2010. Managing Water in River Basins: Hydrology, Economics, and Institutions. Oxford University Press, New Delhi.

Kumar, M.D., 2014a. Of statecraft: managing water, energy and food for long-term national security. In: Kumar, M.D., Bassi, N., Narayanamoorthy, A. (Eds.), Water, Energy, Food Security Nexus: Lessons From India for Development. Routledge/Earthscan, London, UK, pp. 211–220.

Kumar, M.D., 2014b. Thirsty Cities: How Indian Cities Can Manage Their Water Needs. Oxford University Press, New Delhi.

Kumar, M.D., 2017. Market analysis: desalinated water for irrigation and domestic use in India. In: Prepared for Securing Water for Food: A Grand Challenge for Development in the Center for Development Innovation. U.S. Global Development Lab Submitted by DAI Professional Management Services.

Kumar, M.D., Bassi, N., Venkatachalam, L., Sivamohan, M.V.K., Vedantam, N., 2012. Capacity Building in Water Resources Sector of India. Occasional Paper # 4. Institute for Resource Analysis and Policy, Hyderabad, India.

Kumar, M.D., Pandit, C., 2016. India's water management debate: is the civil society making it everlasting? Int. J. Water Resour. Dev. https://doi.org/10.1080/07900627.2016.1204536.

Kumar, M.D., Patel, A., Ravindranath, R., Singh, O.P., 2008a. Chasing a mirage: water harvesting and artificial recharge in naturally water-scarce regions. Econ. Polit. Wkly. 43 (35), 61–71.

Kumar, M.D., Shah, Z., Mukherjee, S., 2008b. In: Kumar, M.D. (Ed.), Water, human development and economic growth: some international perspectives. Managing Water in the Face of Growing Scarcity, Inequity and Diminishing Returns: Exploring Fresh Approaches, proceedings of the 7th Annual Partners' Meet. IWMI-Tata Water Policy Research Program, 2–4 April, 2008, ICRISAT, Patancheru, Andhra Pradesh, pp. 842–858.

Kumar, M.D., Singh, O.P., 2005. Virtual water in global food and water policy making: is there a need for rethinking? Water Resour. Manag. 19, 759–789.

Kumar, M.D., Singh, O.P., Narayanamoorthy, A., Sivamohan, M.V.K., Sharma, M.K., Bassi, N., 2010. Gujarat's Agricultural Growth Story: Exploding Some Myths. Occasional Paper # 2. Institute for Resource Analysis and Policy, Hyderabad.

Malik, R.P.S., Giordano, M., Rathore, M.S., 2016. The negative impact of subsidies on the adoption of drip irrigation in India: evidence from Madhya Pradesh. Int. J. Water Resour. Dev. https://doi.org/10.1080/07900627.2016.1238341.

Meinzen-Dick, R., 1997. Farmer participation in irrigation: 20 years of experience and lessons for the future. Irrig. Drain. Syst. 11 (2), 103–118.

Merrey, D.J., Meinzen-Dick, R., Mollinga, P.P., Karar, E., 2007. Policy and institutional reform: the art of the possible. In: Molden, D. (Ed.), Water for Food-Water for Life: Comprehensive Assessment of Water Management in Agriculture. Earthscan, London, pp. 193–232. (Chapter 5).

Molle, F., 2008. Nirvana concepts, narratives and policy models: insights from the water sector. Water Altern. 1 (1), 131–156.

Molle, F., Mollinga, P., Dick, M.R., 2008. Water, politics and development. Introducing. Water Altern. 1 (1), 1–6.

Mollinga, P.P., Bolding, A., 2004. Research for strategic action. In: Mollinga, P.P., Bolding, A. (Eds.), The Politics of Irrigation Reform, Contested Policy Formulation and Implementation in Asia, Africa and Latin America. Ashgate, Aldershot, UK, pp. 291–318.

Mukherji, A., Shah, T., Banerjee, P., 2012. Kick-starting a second green revolution in Bengal, commentary. Econ. Polit. Wkly. 47 (18), 27–30.

Noll, R., Shirley, M.M., Cowan, S., 2000. Reforming urban water systems in developing countries. In: Anne, O.K. (Ed.), Economic Policy Reform: The Second Stage. University of Chicago Press, Chicago, IL.

OECD, 2015. Scientific Advice for Policy Making: The Role and Responsibility of Expert Bodies and Individual Scientists. OECD Science. Technology and Industry Policy Papers, No. 21. OECD Publishing, Paris. https://doi.org/10.1787/5js33l1jcpwb-en.

Parthasarathy Committee Report, 2006. From Hariyali to Neeranchal: Report of the Technical Committee on Watershed Programmes in India. Department of Land Resources, Ministry of Rural Development, Government of India, New Delhi.

Ranade, R., 2005. Out of sight, out of mind. Econ. Polit. Wkly. 40 (21).

Reserve Bank of India, 2012. State Finances: A Study of Budgets of 2011–12. Reserve Bank of India.

Roth, D., Boelens, R., Zwarteveen, M., 2005. Liquid Relations: Contested Water Rights and Legal Complexity. Rutgers University Press, New York.

Saleth, R.M., 1997. Power tariff policy for groundwater regulation: efficiency, equity and sustainability. Artha Vijnana 39 (3), 312–322.

Saleth, R.M., Dinar, A., 1999. Water Challenge and Institutional Responses (A Cross Country Perspective). Policy Research Working Paper Series 2045. World Bank, Washington, DC.

Shah, M., 2012. Report of the Committee for Revision of MGNREGA Guidelines. Submitted by Dr Mihir Shah, Chairperson, Committee for Revision of MGNREGA Operational Guidelines to Ministry of Rural Development. New Delhi, Government of India.

Shah, M., 2013. Water: towards a paradigm shift in the twelfth plan. Econ. Polit. Wkly. XLVIII (3), 40–52.

Shah, M., 2016a. A 21st Century Institutional Architecture for Water Reforms in India. Final Report Submitted to the Ministry of Water Resources, River Development & Ganga Rejuvenation. Government of India, New Delhi.

Shah, M., 2016b. Paradoxes, bottlenecks and solutions: eliminating poverty in Bihar. Econ. Polit. Wkly. 51 (6), 56–65.

Shah, T., 2009. Taming the Anarchy: Groundwater Governance in Asia. Resources for Future, Washington D.C.

Shah, T., 2011. Past, Present and the Future of Canal Irrigation in India. India Infrastructure Report 2011. Infrastructure Development Finance Corporation, Mumbai.

Shah, T., 2014. Groundwater Governance and Irrigated Agriculture. TEC Background Papers No. 19. Global Water Partnership Technical Committee, Stockholm.

Shah, T., Durga, N., Verma, S., Rathod, R., 2016a. Solar Power as Remunerative Crop. Water Policy Research Highlight # 10. IWMI-Tata Water Policy Program, Anand, Gujarat, India.

Shah, T., Gulati, A., Padhiary, H., Sreedhar, G., Jain, R.C., 2009. Secret of Gujarat's Agrarian Miracle after 2000. Econ. Polit. Wkly. 44 (52), 45–55.

Shah, T., Mishra, G., Kela, P., Chinnasamy, P., 2016b. Har Khet Ko Pani? Madhya Pradesh's irrigation reform as a model, commentary. Econ. Polit. Wkly. 51 (6), 19–24.

Shah, T., Pradhan, P., Rasul, G., 2016c. Water challenges of the ganga basin: an agenda for accelerated reform. In: Bharati, L., Sharma, B.R., Smakhtin, V. (Eds.), The Ganges Basin: Status and Challenges in Water, Environment and Livelihoods. Earthscan/Routledge, London, United Kingdom.

Shah, T., Scott, C.A., Kishore, A., Sharma, A., 2004. Energy Irrigation Nexus in South Asia: Improving Groundwater Conservation and Power Sector Viability. Research Report No. 70. International Water Management Institute, Colombo, Sri Lanka.

Shah, T., Verma, S., 2008. Co-management of electricity and groundwater: an assessment of Gujarat's Jyotirgram scheme. Econ. Polit. Wkly. (Special Article February 16, 2008).

Shah, Z., Kumar, M.D., 2008. In the midst of the large dam controversy: objectives, criteria for assessing large water storages in developing world. Water Res. Manag. 22, 1799–1824.

United Nations Educational, Scientific and Cultural Organization, 2009. Integrated Water Resources Management in Action. United Nations World Water Assessment Programme. Jointly Prepared by DHI Water Policy and UNEP-DHI Centre for Water and Environment, Paris, France.

Vashishtha, P.S., 2006a. Input Subsidy in Andhra Pradesh Agriculture. Unpublished manuscript.

Vashishtha, P.S., 2006b. Input Subsidy in Punjab Agriculture. Unpublished manuscript.

Warner, J. (Ed.), 2008. Multi-Stakeholder Platforms for Integrated Water Management. Ashgate Studies in Environmental Policy and Practice, Ashgate, Aldershot, UK.

Whitty, C., Dercon, S., 2013. The Evidence Debate Continues, Chris Whitty and Stefan Dercon Respond from DFID. People, Spaces, Deliberation. Available at http://blogs.worldbank.org/publicsphere.

World Bank, 2011. Social Protection for a Changing India Volume II, Chapter 4. World Bank, Washington, DC, p. 84.

World Economic Forum, 2012. Energy for Economic Growth Energy Vision Update 2012. Prepared in Partnership with HIS CERA. Industry Agenda, World Economic Forum.

Chapter 2

Assessing the Importance of Water Infrastructure and Institutions for Water Security

2.1 THE DEBATE

The debate on the linkage between water, growth, and development is mounting internationally. The general view of international scholars, who support large water resource projects, is that increased investment in water projects such as irrigation, hydropower, and water supply and sanitation acts as engines of growth in the economy while supporting progress in human development (for instance, see Briscoe, 2005; Braga et al., 1998; HDR, 2006). They harp on the need for investment in water infrastructure and institutions (Kumar et al., 2016).

Grey and Sadoff (2007) suggests that there is a minimum platform of water security, achieved through the right combination of investment in water infrastructure and institutions and governance, which is essential if poor countries are to use water resources effectively and efficiently to achieve rapid economic growth to benefit vast numbers of their population. They suggest an S-curve for growth impacts of investment in water infrastructure and institutions in which returns continue to be nil for early investments. They argue that for poor countries, which experience highly variable climates, the level of investment required to reach the tipping point of water security would be much higher as compared with countries that fall in temperate climate with low variability. But they suggest that for developing countries, the returns on investment in infrastructure would be higher than in management and vice versa in the case of developed countries.

Many environmental groups, on the other hand, advocate small water projects that, according to them, can be managed by the communities. The solutions advocated are watershed management, small water harvesting interventions, community-based water supply systems, and microhydroelectric projects (see, e.g., Dharmadhikary, 2005; D'Souza, 2002; Iyer, 2008, 2011). During the past two decades or so, they have smartly and extensively used their media power and articulation skills to lobby strongly with multilateral donors and philanthropic organizations, through representation in print media (see, e.g., Iyer, 2008, 2011).

Water Policy Science and Politics. https://doi.org/10.1016/B978-0-12-814903-4.00002-6

International literature provides many clues to the fact that water security has the potential to promote inclusive growth. For instance, access to safe water and sanitation can partly determine income. The marginal productivity per unit of water, measured in terms of good health, longevity, or income, is much greater for the poor than for the rich (World Water Council, 2000; Jha, 2010; Van Koppen et al., 2009). Secured water for irrigation, while enabling the poor landholding communities to grow food for their own consumption (Jagadeesan and Kumar, 2015), would generate sufficient employment in rural areas and lower cereal prices while supplying cereals to the markets (Perry, 2001). In poor and developing countries, where large section of the population depends on agriculture and rural wage labor, this is likely to have significant distributional effects on income. The impressive agricultural growth clocked by the state of Madhya Pradesh in the recent past and the recovery in agricultural production attained by the state of Gujarat after two consecutive years of droughts and the role played by large irrigation infrastructure in achieving the same are something that cannot be overlooked (Kumar et al., 2010 for Gujarat and Chapter 8 of this book for Madhya Pradesh).

2.2 THE EVIDENCE

The theoretical discussion on the returns on investment by countries in water infrastructure and institutions is abundant (Grey and Sadoff, 2007; HDR, 2006). The evidence available internationally to the effect that water security can catalyze human development and growth is quite robust (see World Bank, 2004, 2006a; Briscoe, 2005). However, the number of regions for which these are available is not large enough for evolving a global consensus on this complex issue. Till recently, there was no comprehensive database on various factors influencing water security for sufficient number of countries that are at different stages of human development path and economic growth. This contributed to the complexity of the debate. The water poverty index (WPI), conceived and developed for countries by Sullivan (2002), and the international comparisons available from the work by Laurence et al. (2003) for 145 countries enable us to provide an empirical basis for enriching the debate.

The WPI is a composite index consisting of five subindexes, namely, water access index, water use index, water endowment index, water environment index, and institutional capacities in water sector (Sullivan, 2002). In order to realistically assess the water situation of a country, a new index called sustainable water use index (SWUI) was derived from WPI Shah and Kumar (2008) and Kumar (2009).

2.3 THE RESEARCH STUDIES

The Institute for Resource Analysis and Policy (IRAP) set out to carry out empirical research, to address the question of whether water security drives economic growth or vice versa. The study used empirical analysis using global database on SWUI. The paper provides the key results from the empirical analysis using global database on SWUI and many other water and development indicators such as global hunger index (GHI), human poverty index (HPI), and income inequality index to enrich the debate "how water security is linked to human development, poverty reduction, and inclusive growth." The detailed analysis and outputs are available in Kumar et al. (2016). It is supported by recent empirical analysis of the economic and social impacts of one of the multipurpose projects in India, that is, the Sardar Sarovar Project (see Jagadeesan and Kumar, 2015).

We began with three propositions. First, improving the water situation through investments in water infrastructure, institutions, and policies would help ensure economic growth through the human development route. Second, nations can achieve reasonable progress in human development even at low levels of economic growth, through investment in water infrastructure and welfare policies. Third, countries need to invest in building large water storages to support economic prosperity and ensure water security for social advancements. The hypotheses are as follows: (1) improved water situation supports economic growth through the human development route, and (2) countries, which are in tropical climates with aridity, can support their economic growth through enhancing per capita reservoir storage that improves their water security (Kumar et al., 2016).[1]

The values of SWUI were calculated by adding up the values of four of the subindexes of WPI, namely, water access index, water use index, water environment index, and water capacity index. All the subindexes have values ranging from 0 to 20. The maximum value of SWUI for a country therefore is 80.

The results from an intensive study on Sardar Sarovar Project, a multipurpose project in Gujarat state of western India, were used to illustrate how in semiarid and arid region water from large water systems for irrigation can ensure enhanced farm outputs and increased income, directly of canal irrigators and indirectly of well owners, which in turn increases the household expenditure on health and education, enhanced income of agricultural wage laborers, and improved drinking water supplies in villages, thereby affecting all round and inclusive growth.

1. The first hypothesis was tested using a regression of global data on SWUI and data on per capita GDP (ppp adjusted), SWUI and GHI, and SWUI and HDI. GHI is an indicator of the proportion of the population living in undernourished conditions and the child mortality rate. The causality of whether SWUI influences GDP growth or vice versa was tested by running a two-stage least square method, with SWUI as the predictor variable, HDI as the instrumental variable, and per capita GDP as the dependent variable. The underlying premise is that if economic growth drives water situation, then it should change the indicators of human development that are independent of income levels, such as health and education, and that that are interrelated with water situation. The second hypothesis was tested by analyzing the link between per capita GDP (ppp adjusted) and per capita dam storage (m^3/annum) of 22 selected countries falling in hot and arid tropical climate (Kumar et al., 2016).

2.4 WATER AND INCLUSIVE GROWTH

A work by Kellee Institute of Hydrology and Ecology that came out with international comparisons on water poverty of nations had used five indexes, namely, water resources endowment, water access, water use, capacity building in water sector, and water environment, to develop a composite index of water poverty (see Laurence et al., 2003). Among these five indexes, we chose four indexes as important determinants of water situation of a country, and the only subindex we excluded is the water resources endowment. We consider that this subindex is more or less redundant, as three other subindexes, namely, water access, water use, and water environment, take care of what the resource endowment is expected to provide (see Kumar, 2010, for details). That said, all the four subindexes we chose have significant implications for socioeconomic conditions and are influenced by institutional and policy environment and therefore have human element in them. Hence, such a parameter will be appropriate to analyze the effect of institutional interventions in water sector on economy.

It is being hypothesized that that the overall water situation of a country (or SWUI) has a strong influence on its economic growth performance.

Worldwide, experiences show that improved water situation (in terms of its access to water, levels of use of water, the overall health of water environment, and enhancing the technological and institutional capacities to deal with sectoral challenges) leads to better human health and environmental sanitation, food security and nutrition, livelihoods, and greater access to education for the poor (HDR, 2006). This aggregate impact can be segregated with irrigation having direct impact on income of farmers and wage laborers (Narayanamoorthy and Deshpande, 2003; Jagadeesan and Kumar, 2015) and rural poverty (Hussain and Hanjra, 2003); irrigation having impact on food security, livelihoods, and nutrition (Hussain and Hanjra, 2003), with positive effects on productive workforce; and domestic water security having positive connotations for human health (Hutton and Bartram, 2008) and environmental sanitation, with spin-off effects on livelihoods and nutrition (positive), school dropout rates (negative), and productive workforce.

The impact of irrigation on rural wage employment in agrarian societies is very significant and cannot be understated. This impact includes the increase in wage employment and rise in wage rate of farm laborers, with the demand for farm labor pushing wage rates (Narayanamoorthy and Deshpande, 2003). Hence, the impact goes far beyond the farmers who receive canal water providing more inclusive growth of the rural economy. In the villages falling in the command area of SSP, the increase in wage employment (in number of days for wage laborer) post introduction of canal water for irrigation of farm laborers has been quite significant, and highest increase was observed during the winter season. Table 2.1 shows the percentage changes in wage employment in agriculture due to canal irrigation in four districts, which receive gravity irrigation from SSP. Table 2.2 shows the average change in wage rate in real terms, post introduction of Narmada water in the area.

TABLE 2.1 Changes in Wage Employment in Agriculture Due to Canal Irrigation (Days)

Location	Male			Female		
	Monsoon	Winter	Summer	Monsoon	Winter	Summer
Before Narmada						
Panchmahals	36.9	49.7	28.1	38.0	48.5	23.8
Bharuch	55.7	76.6	59.0	55.9	77.6	59.5
Narmada	48.7	45.6	25.5	46.6	45.2	24.4
Vadodara	42.6	65.9	35.3	40.6	57.8	33.1
After Narmada						
Panchmahals	47.5	60.0	34.7	46.5	57.0	28.1
Bharuch	63.4	90.2	61.8	65.0	90.5	61.7
Narmada	58.5	63.0	36.0	57.9	60.5	34.7
Vadodara	50.4	78.6	44.4	47.2	72.0	36.3
Increase in no. of days of employment						
Panchmahals	10.6	10.3	6.6	8.5	8.5	4.3
Bharuch	7.7	13.6	2.8	9.1	12.9	2.2
Narmada	9.8	17.4	10.5	11.3	15.3	10.3
Vadodara	7.8	12.7	9.1	6.6	14.2	3.2

Source: Based on Jagadeesan, S., Kumar, M.D., 2015. The Sardar Sarovar Project: Assessing Economic and Social Impacts. Sage Publications, New Delhi.

TABLE 2.2 Changes in Wage Rate (INR/Day) of Farm Laborers in the Five Locations Over Time Across Seasons for Male and Female Agricultural Laborers

Name of Location	Change in Wage Rates for Men			Change in Wage Rate for Women		
	Monsoon	Winter	Summer	Monsoon	Winter	Summer
1. Ahmedabad	44.3	43.2	54.2	35.1	40.5	59.2
2. Panchmahals	21.87	21.9	21.9	17.5	17.5	22.5
3. Bharuch	56	56	56	54.7	54.7	54.7
4. Mehsana	38.5	38.1	33.8	36.3	36.3	24.5
5. Narmada	34.7	33.7	30.1	24.8	28.2	36.9
6. Vadodara	20.5	22.5	22.5	18.6	28.5	13.5

Source: Based on Jagadeesan, S., Kumar, M.D., 2015. The Sardar Sarovar Project: Assessing Economic and Social Impacts. Sage Publications, New Delhi.

The strong inverse relationship between SWUI and the GHI, developed by IFPRI for 117 countries, provides a broader empirical support for some of the phenomena discussed above. In addition to these 117 countries for which data on GHI are available, we have included 18 developed countries. For these countries, we have considered zero values, assuming that these countries do not face problems of hunger. The estimated *R*-squared value for the regression between SWUI and GHI is 0.60. The coefficient is also significant at 1% level. It shows that with improved water situation, the incidence of infant mortality (below 5 years of age) and impoverishment reduces. In that case, improved water situation should improve the value of human development index, which captures three key spheres of human development such as health, education, and income status.

Thus, all the subindexes of HDI have strong potential to trigger growth in a country's economy, be it educational status, life expectancy, or income levels. When all these factors improve, they would have a synergetic effect on the economic growth. The growth, which occurs from human development, would also be "broad-based" and inclusive. Hence, the "causality" of water as a prime driver for economic growth can be tested if we are able to establish correlation between water situation and HDI.

We first examined how water situation and economic growth of nations are correlated. Regression between SWUI and ppp adjusted per capita GDP for the set of 145 countries shows that it explains level of economic development to an extent of 69%. The coefficient is significant at 1% level. The relation between SWUI and per capita GDP is a power function. Any improvement in water situation beyond a level of 50 in SWUI leads to exponential growth in per capita GDP.

This only means that for countries to be on the track of sustainable growth, the following steps are required: (1) investment in infrastructure and institutional mechanisms and policies to (a) improve access to water for all sectors of use and across the board, (b) enhance the overall level of use of water in different sectors, and (c) regulate the use of water, reduce pollution, and provide water for ecological services, and (2) investment in building human resources and technological capabilities in water sector to tackle new challenges in the sector. Major variations in economic conditions of countries having same levels of SWU can be explained by the economic policies of which the country pursues (Kumar et al., 2016).

2.4.1 How Water Security Impacts Growth?

Debate on whether money or policy reform is more crucial for progress in human development is highly polarized on the international scene (various authors as cited in HDR, 2006, p. 66). Two possible "causal chains" have been discussed by scholars, one that runs between economic growth and human development and the other that runs between human development and economic growth (Ranis, 2004). The causality in the first case occurs when resources from national income are allocated to activities that contribute to human development.

Ranis (2004) argued that low level of economic development would result in a vicious cycle of low levels of human development and high level of economic development would result in the virtuous cycle of high level of human development, whereas in the second case, as indicated by several evidences, better health and nutrition lead to better productivity of labor force (Cornia and Stewart, 1995). Education opens up new economic opportunities in agriculture (Rosenzweig, 1995), impacts on the nature and growth of exports (Wood, 1994), and results in greater income equality, which in itself results in economic growth (Ranis, 2004).

If the stage of economic development determines a country's water situation rather than the vice versa, the variation of human development index should be explained by variation in per capita GDP, rather than water situation in orders of magnitude. Data for 145 countries were used to closely examine this, and analysis was carried out using decomposed values of HDI (which comprises education index and life expectancy index and was run against per capita GDP). The subindex for per capita income was removed to avoid the problem of collinearity. The regression value was 0.75. What is more striking is the fact that 21 countries having per capita income below $2000 per annum have medium levels of decomposed index. Again, 50 countries having per capita GDP (ppp adjusted) less than $5000 per annum have medium levels of decomposed HDI. Significant improvements in HDI values (0.30–0.9) occur within the small range in per capita GDP. The remarkable improvement in HDI values with minor improvements in economic conditions and then "plateauing" means that improvement in HDI is determined more by factors other than economic growth.

The remarkable variation in HDI of countries belonging to the low-income group is because of the quality of governance in these countries, that is, whether good or poor. Many countries that show high HDI also have good governance systems and practices and institutional structures to ensure good literacy and public health (Kumar et al., 2016).[2] Incidentally, many countries with extremely low HDI have highly volatile political systems and ineffective governance and corruption. The investments in building and maintenance of water infrastructure are consequently very poor in these countries (Shah and Kumar, 2008) in spite of huge external aid.

2.4.2 Linkage Between Human Development and Water Security

Regression between SWUI and HDI had shown that variation in human development index can be better explained by *water situation* in a country,

2. For instance, Hungary in eastern Europe; some countries of Latin America, namely, Uruguay, Guatemala, Paraguay, Nicaragua, and Bolivia; and countries of erstwhile Soviet Union, namely, Turkmenistan, Kyrgyzstan, and Armenia have welfare-oriented policies. They made substantial investment in water, health, and educational infrastructure (Kumar, 2009).

expressed in terms of SWUI, than the ppp adjusted per capita GDP (Kumar, 2009; Kumar et al., 2016). Now, such a strong linear relationship between SWUI and HDI explains the exponential relationship between SWUI and per capita GDP, as the improvements in subindexes of HDI contribute to economic growth in its own way (i.e., per capita GDP $= f\{EI, HI\}$; here, EI is the education index, and HI is the health index) by creating a virtuous cycle as argued by Ranis (2004).

The exponential relationship between per capita GDP (ppp adjusted) and HDI (for the year 2007) of nations further reinforces this. Higher levels of HDI result in much higher levels of income. There are some countries that have medium level of development but show very high income. South Africa is one example. They fall off the trajectory that majority of the countries follow. It is important to remember here that this country suffers from high levels of income inequity.

In the absence of time series data on investments in water sector, water access, and water use, changes in water environment, and economic condition of nations, to analyze the impact of a country's water situation on its economic growth performance, the best alternative is to analyze the impact of natural water endowment, that is, rainfall, on economic growth in a situation where investments in infrastructure and institutions and water governance are poor (Kumar, 2010; Kumar et al., 2016). The most ideal place that illustrated this effect is sub-Saharan Africa.[3]

We further tested the causality of SWUI acting as a determinant of economic growth by running a two-stage least square method with HDI as the instrumental variable, SWI as the predictor variable, and per capita GDP (ppp adjusted) as the dependent variable. Here, it was assumed that SWUI will have a major bearing on HDI. The link has also been established through empirical analysis. The results showed an R-squared value of 0.50, at 1% level of significance. The beta coefficient was 1192, and the constant was −44,636.0. The analyses suggested that improving the water situation of a country is of paramount importance if we need to achieve sustained growth. While the natural water endowment in both qualitative and quantitative terms cannot be improved through ordinary measures, the *water situation* can be improved through economically efficient, just, and ecologically sound development and use of water in river basins (Kumar et al., 2016).

3. Barrios et al. (2004) showed a strong correlation between rainfall trend since the 1960s and GDP growth rates in the region during the same period, to argue that the low economic growth performance could be attributed to long-term decline in rainfall in the region. Such a dramatic outcome of rainfall failure can be explained partly by the failure of the governments to build adequate water infrastructure. The region has smallest proportion of its cultivated area (<3%) under irrigation (HDR, 2006). Due to this reason, reduction in rainfall leads to decline in agricultural production, food insecurity, malnutrition, loss of employment opportunities, and an overall drop in economic growth in rural areas (Kumar, 2009).

2.4.3 Impact of Water Security on Income Inequality and Poverty

The causality of water security as a driver of economic growth was tested by establishing the strong linkage between water security and human development (R^2=0.8). Scholars have earlier shown the negative impact of education on income inequality based on empirical evidence from selected countries around the world (Source: based on Alejandro et al., 1997). With the availability of data on income inequality for a large number of countries (for the year 2007) from the Human Development Report of 2009, the impact of improved water security on inclusiveness of economic growth could be empirically tested by analyzing the link between HDI and income inequity existing in the countries. For this, the data on income inequality, expressed in terms of Gini coefficient of income, were obtained for 125 countries from the Human Development Report of 2009 (HDR, 2009).

Analysis showed that HDI had a direct positive impact on income equality, when countries with high levels of human development (above 0.65 and up to 0.971) are considered for the analysis. There was an inverse exponential relationship between HDI and income inequality, with an R^2 value of 0.29. But such a relationship did not emerge when the analysis was performed after considering data from all the low-HDI and high-HDI countries. The reason is some of the low-HDI countries have low income inequality.

But countries with low HDI have very low per capita GDP, though the vice versa was not true (Kumar et al., 2016). The analysis meant that when the human development of a country crosses a particular threshold level, the national wealth gets better distributed. Since water security influences HDI positively, the type of relationship that will emerge between water security and income inequality is likely to be same as the relationship found between HDI and income inequality. This means improving water security of a nation would be a necessary and sufficient condition for achieving high levels of development and inclusive growth. The analysis using SWI for 79 countries (with SWUI values exceeding 45) and the values of income inequality for those countries showed an inverse correlation, meaning variation in SWUI explains variations in income inequality by 22%. The relationship was rather weak, when regression was run for all the 125 countries for which data on both SWI and income inequality are available. This means that the distributional effect of national income gets affected when water security and therefore human development cross a certain threshold.

Improved water security results in better economic growth conditions, and when water security reaches certain threshold, it results in better distribution of income. To find out whether this has a real impact on poverty reduction, analysis was carried out using the data of HPI-1 (available for 113 countries only) and SWUI. The analysis shows a strong inverse correlation between the two, with countries having higher values of SWUI showing lower poverty rates

(Kumar et al., 2016). While SWUI also suggests better access to water for irrigation, this result validates the earlier work on poverty reduction impact of irrigation (Hussain and Hanjra, 2003).

2.4.4 Importance of Storage Development for Economic Growth in Arid Tropical Countries

The crucial role of reservoir storage in improving the access to water is evident from the direct logarithmic relationship between storage development and water security (expressed in terms of SWUI), with an estimated *R*-squared value of 0.39. Major improvements in water security happened within the range of 0–1500 m^3 per capita per annum of reservoir storage and leveled off thereafter (Kumar et al., 2016).

However, the amount of storage needed to improve access to and use of water depends on the type of climatic conditions and amount of cultivable land. In temperate and cold climates, the demand of irrigation water would be considerably small as compared with tropical and hot climates. Hence, the storage requirements in such regions would be much lower and would be mainly limited to that for meeting domestic/municipal water needs and water for manufacturing. Therefore, it would be easier to explore the linkage between storage development for meeting various human needs and economic growth in the case of tropical and hot climates (Kumar, 2009; Kumar et al., 2016).

But the sheer scale of water infrastructure in rich countries is not widely recognized and appreciated (HDR, 2006, p. 155). Many developed countries of the world that experience tropical climates had high water storage in per capita terms. The United States, for instance, had created a per capita storage capacity of nearly 6000 m^3. In Australia, the 447 large dams alone create a total storage of 79,000 MCM per annum, providing per capita water storage of nearly 3808 m^3 per annum. Aquifers supply another 4000 MCM per annum. China, the fastest-growing economy in the world, has a per capita reservoir storage capacity of 2000 m^3 per annum through dams and an actual storage of nearly 360 m^3 per capita (Kumar, 2009).

When compared with these figures, India has a per capita storage of only 220 m^3 per annum. Though a much higher level of withdrawal of nearly 600 m^3 per capita per annum is maintained by the country, a large percentage of this (240 BCM per annum or nearly 217 m^3 per capita per annum) comes from groundwater draft, and there is increasing evidence to suggest that this won't be sustainable (Government of India, 2014).

Regression analysis of per capita water storage and the per capita GDP (ppp adjusted) for a group of 24 countries falling in the arid and semiarid tropics showed a strong relationship between level of storage development and country's economic prosperity. The *R*-squared value is 0.55, and the coefficient is significant at 1% level (Kumar et al., 2016). The strong relationship is explained in the following way. Storage reservoirs reduce risks and improve water security. In many regions, investments in hydraulic infrastructure had supported economic prosperity and social progress, though in some regions had caused environmental damage (Grey and Sadoff, 2007; HDR, 2006, based on various authors, p. 140).

The returns on investments in building water storages were quite visible in India. The impact of the yet-to-be-completed Sardar Sarovar Project (SSP) in reviving the agricultural production in Gujarat state of western India, after it experienced a major dip following two consecutive years of drought (1999 and 2000), has been remarkable (Kumar et al., 2010). The project, which brings water from the water-rich South Gujarat to the water-scarce regions of North Gujarat, Saurashtra, and Kachchh, reduces the regional imbalances in water availability and demand (Jagadeesan and Kumar, 2015).

Table 2.3 illustrates the economic impact of providing gravity irrigation farmers in the command area of SSP in four districts of Gujarat. The values of different variables that were used to estimate the increase in farm surplus per hectare of land of gravity irrigators and the final value of farm surplus for the four districts are also provided in Table 2.3 (last row).The farm surplus per hectare of gross cropped area ranged from INR 24903 in the case of Panchmahals to INR 48,348 in the case of Bharuch. It can be seen from Table 2.3 that the increase in average net income of farmers in different locations of Narmada command per hectare of land is even higher than the average net income prior to Narmada water introduction. Such a dramatic change has happened because of the slight change in cropping pattern, shift from rainfed to irrigated crops, introduction of high-yielding varieties, and intensive farming using better inputs such as fertilizers and pesticides. This alone was sufficient to increase the farm surplus manifold without any changes in the gross cropped area. But, over and above this, there has been increase in income from dairy production as well, indicated by increase in average net income from unit of livestock and increase in average number of livestock per unit of land.

The study showed that the income surplus owing to irrigated farming spilled over to household expenditure on food and nutrition, education, health, and asset building. While there was a marked increase in annual family expenditure of canal irrigators after the introduction of Narmada waters, the expenditure on education increased not only in absolute terms but also in terms of percentage (7.3%–14.8%), whereas the expenditure on health reduced marginally in percentage terms (from 10% to 9.4%) but increased significantly in absolute terms (Jagadeesan and Kumar, 2015).

The potential positive impact of water infrastructure on economic growth in regions that experience seasonal climates, rainfall variability, and floods and droughts can be better demonstrated by the economic losses that water-related natural disasters cause in the regions that lack them (Kumar, 2009). For instance, in the Indian state of Gujarat, the value of agricultural output dropped from Rs. 268.37 billion in 1998–99 to Rs. 189.0 billion in 2000–01 following the droughts in 1999 and 2000 (Kumar et al., 2010: Fig. 1, p. 4).

TABLE 2.3 Overall Farm Surplus of Canal Irrigators

	Impact Variables	Farm Surplus Estimates For			
		Bharuch	Narmada	Panchmahals	Vadodara
1	Current average GCA (ha/farmer)	5.16	0.85	6.02	5.64
2	Pre-Narmada Average GCA (ha/farmer)	5.21	0.81	5.09	5.20
3	Average increase in GCA (ha/farmer)	−0.05	0.04	0.93	0.44
4	Current average GIA from canals (ha/farmer) (as per respondent survey)	5.09	0.72	0.98	1.01
5	NIA from SSP in the region (ha)				
6	Average net income from crops (pre-Narmada) (INR/ha)	6885.4	5484.8	13,649	12,707.9
7	Average increase in net income crops (INR/ha)	47,043.2	27,661.5	21,029.2	42,463.2
8	Average livestock holding (present) (no./ha of GCA)	0.14	1.68	0.20	0.52
9	Increase in livestock holding (no./ha of GCA)	0.03	0.56	0.03	0.00
10	Average net income from livestock (pre-Narmada) (INR/livestock)	2380.0	7054.4	5018.4	5788.8
11	Increase in average net income, livestock (INR/livestock)	9288.3	5717.4	8076.6	8982.8
12	Farm surplus per hectare of GCA	**48,348.2**	**41,475.3**	**24,903.6**	**48,125.7**

Source: Based on Jagadeesan, S., Kumar, M.D., 2015. The Sardar Sarovar Project: Assessing Economic and Social Impacts. Sage Publications, New Delhi.

Nevertheless, the overall economic growth impact of water storage would depend on the nature of uses for which the resources are developed, the effectiveness of the institutions that are created to allocate the resource, and the nature of institutional and policy regimes that govern the use of the resource. Though the per capita water storage in Israel is quite low (nearly $150\,m^3$ per annum), the efficiency with which water is used in different sectors is extremely high. Nearly 90% of the country's irrigated area is under micro irrigation systems. A large portion of the water used in urban areas is recycled and put back to use for irrigation. Not only water is priced on volumetric basis, but also its allocation for irrigation is rationed (Kumar, 2009). The impact of water storage on the economy will also be decided by the structure of economy. In countries that have less amount of agriculture, like some of the highly industrialized countries (the UAE, Japan, Denmark, and Singapore), the per capita water requirement for all uses will be much less as compared with those that are agrarian. In these countries, every unit of water consumed would ideally generate much greater economic output than in countries that are agrarian as most of the water gets consumed in high-value sectors.

What is least appreciated, though, is the multiple-use benefits of large reservoir projects. In the case of SSP, the irrigation return flows from gravity-irrigated fields recharge the farm wells tapping the shallow aquifers in the command area. This reduced the energy use per cubic meter of water pumped and the average annual cost of well deepening and increased the average area irrigated by wells and the intensity of water application per unit irrigated area and livestock holding, all resulting in substantial increase in farm surplus of the well irrigators (Jagadeesan and Kumar, 2015). The values of different variables that are required to estimate the indirect impact of canal irrigation on well irrigation per hectare of gross well-irrigated area and the incremental farm surplus of the well irrigators are provided in Table 2.4.

It can be seen from Table 2.4 that while the average area under well irrigation had declined in four out of the six cases, the average net income from crop production substantially increased, with the average incremental income often exceeding the average net income prior to Narmada waters. The shift in cropping pattern towards high-valued, irrigated crops influenced the overall net income per unit area of the crops. On the other hand, dairying also influenced farm surplus. There has been a marginal increase in the number of livestock and a substantial increase in average net income from dairy production per unit of livestock kept by the well irrigators, post Narmada water introduction.

Gravity irrigation also improved the sustainability of drinking water wells in the areas that are served by the SSP canal network, through the following ways: (1) rise in water levels in the wells, which reduced the cost of pumping groundwater, thereby reducing the operational costs of the village panchayats running these schemes; (2) reduced incidence of well failures and therefore reduced investment for well deepening, etc.; (3) improved the yield of wells, resulting in increased supply of water, especially during summer months, thereby avoiding costly tanker

TABLE 2.4 Changes in Farm Surplus of Well Irrigators in Canal Command Areas

	Ahmedabad	Bharuch	Mehsana	Narmada	Panchmahals	Vadodara
Current average gross irrigated area (ha/farmer)	0.17	4.47	3.80	0.87	3.50	3.05
Pre-Narmada average gross irrigated area (ha/farmer)	0.17	5.67	4.28	1.01	3.32	2.95
Average increase in gross irrigated area (ha/farmer)	−0.01	−1.20	−0.48	−0.14	0.18	0.10
Average net income from well-irrigated crop (pre-Narmada)	11,320.3	10,923.2	24,151.4	5852.2	9451.8	17,966.4
Average increase in net income from well-irrigated crops (INR/ha)	9814.1	34,836.7	45,249.5	28,703	18,212.3	36,228
Average livestock holding (present) (no./ha of GWIA)	8.82	0.09	0.23	2.03	0.55	0.77
Average net income per livestock (INR/unit): pre-Narmada	12,071.3	3179.3	14,910	8484.9	5018.4	6478
Increase in average net income per livestock (INR/unit)	11,088.4	7664.5	8292.2	5157.4	8118	8809.8
Average increase in no. of livestock per hectare of GWIA	0.55	−0.04	0.17	1.29	0.31	0.59
Positive externality of canal Irrigation on well Irrigation per hectare of gross well-irrigated area	**113,587.1**	**32,466.9**	**46,640.7**	**49,176.3**	**24,719.0**	**47,422.6**

Source: Based on Jagadeesan, S., Kumar, M.D., 2015. The Sardar Sarovar Project: Assessing Economic and Social Impacts. Sage Publications, New Delhi.

water supply from outside; and (4) dilution of the minerals present in the groundwater, thereby improving the quality of the drinking water supplied by well-based schemes. The first three impacts were quantified in value terms and were estimated to be Rs. 1069.9 million (app. US $16 million) per annum, for a population of 5 million people hit by water scarcity in that region. The economic benefit due to dilution of minerals in groundwater was not quantified as the groundwater contamination due to minerals was not serious prior to introduction of canal water in that region (Jagadeesan and Kumar, 2015). This is one way of looking at the impacts. But if we consider a scenario of the government not responding to the water crisis in the region adequately with negative public health consequences, the actual positive impacts of this intervention social terms could be far greater.

One could as well argue that access to water could be better improved through local water resources development intervention including small water harvesting structures or through groundwater development. Small water harvesting systems had been suggested for water-scarce regions of India (Iyer, 2008) and the poor countries of sub-Saharan Africa (Rockström et al., 2002). But recent evidence suggest that they cannot make any significant dent in increasing water supplies in countries like India due to the unique hydrologic regimes and can also prove to be prohibitively expensive in many situations (Glendenning and Vervoort, 2011; Kumar et al., 2008). Also, to meet large concentrated demands in urban and industrial areas, several thousands of small water harvesting systems would be required. The type of engineering interventions and the economic viability of doing the same are open to question. Evidence also suggests that small reservoirs get silted up much faster than the large ones (Vora, 1994), a problem for which large dams are criticized world over (see McCully, 1996).

As regards groundwater development, intensive use of groundwater resources for agricultural production is proving to be catastrophic in many semiarid and arid regions of the world, including developed and developing countries (Food and Agriculture Organisation of the United Nations (FAO), 2011), though some of the developed countries have achieved some degree of success in controlling it through establishment of management regimes (Kumar, 2007) with physical and institutional interventions like in the western United States (Rosegrant and Sinclair, 1994; Rosegrant and Binswanger, 1994) or through institutional interventions such as formal water markets like in the Murray-Darling basin of Australia (National Water Commission, 2010) or through physical interventions alone like in Israel.

But in the basins facing problems of environmental water scarcity and degradation (see Smakhtin et al., 2004; FAO, 2011) due to large water projects, river flows are appropriated and transferred for various consumptive needs. Some of these basins are the Colorado river basin in the western United States; Yellow river basin in northern China; Aral sea basins, namely, Amu Darya and Syr Darya in Central Asia; Indus basin in Pakistan and India; basins of northern Spain; Nile basin in northern Africa; basins of Euphrates and Tigris; the Jordan river; Cauvery, Krishna, and Pennar basins of peninsular India; and river basins of western India including Sabarmati, Banas, and Narmada, located in Gujarat, Rajasthan, and Madhya Pradesh in India. The diverted water mostly meets irrigation water demands of these regions that have become agriculturally prosperous (Kumar et al., 2016). These basins not only meet the food requirements of the region but also export significant proportion of the food produced to other regions of the world, including some of the water-rich regions within the country's territory (Amarasinghe et al., 2004, for Indus basin and peninsular region in India; Kumar and Singh, 2005, for many water-scarce countries of the world; and Jia et al., 2016 for China).

Strikingly, wherever aquifers are available for exploitation, these regions had experienced problems of groundwater overdraft, though some developed countries had developed the science to deal with it. The most glaring examples are aquifers in the western United States; aquifers in the countries of the Middle East including Yemen, Iran, and Iraq (Voss et al., 2013); aquifers in Mexico; aquifers in north China plains (Feng et al., 2013); alluvial aquifers of Indus basin areas in India; hard-rock aquifers of Peninsular India; and aquifers in western and central India (GOI, 2014).

Without these large surface-water projects, agricultural growth might have caused far more serious negative impact on groundwater resources in these regions. In fact, it is this surface-water availability that to a great extent helps reduce dependence of farmers and cities on groundwater (Kumar, 2009). For instance, imported water from Indus basin through canal in Indian and Pakistan Punjab sustains intensive groundwater use in the regions, through continuously providing replenishment through return flows from surface irrigation (Qureshi et al., 2009). In India, water imported from a large reservoir named Sardar Sarovar in Narmada basin had started supplying water to rejuvenate the rivers in environmentally stressed basins of North Gujarat (Jagadeesan and Kumar, 2015).

2.5 IMPACT OF STORAGE DEVELOPMENT ON MALNUTRITION AND CHILD MORTALITY

Storage development is found to have a direct impact on malnutrition and infant mortality, when we considered zero values of GHI for developed countries, namely, the United States, Australia, and Spain for which data on GHI are not available (Kumar, 2009). Regression showed an R-squared value of 0.59, with an inverse, logarithmic relationship. Greater water storage reduced the chances of human hunger. This inverse relationship can be explained this way. For the sample countries, the ability to cultivate the available arable land intensively would increase with the amount of water storage facilities available. Increased availability of irrigation water reduces the risk of crop failure, enhances the ability of farmers to produce more crops to improve their own domestic consumption of food, and takes care of the cash needs (Kumar et al., 2016). The critical role of large reservoirs in drought-prone countries like India cannot be overstated as they become the only source of water for meeting the basic survival needs of drought-hit areas and other local sources dry up no matter how far they are located from such areas (Chapter 4). Also, increased irrigated production improves food and nutritional security of the population at large by lowering cereal prices in the region in question as the gap between cereal demand and supplies is reduced (Hussain and Hanjra, 2003).

Irrigation expansion through large storages had contributed nearly 47 million tons of additional cereals to India's bread basket (Perry, 2001, p. 104). The most illustrious example of the impact of irrigation from the recent times is of the Sardar Sarovar Project on food production and agricultural growth in Gujarat (Jagadeesan and Kumar, 2015; Kumar et al., 2010). Availability of surface water through canals had motivated farmers in south and central Gujarat to take up paddy and wheat and achieve bumper food-grain production in the recent years, also resulting in substantial increase in income of farmers (Jagadeesan and Kumar, 2015). Shah and Kumar (2008) made a rough estimate of the positive externality it created in terms of lowering food prices for the consumers in India as US $20 per ton of cereals. One could also argue that rich countries could afford to import food. But what is important is that water had played a big role for these countries to achieve a certain level of economic growth and prosperity, by virtue of which they can now afford to import food instead of resorting to domestic production (Kumar et al., 2016). Equally important is the fact that in countries with very large population base like India, China, the United States, and Indonesia, large-scale cereal import cannot be a viable option, no matter how their economic conditions are because of the implication that it will have on international cereal trade.

2.6 THE FINDINGS

The results presented in this chapter based on earlier works led by the authors (Kumar, 2009; Kumar et al., 2016) and recent microlevel evidence available from vast amount of primary research (Jagadeesan and Kumar, 2015; Kumar et al., 2010) have shown that improving the water situation can trigger economic growth of a nation. Such growth would be inclusive as this occurs through the human development route. This strong linkage can be partly explained by the reduction in malnutrition and infant mortality, with improvement in water situation. Further, nations could achieve good indicators of development even at low levels of economic growth, through investment in water infrastructure and welfare-oriented policies.

Analysis has also shown that improving water security also promotes better distribution of national income through the human development route. Improved water security also reduces rural poverty through various routes. Public investments in surface irrigation, for instance, increase the income of not only farmers who get canal water but also well irrigators in the command and rural wage laborers (Jagadeesan and Kumar, 2015). Countries that fall in tropical semiarid and arid climate can improve their economic conditions by enhancing their reservoir storage. Greater storage provides increased water security, which reduces the risks associated with droughts and floods. Such natural calamities, which cause huge economic losses, are characteristic of these countries. Nevertheless, the impact of storage could depend on the nature of uses for which the resources are developed, the effectiveness of the institutions that allocate the resources, and the nature of institutional and policy regimes that govern their use (Kumar et al., 2016). For countries having large agriculture-based rural population like India, Pakistan, Afghanistan, Iran, and most sub-Saharan African countries, it is important that water allocation decisions give highest priority to irrigation, after enough water is kept for basic needs.

The indirect benefits of allocating large volumes of water for surface irrigation are often very large, resulting from increased farm income of well irrigators, improved yield of drinking water supply sources in the command area, and enhanced wage rates of rural laborers, in addition to reduced subsidy burden on the state power utilities, as demonstrated by the study of Sardar Sarovar Project (Jagadeesan and Kumar, 2015).

Poor countries should start investing in building water infrastructure, create institutions, and introduce policy reforms in water sector that could lead to sustainable water use. Hot and arid tropical countries should invest in large water resource systems to raise the per capita storage (Kumar, 2009). India, which belongs to this category, has per capita reservoir storage of only 220 m^3. With population rise and silting up of reservoirs, this capacity will reduce over a period of time. Countries that are characterized by huge regional variations in water endowment such as India will have to have the infrastructure to transfer the stored water to regions of deficit. The Sardar Sarovar Project amply demonstrates the importance of such water transfers by transporting several hundred million cubic meter of water from a relatively water-rich South Gujarat to absolutely water-scarce North Gujarat annually (Jagadeesan and Kumar, 2015). This will help them fight hunger and poverty, malnutrition, and infant mortality and reduce the incidence of water-related disasters. We must remember that when droughts hit large regions, it is always the water from big manmade reservoirs, transported to villages and towns, that protects millions of people from complete desiccation.

2.7 THE LAST WORD

There have been determined attempts from some quarters in the recent past to misinterpret the concept of water sector reforms in order to move away from infrastructure development and serious institutional change processes and policy measures and merely parade vacuous slogans such as participatory management, "community management," "making water everybody's business," etc. and label them as "paradigm shift" in water management, without suggesting any concrete institutional reforms that are critical to achieving the goals of equity in access to water and efficiency and sustainability in water use (see, e.g., Shah, 2013, 2016). Such attempts are not driven by any serious considerations about the nature of water problems that we are confronted with but a pathological hatred toward good water science, professional engineers, and hydrologists (Kumar and Pandit, 2016).

There is an imminent danger that politicians and policy makers get carried away by such catchy slogans and even use them to masquerade their lack of ability to fix the problems in the underperforming sector. India has been a witness to this for several years now. The social welfare schemes such as the Mahatma Gandhi National Rural Employment Guarantee Scheme (MGNREGS) became epitomes of the lack of transparency and accountability and rampant corruption, resulting from poor planning, the lack of supervision, and inadequate work monitoring (World Bank, 2011). Recent analysis by a comptroller and auditor general (CAG) of India based on the report of CAG and analysis presented elsewhere make a case that though the program has been there for 11 years now, with a total expenditure of Rs. 300,000 crore (US $46 billion), it has achieved too little in terms of reducing rural poverty, one major aim of this social welfare scheme, with the lack of correction between the number of poor people in different states and the number of people getting employment benefit under the schemes remaining as a major concern (Bhattacharjee, 2017). As correctly argued by Bhattacharjee (2016), poverty can be overcome only through education and health, and acquisition of employable skills and programs like the MGNREGS does not facilitate any of these and merely provides some aid, which is sufficient for subsistence.

The fact is that nearly 60% of the work under MGNREGS goes for water conservation and drought-proofing-related works. Because of the poor conceptualization and flawed design, the available evidence does not suggest any impact of the program in improving the water security of the people in rural areas through enhanced irrigation and drinking water supplies or groundwater recharge at a regional scale (Bassi and Kumar, 2010). The country has paid a huge price for this, with hundreds of thousands of wells going dry every year, as farmers invest their scarce resources for drilling deeper to mine the overexploited aquifers.

Strangely, even when politicians make attempts to modify such schemes, they have to face fierce resistance from the members of the civil society with deep vested interests. The recent example is of several economists and social activists making an appeal to the prime minister of India to continue the MGNREGS in the format in which it was when the new government took over in May 2014, arguing "Despite numerous hurdles, the NREGA has achieved significant results. At a relatively small cost (currently 0.3% of India's GDP), about 50 million households are getting some employment at NREGA worksites every year. A majority of NREGA workers are women, and close to half are Dalits or Adivasis. A large body of research shows that the NREGA has wide-ranging social benefits, including the creation of productive assets" (The Hindu, 2014). The political dispensation will have to manage such pressures from the civil society to carry on with their decisions.

REFERENCES

Alejandro, R., Ranis, G., Stewart, F., 1997. Economic Growth and Human Development. Centre Discussion Paper # 787. Economic Growth Centre, Yale University, New York.

Amarasinghe, U., Sharma, B.R., Noel, A., Christopher, S., Vladimir, S., de Fraiture, C., Sinha, A.K., Shukla, A.K., 2004. Spatial Variation in Water Supply and Demand across River Basins of India. IWMI Research Report 83. Colombo, Sri Lanka.

Barrios, S., Bertinelli, L., Strobl, E., 2004. Rainfall and Africa's Growth Tragedy, 2004 paper.

Bassi, N., Kumar, M.D., 2010. NREGA and Rural Water Management in India: Improving the Welfare Effects. Occasional Paper # 3. Institute for Resource Analysis and Policy, Hyderabad, India.

Bhattacharjee, G., 2016. Treat the disease, not the symptoms. In: Singh, V.V. (Ed.), Indian Economy: A Roadmap Towards Development. Flying Pen, Jaipur.

Bhattacharjee, G., 2017. MGNREGA as distribution of dole. Econ. Polit. Wkly. 52 (25 and 26), 29–33.

Braga, B., Rocha, O., Tundisi, J., 1998. Dams and the environment: the Brazilian experience. Int. J. Water Resour. Dev. 14 (2), 127–140.

Briscoe, J., 2005. In: India's water economy: bracing up for a turbulent future. Key Note Address at the 4th Annual Partners' Meet of IWMI-Tata Water Policy Research Program, Institute of Rural Management Anand, 26–29 February, 2005.

Cornia, G.A., Stewart, F., 1995. Two errors of targeting. In: Stewart, F. (Ed.), Adjustment and Poverty: Options and Choices. Routledge, London, pp. 82–107.

D'Souza, D., 2002. Narmada Dammed: An Enquiry into the Politics of Development. Penguin Books India, New Delhi.

Dharmadhikary, S., 2005. Unraveling Bhakra: Assessing the Temple of Resurgent India. Manthan Adhyayan Kendra, New Delhi.

Feng, W., Zhong, M., Jean-Michel Lemoin, J.-M., Biancale, R., Hsu, H.-T., Xia, J., 2013. Evaluation of groundwater depletion in North China using the gravity recovery and climate experiment (GRACE) data and ground-based measurements. Water Resour. Res. 49 (4), 2110–2118.

Food and Agriculture Organisation of the United Nations (FAO), 2011. The State of the World's Land and Water Resources for Food and Agriculture (SOLAW)—Managing Systems at Risk. Food and Agriculture Organization of the United Nations/Earthscan, Rome/London.

Glendenning, C.J., Vervoort, R.W., 2011. Hydrological impacts of rainwater harvesting (RWH) in the case study catchment: the Arvari River, Rajasthan, India. Part 2. Catchment-scale impacts. Agric. Water Manag. 98, 715–730.

Government of India, 2014. Dynamic Ground Water Resources of India (as on March 31, 2011). Central Ground Water Board, Ministry of Water Resources, River Development and Ganga Rejuvenation Government of India.

Grey, D., Sadoff, C., 2007. Sink or swim: water security for growth and development. Water Policy 9 (6), 541–570.

Human Development Report (HDR), 2006. Human Development Report-2006. United Nations, New York.

Human Development Report (HDR), 2009. Human Development Report-2009. United Nations, New York.

Hussain, I., Hanjra, M., 2003. Does irrigation water matter for rural poverty alleviation? Evidence from South and South East Asia. Water Policy 5 (5), 429–442.

Hutton, G., Bartram, J., 2008. Global costs of attaining the millennium development goal for water supply and sanitation. Bull. World Health Org. 86, 13–19.

Iyer, R.R., 2008. Water: a critique of three basic concepts. Econ. Polit. Wkly. 43 (1), 15–17.

Iyer, R.R., 2011. National water policy: an alternative draft for consideration. Econ. Polit. Wkly. XLVI (26–27).

Jagadeesan, S., Kumar, M.D., 2015. The Sardar Sarovar Project: Assessing Economic and Social Impacts. Sage Publications, New Delhi.
Jha, N., 2010. Access of the Poor to Water and Sanitation in India: Salient Concepts, Issues and Cases. Working Paper 62. International Policy Centre for Inclusive Growth.
Jia, S., Qiubo, L., Wenhua, L., 2016. The fallacious strategy of virtual water trade. Int. J. Water Resour. Dev. https://doi.org/10.1080/07900627.2016.1180591.
van Koppen, B., Smits, B.S., Moriarty, R., Devries, P.F., Mikhail, M., Boelee, E., 2009. Climbing the Water Ladder: Multiple Use Water Services for Poverty Reduction. IRC International Water and Sanitation Centre and International Water Management Institute, The Hague, The Netherlands. TP Series; No. 52, 213.
Kumar, M.D., 2007. Groundwater Management in India: Physical, Institutional and Policy Alternatives. Sage Publications, New Delhi.
Kumar, M.D., 2009. Water Management in India: What Works, What Doesn't. Gyan Books, New Delhi.
Kumar, M.D., 2010. Managing Water in River Basins: Hydrology, Economics, and Institutions. Oxford University Press, New Delhi.
Kumar, M.D., Narayanamoorthy, A., Singh, O.P., Sivamohan, M.V.K., Sharma, M., Bassi, N., 2010. Gujarat's Agricultural Growth Story: Exploding Some Myths. Occasional Paper # 2. Institute for Resource Analysis and Policy, Hyderabad.
Kumar, M.D., Pandit, C., 2016. India's water management debate: is the civil society making it everlasting? Int. J. Water Resour. Dev. https://doi.org/10.1080/07900627.2016.1204536.
Kumar, M.D., Saleth, R.M., Foster, J.D., Vedantam, N., Sivamohan, M.V.K., 2016. Water, human development, inclusive growth, and poverty alleviation: international perspectives. In: Kumar, M.D., James, A.J., Kabir, Y. (Eds.), Rural Water Systems for Multiple Uses and Livelihood Security. Elsevier Science.
Kumar, M.D., Singh, O.P., 2005. Virtual water in global food and water policy making: is there a need for rethinking? Water Resour. Manag. 19, 759–789.
Kumar, M.D., Patel, A., Ravindranath, R., Singh, O.P., 2008. Chasing a mirage: water harvesting and artificial recharge in naturally water-scarce regions. Econ. Political Wkly. 43 (35), 61–71.
Laurence, M., Meigh, J., Sullivan, C., 2003. Water Poverty of Nations: International Comparisons. Kellee University, Wallingford, UK.
McCully, P., 1996. Climate Change Dooms Dams, Silenced Rivers: The Ecology and Politics of Large Dam. Zed Books, London.
Narayanamoorthy, A., Deshpande, R.S., 2003. Irrigation development and agricultural wages: analysis across states. Econ. Polit. Wkly. 38 (35), 3716–3722.
National Water Commission, 2010. The Impacts of Water Trading in the Southern Murray–Darling Basin: An Economic, Social and Environmental Assessment. NWC, Canberra.
Perry, C.J., 2001. World commission on dams: implications for food and irrigation. Irrig. Drain. 50, 101–107.
Qureshi, A.S., McCornick, P.G., Sarwar, A., Sharma, B.R., 2009. Challenges and prospects of sustainable groundwater management in the Indus Basin, Pakistan. Water Resour. Manag. 24 (8), 1551–1569.
Ranis, G., 2004. Human Development and Economic Growth. Yale University Economic Growth Center Discussion Paper No. 887.
Rockström, J., Barron, J., Fox, P., 2002. Rainwater management for improving productivity among small holder farmers in drought prone environments. Phys. Chem. Earth 27 (2002), 949–959.
Rosegrant, M., Binswanger, H.P., 1994. Markets in tradable water rights: potential for efficiency gains in developing country water resource allocation. World Dev. 22 (11), 1613–1625.

Rosegrant, M.W., Sinclair, R.G., 1994. Reforming Water Allocation Policy through Markets in Tradable Water Rights: Lessons from Chile, Mexico, and California. EPTD Discussion paper 2. International Food Policy Research Institute, Washington.

Rosenzweig, M.R., 1995. Why are there returns in schooling? Am. Econ. Rev. 85, 2.

Shah, M., 2013. Water: towards a paradigm shift in the twelfth plan. Econ. Polit. Wkly. XLVIII (3), 40–52.

Shah, M., 2016. A 21st Century Institutional Architecture for Water Reforms in India. Final Report Submitted to the Ministry of Water Resources. River Development & Ganga Rejuvenation, Government of India, New Delhi.

Shah, Z., Kumar, M.D., 2008. In the midst of the large dam controversy: objectives, criteria for assessing large water storages in developing world. Water Resour. Manag. 22, 1799–1824.

Smakhtin, V., Carmen, R., Doll, P., 2004. Taking into Account Environmental Water Requirements in Global-Scale Water Resources Assessments, Comprehensive Assessment Report 2. Comprehensive Assessment Secretariat, Colombo.

Sullivan, C., 2002. Calculating water poverty index. World Dev. 30 (7), 1195–1211.

The Hindu, 2014. Continue With NREGA, Economists Urge Modi. The Hindu.

Vora, B.B., 1994. Major and Medium Dams: Myth and the Reality. Hindu Survey of the Environment. The Hindu, Chennai.

Voss, K.A., Flamigietti, J.S., Lo, M., Linage, C., Rodell, M., Swenson, S.C., 2013. Groundwater depletion in the Middle East from GRACE with implications for transboundary water management in the Tigris-Euphrates-Western Iran region. Water Resour. Res. 49 (2), 904–914.

Wood, A., 1994. North-South Trade, Employment and Inequality: Changing Fortunes in a Skill-Driven World. IDS Development Studies Series. Oxford University Press, Oxford.

World Bank, 2004. Towards a Water Secure Kenya: Water Resources Sector Memorandum. World Bank, Washington, DC.

World Bank, 2006a. Managing Water Resources to Maximize Sustainable Growth: A Country Water Resources Assistance Strategy for Ethiopia. World Bank, Washington, DC.

World Bank, 2011. Social Protection for a Changing India Volume II, Chapter 4, 84. World Bank, Washington, DC.

World Water Council, 2000. 2nd World Water Forum. World Water Council, The Hague, The Netherlands.

Chapter 3

Why India Needs Large Water Resource Projects Involving Interbasin Water Transfers

3.1. INDIA'S WATER CRISIS AND THE DEBATE ON INTERLINKING OF RIVERS

The National River Linking Project of India, unlike what the name suggests, is not a project, but a concept plan for building about 30 river links, with canals and reservoirs for transferring water across several of the Himalayan and Peninsular rivers in India for irrigation, flood control, drought proofing, and hydropower benefits, and the feasibility of each one is being studied separately. These are essentially interbasin water transfer projects. These ideas have evolved over long time periods—from the proposal for a garland canal around the country made by Capt Dastur to the current proposal that the National Water Development Board (NWDB) is working on to develop prefeasibility reports and detailed project reports. That said, many river links already exist in India, some of them for several decades. The most remarkable ones are Periyar-Vaigai link and Beas-Sutlej link. Two important interbasin water transfer projects are Indira Gandhi Nehar Project and Sardar Sarovar Project.[1]

It is generally understood that many river basins in the north and east and one in the south, including Ganges (north), Brahmaputra, Mahanadi (east), and Godavari (south), are water-rich and also highly flood-prone, whereas many river basins in the south, namely, Cauvery, Krishna, and Pennar, are water-scarce. With the increasing water scarcity problems in India, there is a growing chorus from the engineering fraternity, especially those working in the water and electricity sectors in the central and provincial governments, and also from the larger civil society and farmers' organizations that projects should be built to transfer water from the water-rich basins to the water-scarce basins, as permanent solution for droughts and floods. However, since then, there has been fierce opposition to the "project" from many civil society organizations, social activists, and ecologists, well known for their strong views against large dams

1. Indira Gandhi Nehar Project transfers water from the Sutlej River in Punjab to the Thar Desert of Rajasthan. The SSP transfers water from the Narmada River in the relatively water-rich South Gujarat to the North Gujarat, through the 458 km-long Narmada Main Canal.

Water Policy Science and Politics. https://doi.org/10.1016/B978-0-12-814903-4.00003-8

(see Bandyopadhyay and Parveen, 2004; Iyer, 2011). They believe that interlinking of rivers would spell ecological disaster for the country, in spite of the fact that no concrete plans exist today on interbasin water transfers, vis-à-vis the number of links, their alignment, the amount of water to be transferred, the number and size of reservoirs to be built, and their location. The technical, economic, social, financial, and legal viability of such projects is also challenged.

In this chapter, we argue how the unique pattern of growth in India's water demands makes transfer of water from water-rich basins to water-scarce basins essential to mitigate future water scarcity and food insecurity problems, how other water management solutions such as water harvesting and water demand management fail to become alternatives to interbasin transfers, and how India should minimize the negative social and ecological consequences of large-scale water transfer projects.

3.2. WHY INTERLINKING OF RIVERS IS IMPORTANT FOR INDIA

Many proponents of river linking, including the National Water Development Agency,[2] used the argument of "water equalizer," to justify the large-scale transfer of water from water-surplus basins to water-scarce basins. They believe that such projects are in the national interest as they would provide sufficient water to water-scarce regions that are characterized by crops dependent on monsoon rains, which in turn will help the country to enhance its agricultural production for national food security, while insulating the water-surplus regions of devastating floods (www.nwda.gov.in). Arguments such as "protecting the livelihoods of farmers in these areas" were just afterthoughts. This was a colossal mistake. It was fraught with the counterargument by the antidam lobbyists that water-surplus regions could produce more food, with the intensive use of groundwater supported by enabling policies (see Shah, 2013). The historical growth in well irrigation was used to substantiate this argument. Some of the activists also used the virtual water trade argument by saying that the water-surplus regions could produce more food and export to the water-scarce regions, instead of engaging in physical transfer of water (see Iyer, 2008).

2. The National Water Development Agency (NWDA) was setup in July 1982 to carry out scientific studies for optimum utilization of water resources of the Peninsular river system for preparation of feasibility reports and thus to give concrete shape to Peninsular river development component of National Perspective Plan (NPP) prepared by the Central Water Commission and the then Ministry of Irrigation (now MoWR, RD, and GR). In the year 1994, NWDA was also entrusted with the task of Himalayan Component of the National Perspective Plan. In 2006, it was decided that NWDA will explore the feasibility of intrastate links and to take up the work for the preparation of Detailed Project Report (DPR) of river-link proposals under NPP. The functions of NWDA were further modified vide MoWR resolution dated 19th May 2011, to undertake the work of preparation of DPRs of intrastate links. Recently on 7th October 2016, the functions of NWDA were further modified to undertake the implementation of all ILR projects (www.nwda.gov.in).

We would like to argue that there are at least five important reasons why India would require major projects that involve large-scale water transfers across basins. They are as follows: (1) the growing demand-supply gap in water and increasing urban water scarcity, (2) the land-water-food security nexus, (3) the livelihood security of people living in rural areas, (4) the lack of alternatives to water transfers, and (5) the growing social tensions over acute scarcity of water.

3.3. THE DRIVERS OF FUTURE WATER DEMAND

A reality that we cannot wish away is that India's population would continue to grow in the coming few decades, with greater rate of growth in urban areas. This means the demand for domestic water supply in rural and urban areas would increase proportionately. But, with increase in per capita income, the water demand for domestic uses is likely to increase, with greater demand for water for washing transport vehicles, gardening, etc. The impact of a unit increase in urban population on water demand will be much greater than that of rural population, in lieu of the fact that the per capita water demand of people in urban areas is much higher than that of their rural counterparts, in lieu of the need for running sewer systems.

A major contributor to growth in India's future water demand, however, is likely to come from growth in food demand, for cereals, dairy products, fruits and vegetables, and meat products. The food grain demand in 2050 is estimated to be in the range of 420–494 million tons. The calorie intake, which is one of the lowest in the developing world, is going to increase substantially in the coming years, with rising per capita income. The demand for dairy products has been growing exponentially in India—from a lowest of 42 kg/capita/year in 1979–81 to 71 kg in 2005–07 (Alexandratos and Bruinsma, 2012) and is projected to reach 133 kg/capita/year by the year 2022 (Punjabi, 2010)—and this has major implications for water use, as dry regions are producing maximum milk.

Another major contributor to the growth in future water demand is from the cities, to supply water to around 650–900 M people. This alone would require around 42–58 BCM of water, at 150 lpcd, and a loss of around 20%. The National Commission on Integrated Water Resources Development had estimated the urban population in India to be 603 million people by 2025 and 971 by 2050 (as per high rate of urbanization) and 476 million people in 2025 and 646 million people by 2050 (as per low rate of urbanization) (Table 3.1). The urban water demand was estimated to be 48.42 BCM by the year 2025 under a high growth scenario and 38.2 BCM under a low growth scenario. For the year 2050, the demand was estimated to be 78 BCM under a high growth scenario and 51.87 BCM under a low growth scenario (source: based on Government of India, 1999).

The water demand for industrial uses will also increase, including that for thermal power generation. Kumar (2010) had projected the industrial water demand in India to be 10,015 MCM (i.e., 10.015 BCM) by 2025. The estimation did not consider thermal power, which is now emerging as a major

TABLE 3.1 India's Projected Population for 2025 and 2050

	2025		2050	
Population	**High**	**Low**	**High**	**Low**
Rural	730	810	610	700
Urban	603	476	971	646
Total	1333	1286	1581	1346

water-consuming subsector within the industrial sector. Every megawatt hour of thermal power generation takes nearly 3.5 m^3 of water for plants that use evaporation cooking, and the installed capacity of such plants had increased substantially in the recent past.[3] The National Commission estimated the industrial water demand in 2050 to be 70 BCM for the estimated installed capacity for that year, based on the assumption that coal-based thermal power plants would consume 4.48 m^3/h/MW of plant capacity. If we assume the water requirement to drop to 3.5 m^3/h/MW, the requirement would drop to 54.7 BCM.

At the aggregate level, we have total renewable water resources of 2295 BCM, for the country as a whole, including surface runoff (1890 BCM) and renewable groundwater resources (405 BCM). This is far in excess of the projected demand for water even for the year 2050 (1180 BCM) to meet all the needs that include irrigation, domestic water supply in rural areas, municipal water supply, water for livestock, and water for industrial use, including thermal power generation and water for environmental management (GOI, 1999). These figures were conveniently used by some to argue for rainwater harvesting systems, as an alternative to large reservoirs and canals (see Puttaswamy, 2015).

3.4. THE FOOD SECURITY CHALLENGE OF LAND-WATER-FOOD SECURITY NEXUS IN INDIA

The river basins that have a large amount of annual streamflows and groundwater and that mainly contribute to our positive water balance at the aggregate level are the Ganges, Godavari, Brahmaputra, and Meghna basins. These basins (barring Godavari) are not known for water scarcity, and the problem they face is of floods, and hence, rainwater harvesting is of not much consequence for these basins.[4] The amount of water in these basins that can be

3. Within the coal-rich state of Chhattisgarh alone, the total installed capacity of thermal power stations today is around 70,000 MW.

4. While Godavari has plenty of unutilized water resources and there is also unmet demand for water in the basin, especially in Maharashtra and Telangana, the cost of exploiting this water is prohibitively high.

utilized using conventional water resource development technologies is very low (690+342=1032 BCM), due to unfavorable topography, limited arable land that needs irrigation water, and peak floods that occur for very short duration. Large parts of Ganges and Brahmaputra have completely flat terrain, and building of dams and diversion systems is not feasible.

Due to land scarcity, high population density, and ecological constraints, we cannot produce surplus food to meet the future demands in the water-rich basins of Ganges and Brahmaputra. In spite of having high cropping intensity, the per capita cropped area is lowest in the region, as is evident from the data for Bihar, West Bengal, Assam, and Uttar Pradesh available in Fig. 3.1 (Kumar, 2003). These regions are already food insecure (Amarasinghe et al., 2004), and their population is growing faster than in water-scarce regions. Unfortunately, the issue of land crisis in water-rich regions is hardly mentioned in the discourse on food security and food self-sufficiency.

The regions that are currently contributing to India's food security and that have the potential in terms of access to arable land and production technologies are South Indian peninsula, Central Indian belt, northwestern India, and western India (Rada, 2013). They are all in the water-scarce basins. To name a few, there are Krishna, Cauvery, Pennar, Sabarmati, Narmada, Indus, and some western Indian river basins. Their water demand far exceeds the renewable supplies in these regions. They have already run out of water, and there is no extra water in these basins that can be harnessed. The intensity of groundwater use is very high in these regions, and their groundwater resources are also overexploited and mined (Kumar et al., 2012).

Hence, there is a huge mismatch between water availability and water demand with respect to space and time. The regions that have a huge amount of water cannot use it for crop production due to very limited arable land, and regions, which are parched, do not have any extra water to increase the production for future. We need to remember the fact that only water can be physically transferred and not "land." This means water has to be physically transferred

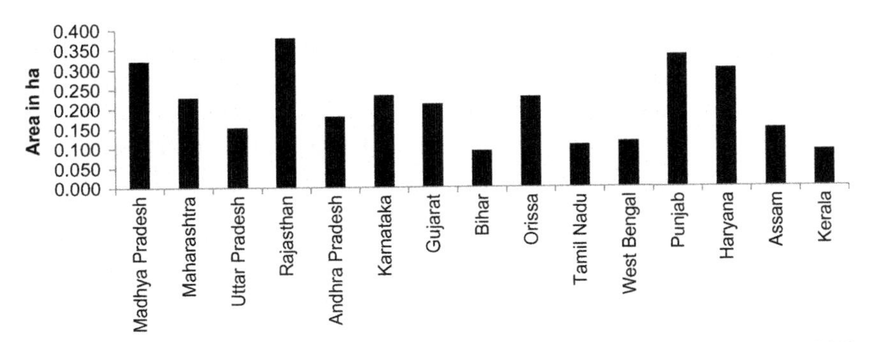

FIG. 3.1 Per capita cropping in different states. *(Adapted from Figure 2 in Kumar, M.D., 2003. Food Security and Sustainable agriculture in India: the Water Management Challenge. Working Paper # 60. International Water Management Institute, Colombo, Sri Lanka).*

from land-scarce areas to land-rich areas. The spatial mismatch between water availability and water demand is evident from the two tables (Tables 3.2 and 3.3) presented below. The basins, namely, Narmada, Sabarmati, Cauvery, Pennar, and Krishna, are water-scarce, and Ganga and Brahmaputra are water-rich basins.

Even today, around 55% of the people in rural area depend on agriculture, and sustaining these livelihoods would require water. If agricultural production in the country is to be sustained to meet the growing demands of the population and livelihoods of poor people in the villages to be protected, the water-scarce basins would require a large amount of exogenous water. Further, most of the fast-growing cities of India today are unfortunately located in the water-scarce regions—Bangalore, Chennai, Hyderabad, Ahmedabad, New Delhi, and Jaipur. They cannot sustain the growth using the water that is currently available. They are already heavily dependent on water from rural areas. Further reallocation of water from reservoirs primarily meant for irrigating rural areas would mean increasing social tension.

3.5. ALTERNATIVES TO RIVER LINKING

As regards alternatives to NRLP, rooftop water harvesting and groundwater use in the urban areas are widely talked about by the civil society organizations (see Srinivasan et al., 2010) against any move by governments to engage in large-scale water transfer projects, for meeting urban water demands. As regards roof water harvesting, it can add too little to augment the overall water supply, but would be prohibitively expensive also (Kumar, 2009).

Interestingly, many of the cities and towns that had earlier used groundwater to meet their municipal demands have slowly increased the proportion of imported (surface) water as their population grew and demand increased. The fact is that the local groundwater is simply far less than sufficient to meet their high water demands, more so in the hard-rock regions of peninsular and central India (Kumar, 2014; Mukherjee et al., 2010). As it is evident today, many metropolitan cities of southern and western India with 1 million plus population that are rapidly growing, including Hyderabad, Chennai, Bangalore, Mysore, Ahmedabad, Delhi, Jodhpur, Mangalore, and Visakhapatnam, are heavily dependent on water imported from long distance from outside basins. Hyderabad gets water supply from sources located several hundred kilometers away from the Krishna and Godavari basins.

Small-scale rainwater harvesting systems have very limited potential in water-scarce regions of India, and this potential is already exploited. Not only they do not produce hydrologic (Glendenning and Vervoort, 2011; Kumar et al., 2006; Kumar et al., 2008a) and economic benefits at the basin scale, but also they seem to cause ecological destruction, with drying up of wetlands (tanks and lakes) and man-made reservoirs (Kumar et al., 2008a, 2011).

TABLE 3.2 Average Evapotranspiration and Effective Annual Water Resources in Selected River Basins in Water-Scarce Regions

Name of the Basin	Mean Annual Rainfall (mm)		Average Annual Water Resources (mm)	Effective Annual Water Resources (mm)	Evapotranspiration (mm)	
	Upper Basin Area	Lower Basin Area			Upper Basin Area	Lower Basin Area
Narmada	1352.00	792.00	444.70	937.60	1639.00	2127.00
Sabarmati	643.00	821.00	222.84	309.61	1263.00	1788.80
Cauvery	3283.00	1337.00	316.15	682.80	1586.90	1852.90
Pennar	900.00	567.00	193.90	467.80	1783.00	1888.00
Krishna	2100.00	1029.00	249.16	489.15	1637.00	1785.90

Source: Kumar, M.D., Patel, A., Ravindranath, R., Singh, OP., 2008a. Chasing a mirage: water harvesting and artificial recharge in naturally water-scarce regions. Econ. Polit. Wkly. 43(35), 61–71.

TABLE 3.3 Per capita Renewable Water Resources and Per Capita Water Demand in Agriculture in Two River Basins

Name of the Basin	Average Annual Rainfall in the basin (mm)		Average Renewable Water Resources (m^3/capita/annum)	Average Effective Water Resources (m^3/capita/annum)	Mean Annual Reference Evapotranspiration (mm)		Water Demand for Agriculture (m^3/capita/annum)
	Upper	Lower			Upper Catchment	Lower Catchment	
Ganga	1675.0	1449.0	1179.9	1399.4	710.0	1397.0	721.5
Brahmaputra	2359.0	2641.0	1737.1	2052.8	1064.0	1205.0	1180.9

Source: authors' own estimates based on ET_0 values estimated from FAO CROPWAT and population and renewable water availability figure obtained from Government of India, 1999. Integrated Water Resources Development: A Plan for Action, Report of the National Commission on Integrated Water Resources Development, vol. 1. Ministry of Water Resources, Govt. of India, New Delhi.

Water demand management as an alternative to large water transfers is being advocated by a few authors in the Indian context (Iyer, 2008, 2011). Obviously, agriculture is a high-priority sector to be targeted to achieve water demand management, as it consumes more than 80% of the "managed water" (Kumar, 2010). An important point we need to remember is that there is very little water saving we can achieve through improvement in water-use efficiency (WUE) in agriculture, using technologies such as micro irrigation. We have estimated it to be 44 BCM, for the existing cropping pattern, considering the crops that are amenable to water-saving micro irrigation systems in water-scarce basins. The reason is that area under crops that are amenable to water-saving micro irrigation systems and that are under the traditional method of irrigation is very limited (Kumar et al., 2008b).

More importantly, the efficiency of water use in crop production is not very low at the basin scale, particularly those located in the naturally water-scarce regions, due to recycling and reuse of water within the system. Therefore, the amount of water that can be saved through efficiency improvements is unlikely to be very significant (Kumar, 2017a). While there are approaches other than micro irrigation system that can help enhance WUE (such as deficit irrigation (Deng et al., 2006) and shift to cultivars that yield higher WUE in relation to ET (Geert and Dirk, 2009), the current institutional and policy environment in terms of pricing of water, pricing of electricity, and regulations on water use is not favorable (Kumar and van Dam, 2013)).

There is also a scope for improving WUE in economic terms (dollar per cubic meter of water), through crop shifts in the water-scarce regions. Some of the crops that currently yield high water productivity in economic terms and that can grow well under these climates are spices such as cumin and fennel; vegetables such as chili, tomato, capsicum, cauliflower, and brinjal; flowers such as marigold; and fruits such as pomegranate, watermelon, berry, and mandarin orange.

Already, there is a significant area under high-value crops in these regions encompassing Gujarat, Rajasthan, Telangana, Andhra Pradesh, Karnataka, and Haryana. Their total area is estimated to be 8.1 M ha (Kumar, 2017b). A large-scale shift in the cropping pattern toward high-value crops, with the replacement of traditional cereal crops such as wheat and paddy, is unlikely as such a shift will have implications for farming system resilience and national food security.

In the basins where excessive use of water in irrigation systems is significant (mainly the Ganges), water is relatively abundant, and therefore, the economic value of water is comparatively low. Technological measures for WUE improvement (such as drip and sprinkler irrigation) may not make much sense either in physical terms or in economic terms. The physical benefit will be less because the real water saving will be negligible in such regions owing to shallow groundwater table and temperate climate. The economic benefit will be negligible because of the low economic value of the saved water.

The physical infrastructure, which is necessary for the use of micro irrigation systems, such as independent pressurized irrigation sources and quality electricity supply, is accessible to only a limited number of farmers in the region encompassing the states of Uttar Pradesh, Bihar, West Bengal, and Assam, and most farmers use either diesel pumps or purchased water from neighboring well owners (Mukherji et al., 2012).

Rainfed area produces much less in terms of value of output per unit volume of water, as shown by an analysis in Narmada river basin, and the scope for increasing area under rainfed crops also does not exist, given the fact that land use intensity is already very high in the region.

The real demand reduction will have to come from the use of market instruments, with an allocation of water rights or water entitlements (in public irrigation systems and groundwater), and efficient pricing of water in urban areas and pricing of water and energy in agriculture (Kumar, 2010, 2017a). Once this is done, farmers and other water users would adopt efficient and water-saving use practices. In agricultural sector, this can be in the form of drip irrigation, mulching, deficit irrigation, allocation of water to economic efficient crops, etc. This is different from the technological approach in which farmers are incentivized to use water-efficient irrigation system through capital subsidies. In the latter case, the farmers can expand the area under irrigation (called the "rebound effect"), after saving some water per unit of land cropped, and with the result, there would be no water saving. We will deal with the issue of "rebound effect" in the chapter on drip irrigation systems.

But, there is too little progress in this direction. Most of the decisions are in the political domain. The ruling political class is generally averse to putting a price for water that signals its scarcity value, fearing backlash (Kumar and Pandit, 2016). There are very few states in India, where farmers are paying for electricity used for well irrigation on the basis of consumption, and there again, it is heavily subsidized (see Chapter 15 for details). In the rest of the states, the electricity charges in the farm sector are regressive as it is linked to the connected load, and therefore, the cost of irrigation as a percentage of gross farm income decreases with an increase in farm size (Gulati and Pahuja, 2015). Canal water is heavily subsidized in all states, and water charges are not based on volumetric water delivery and instead are based on crop type and area cultivated. While the agencies of the Union Ministry of Water Resources advocate technological and institutional measures for achieving water demand management in agriculture (Central Water Commission, 2014), more and more governments are keen to offer free water and free electricity to farmers as the easy way to garner votes. Recently, the government in the newly formed state of Telangana declared that it would provide 24×7 power supply to rural areas free of charge. Its impact on electricity use in the farm sector and sustainability of groundwater resource is anyone's guess.

3.6. THE SOCIAL COST OF NOT DOING INTER-BASIN WATER TRANSFERS

The civil society organizations, which oppose ideas like the National River Linking Project (NRLP) tooth and nail, do not talk about the "social cost" of not providing water to water-scarce regions and the droughts and floods. Already, there is a great degree of poverty, socioeconomic deprivation, and social tension in many drought- and flood-prone regions of India, due to the lack of sustainable livelihoods. Out-migration and farmer suicides are on the rise. While the importance of ecology cannot be negated, great rivers in their pristine conditions would mean nothing if the people living in their flood plains do not have one square meal a day; farmers in the drought prone areas look for rains to grow crops, and there is no clean water for poor people living in the cities and towns.

Conflicts over sharing of water are on the rise, not only between states located in different river basins but also between states that share the same river basin. The latest examples are conflict between Punjab and Haryana and between Punjab and Delhi over sharing of water from Sutlej River in the Indus basin, conflict between Tamil Nadu and Karnataka over sharing of water from Cauvery basin, brewing tension between Odisha and Chhattisgarh over sharing of water from Mahanadi basin, and brewing tension between Andhra Pradesh and Telangana over sharing of water from Krishna and Godavari. Increasing scarcity of water, including localized ones, is the main cause of these tensions. These conflicts escalate during droughts, as the recent incidents in Tamil Nadu and Karnataka had demonstrated. In many of these situations mentioned above, the only way to reduce the growing tension is to transfer additional water to the basins, and there are no alternatives in sight, though there is an equal need for new legal instruments (Cullet, 2016) and effective institutions (Economic and Political Weekly, 2016) to resolve interstate water disputes.

That being the case, it is also true that India still hasn't perfected the science and art of building water systems with least social, ecological, and environmental consequences. As YK Alagh, a renowned economist and an ardent supporter of public irrigation systems, had commented more than a decade ago, in an article, "India had built some of the fanciest dams, but the lousiest delivery systems." Since then, we have made some progress by planning large water projects better in a way that result in minimum submergence and human displacement, as our awareness of the related issues had increased with experience gained from already existing projects.

We have also made progress in terms of working out rehabilitation and resettlement package for the oustees of water projects, in a way that it minimizes the hardship and psychological trauma of oustees (Jagadeesan and Kumar, 2015; Patel, 2012). New models of benefit sharing with project-affected people in the case of hydropower projects are now in use in Bhutan under cooperation between India and Bhutan (Biswas and Tortajada, 2017) and in the northeast

(Sikkim) (Rao, 2014). We need to extend this to other large water projects too, by evolving appropriate institutional mechanisms through which the people who benefit from the project transfer a share of the benefits to those who bear the costs in the form of displacement of habitations and loss of productive land that serve as their source of livelihood. To begin with, we need to assess the cost of displacement and submergence realistically, include them in the project costs at the planning stage itself, and generate revenues by charging adequately for the services to project beneficiaries to cover these costs (Shah and Kumar, 2008).

3.7. CONCLUSIONS

If India has to increase agricultural production to sustain the supplies of cereals, fibers, oil, sugar, fish and meat, and fruits and vegetable for its growing population, many of the agriculturally productive regions that are currently facing water shortage and groundwater depletion but having plenty of arable land will have to be supplied with additional water from outside their basins (Kumar et al., 2012; Sarkar, 2012). This is because the water-rich regions do not have a sufficient amount of arable land that can be used for expanding irrigated area and enhancing agricultural production (see Chapter 12, this book), a point that has been terribly missed by the advocates of large water transfer projects. This water required by water-scarce regions can come only through interbasin water transfers. Water is also required for meeting the future manufacturing needs and water needs of the rapidly growing urban population. Most of this additional demand would come from water-scarce regions, which have most of the large cities and manufacturing hubs.

Unfortunately, the major critiques to the idea of "interbasin water transfers" are largely built on the semantics of "water-scarce basins," "water-surplus basins," "river linking," etc. The view of those who opposed the idea of river linking is that the concepts of "water scarcity" and "water surplus" are fallacious as they are defined in an unscientific way, river linking is economically unviable (Bandyopadhyay and Parveen, 2004), rivers cannot be "linked", and those who talk about river linking essentially do not understand the "functions of rivers" (Iyer, 2011).

Concerns are raised about ecological health of rivers and larger environmental impacts of large-scale water transfer including coastal seawater intrusion, damage to estuarine fisheries, forest submergence, and threat to wild life. While they are serious and cannot be wished away, the existing institutional regimes at the state and central level, consisting of statutory bodies for environmental clearance of the projects and the Judiciary, are capable enough to address these concerns. Unfortunately, as noted by Verghese (2003), while a project like interlinking of rivers should be considered as a project with 50- to 100-year time frame, with sufficient time for technical feasibility studies, preparation of detailed project reports, design of structural components and execution, certain civil society groups and academics expressed exaggerated fear by variously describing it as "frighteningly grandiose," a "misapplied vision," "extravagantly stupid," a case of putting the "cart before the horse," a "subcontinental fiasco," "a flood of nonsense," a "dangerous delusion," or a case of "hydro hubris" (Verghese, 2003). Notably, the opponents have not yet offered any viable alternatives to interbasin water transfer to solve the growing water management problems in the country.

In the case of large water projects, we need to make significant improvements in the method of working out the project costs so as to include the real cost of displacement and submergence and evolve appropriate institutional mechanisms wherein the project-affected people are adequately compensated or get a share of the project benefits and the beneficiaries of the project share part of the costs. Only then, such projects will become socially viable. But in the author's opinion, it is the inability to quantify some of the real benefits that are indirect in nature, but significant (Cestti and Malik, 2012), which prevents the planners of such large water projects from either underestimating the direct and indirect costs or altogether neglecting them as they fear of not being able to justify the investments in terms of B-C ratio. Hence, improvements are needed in the methodologies used for evaluating project benefits, especially the positive externalities.

REFERENCES

Alexandratos, N., Bruinsma, J., 2012. World agriculture Towards 2030/2050: The 2012 Revision. ESA Working Paper No. 12-03, Agricultural Development Economics Division, FAO, Rome.

Amarasinghe, U.A., Sharma, B.R., Aloysius, N., Scott, C.A., Smakhtin, V., de Fraiture, C., Sinha A.K., Shukla, A.K., 2004. Spatial Variation in Water Supply and Demand across River Basins of India. IWMI Research Report 83. Colombo, Sri Lanka.

Bandyopadhyay, J., Parveen, S., 2004. Interlinking of rivers in India. Econ. Polit. Wkly. 39 (50).

Biswas, A.K., Tortajada, C., 2017. Cooperation on Shared Waters. Kathmandu Post.

Central Water Commission, 2014. Guidelines for Improving Water Use Efficiency in Irrigation, Domestic and Industrial Sectors. Performance Overview and Management Improvement Organization, Central Water Commission, Ministry of Water Resources, Govt. of India, New Delhi.

Cestti, R., Malik, R.P.S., 2012. Indirect economic impact of dams. In: Tortajada, C., et al. (Eds.), Impacts of Large Dams: A Global Assessment, Water Resources Development and Management. Springer-Verlag, Berlin Heidelberg, https://doi.org/10.1007/978-3-642-23571-9_2.

Cullet, P., 2016. Governing water to foster equity and conservation: need for new legal instruments. Econ. Polit. Wkly. 51 (53).

Deng, X., Lun, S., Zhang, H., Turner, N.C., 2006. Improving agricultural water use efficiency in semi-arid and arid areas of China. Agric. Water Manag. 80 (1–3), 23–40.

Economic and Political Weekly, 2016. Water wars. Editorial. Econ. Polit. Wkly. 51 (38).

Geert, S., Dirk, R., 2009. Deficit irrigation as an on-farm strategy to maximize crop water productivity in dry areas. Agric. Water Manag. 96 (2009), 1275–1284.

Glendenning, C.J., Vervoort, R.W., 2011. Hydrological impacts of rainwater harvesting (RWH) in the case study catchment: The Arvari River, Rajasthan, India. Part 2. Catchment-scale impacts. Agric. Water Manag. 98, 715–730.

Government of India, 1999. Integrated Water Resources Development: A Plan for Action, Report of the National Commission on Integrated Water Resources Development. vol. 1. Ministry of Water Resources, Govt. of India, New Delhi.

Gulati, M., Pahuja, S., 2015. Direct delivery of power subsidy to manage energy–ground water-agriculture nexus. Aquat. Procedia 5 (2015), 22–30.

Iyer, R.R., 2008. Water: a critique of three basic concepts. Econ. Polit. Wkly. 43 (1), 15–17.

Iyer, R.R., 2011. National Water Policy: an alternative draft for consideration. Econ. Polit. Wkly. XLVI (26 and 27), 201–214.

Jagadeesan, S., Kumar, M.D., 2015. The Sardar Sarovar Project: Assessing Economic and Social Impacts. Sage Publications, New Delhi.

Kumar, M.D., 2003. Food Security and Sustainable agriculture in India: the Water Management Challenge. Working Paper # 60. International Water Management Institute, Colombo, Sri Lanka.

Kumar, M.D., 2009. Water Management in India: What Works, What Doesn't. Gyan Books, New Delhi.

Kumar, M.D., 2010. Managing Water in River Basins: Hydrology, Economics, and Institutions. Oxford University Press, New Delhi.

Kumar, M.D., 2014. Thirsty Cities: How Indian Cities can Manage their Water Needs. Oxford University Press, New Delhi.

Kumar, M.D., 2017a. Proposing a solution to India's water crisis: paradigm shift or pushing outdated concepts? Int. J. Water Resour. Dev. https://doi.org/10.1080/07900627.2016.1253545.

Kumar, M.D., 2017b. Market Analysis: Desalinated Water for Irrigation and Domestic Use in India. Prepared for Securing Water for Food: A Grand Challenge for Development. Center for Development Innovation U.S. Global Development Lab Submitted by DAI Professional Management Services.

Kumar, M.D., Ghosh, S., Patel, A., Singh, O.P., Ravindranath, R., 2006. Rainwater harvesting in India: some critical issues for basin planning and research. Land Use Water Res. Res 6 (2006), 1–17.

Kumar, M.D., Bassi, N., Sivamohan, M.V.K., Vedantam, N., 2011. Employment guarantee and environmental benefits: are the claims valid? Econ. Polit. Wkly. 46 (34), 49–51.

Kumar, M.D., Hugh, T., Sharma, B., Upali, A., Singh, O.P., , 2008b. In: Kumar, M.D. (Ed.), Water saving and yield enhancing micro irrigation technologies in india: when do they become best bet technologies? Managing Water in the Face of Growing Scarcity, Inequity and Declining Returns: Exploring Fresh Approaches, Vol. 1. Proceedings of the 7th Annual Partners' Meet of IWMI-Tata Water Policy Research program, ICRISAT, Hyderabad, pp. 13–36.

Kumar, M.D., Pandit, C., 2016. India's water management debate: is the "Civil Society" making it everlasting? Int. J. Water Resour. Dev. https://doi.org/10.1080/07900627.2016.1204536.

Kumar, M.D., Patel, A., Ravindranath, R., Singh, O.P., 2008a. Chasing a mirage: water harvesting and artificial recharge in naturally water-scarce regions. Econ. Polit. Wkly. 43 (35), 61–71.

Kumar, M.D., Sivamohan, M.V.K., Narayanamoorthy, A., 2012. The food security challenge of the food-land-water nexus in India. Food Security 4 (4), 539–556.

Kumar, M.D., van Dam, J., 2013. Drivers of change in agricultural water productivity and its improvement at basin scale in developing economies. Water Int. 38 (3), 312–325.

Mukherjee, S., Shah, Z., Kumar, M.D., 2010. Sustaining urban water supplies in India: increasing role of large reservoirs. Water Res. Manag. 24 (10), 2035–2055.

Mukherji, A., Shah, T., Banerjee, P.S., 2012. Kick-starting a second green revolution in Bengal, commentary. Econ. Polit. Wkly. XLVII (18), 27–30.

Patel, A., 2012. What do the Narmada valley tribals want? In: Parthasarathy, R., Dholakia, R. (Eds.), History of Rehabilitation and Implementation, Vol. 2, Sardar Sarovar Project on the River Narmada. CEPT University Press, Ahmedabad.

Punjabi, M., 2010. India: increasing demand challenges the dairy sector. In: Morgan, N. (Ed.), Smallholder Livestock Development: Lessons Learned From Asia. Animal Production and Health Commission for Asia and the Pacific, Food and Agriculture Organization of the United Nations, Bangkok.

Puttaswamy, H.J., 2015. Response to the opinion article entitled "Environmental over enthusiasm" by Pandit, C. Int. J. Water Resour. Dev. 31 (4), 780–784.

Rada, N.E., 2013. In: Agricultural growth in India: examining the post green revolution transition. Selected Paper Prepared for Presentation at the Agricultural & Applied Economics Association's 2013 AAEA & CAES Joint Annual Meeting, Washington, D.C., August 4–6, 2013.

Rao, N., 2014. Benefit sharing mechanism for hydropower projects: pointers for North East India. In: Kumar, M.D., Bassi, N., Narayanamoorthy, A., Sivamohan, M.V.K. (Eds.), Water, Energy and Food Security Nexus: Lessons From India for Development. Routledge/Earthscan, London, UK.

Sarkar, A., 2012. Equity in access to irrigation water: a comparative analysis of tube-well irrigation system and conjunctive irrigation system. In: Kumar, M. (Ed.), Problems, Perspectives and Challenges of Agricultural Water Management. Intech Publishers.

Shah, M., 2013. Water: towards a paradigm shift in the twelfth plan. Econ Political Wkly. 48 (3), 40–52.

Shah, Z., Kumar, M.D., 2008. In the midst of the large dam controversy: objectives, criteria for assessing large water storages in the developing world. Water Resour. Manag. 22 (12), 1799–1824.

Srinivasan, V., Steven, M.G., Lawrence, G., 2010. Sustainable urban water supply in south India: desalination, efficiency improvement, or rainwater harvesting? Water Resour. Res. 46 (W10504), 2010. https://doi.org/10.1029/2009WR008698.

Verghese, B.G., 2003. Exaggerated Fears on Linking Rivers. http://www.himalmag.com/2003/.

FURTHER READING

Sacchidananda, M., Shah, Z., Kumar, M.D., 2010. Sustaining urban water supplies in India: increasing role of large reservoirs. Water Resour. Manag. 24 (10), 2035–2055.

Chapter 4

Recent Droughts in India: Nature's Fury or Poor Statecraft?

4.1. MANAGING NATURAL RESOURCES DURING CRISIS

In today's world, when natural resources are becoming scarce, managing water, energy, and food requires sound knowledge of statecraft and economic management. For a country like India, the ability to maintain good economic growth rates and move people out of poverty depends heavily on how water and energy resources are managed for sustainable water supplies and energy and food security. This requires solving several complex problems relating to demand and supply of water, energy, and food. We need to know where the basic resources are available (say, for instance, water), which technologies to be used to appropriate these resources to access it in a cost-effective manner, and which are the high priority demands that need to be catered to so as to minimize conflicts over their allocation.

Along with good technology and human capital, finding solutions to such complex problems would more often require hard decisions on the part of the state bureaucracies dealing with water, energy, and agriculture, supported by strong political leadership. There can be no better situation than a "drought" to illustrate the importance of this. Normally, such decisions taken during droughts include water utilities putting ban on use of water supplied by them for watering gardens and lawns in urban areas, release of water from reservoirs for agricultural uses, and sealing of agrowells in rural areas.

There is widespread belief that for democratically elected government to survive in (developing) economies, the decisions concerning the management and use of resources, which directly affect the livelihoods of the millions of ordinary people, should be driven by popular concerns rather than the macroeconomic interests, more so for fledgling democracies like India with a significant share of the population that is illiterate or with low level of formal education. This is because of a growing chorus that these decisions have to be "politically correct." The underlying assumption is that hard decisions to improve economy give dividends only in the long run and are bound to be in direct conflict with measures that can appease ordinary people and that the latter better suit the political class as they produce immediate electoral gains. A related assumption is that the ordinary people are only interested in measures that produce immediate gains and are not concerned with long-term economic management.

Water Policy Science and Politics. https://doi.org/10.1016/B978-0-12-814903-4.00004-X

This chapter discusses how the changing political economic landscape of India is changing the narrative for the governments for making policies for water and other related sectors in India, how the academics-bureaucracy-politician nexus ensures that certain populist policies and schemes are continued by the government in spite of increasing evidence of their ineffectiveness and social costs associated with such measures, and how media power is used by civil society groups to convince the government of pursuing an agenda that is not very much aligned with national goals.

4.2. CHANGING POLITICAL ECONOMIC LANDSCAPE OF INDIA

To a great extent, this is true as a large proportion of the people living in villages and towns are concerned with their immediate survival needs. However, it is also true that there is a large proportion of the population, the middle class, which is quite concerned with the long-term economic growth needs. They are concerned about the implications of the short-term populist measures such as subsidized or free electricity and free water on the growth of individual sectors, the state and national economies, and their own household economy. Their size is growing, a fact many of the people in the development sector refuse to acknowledge. This class does not benefit from farm power and irrigation subsidies. They also do not benefit much from the huge diesel subsidies. Good politics also imply that this segment of the population is not simply ignored for the sake of a larger chunk of the population, which benefit from such populist measures. In a recent public speech, the prime minister of India echoed this growing concern of the middle class by saying "today, our middle class are over-burdened by taxes, which are levied by the government to reduce the hardship of the poor; are under obligation to respect the many laws and rules that exist in our society, and to reduce the stress on them, it is important that we make quick progress in eradicating poverty in our country."

The most recent example that validates the hypothesis about changing political economic landscape in India is that the political party that came to power in the state of Delhi in 2013 with a thumping majority lost many elections within 4 years, including the Delhi municipal election, in spite of the fact that a significant proportion of the people in the capital city get heavily subsidized electricity for domestic uses and free municipal water supply. The stark reality is that the attempt by the political parties in power to appease one section of the population through subsidies, cash doles, and loan waivers can annoy many other segments of the society that do not get these benefits.

4.3. SOCIAL COST OF POPULIST MEASURES

But as the aftermath of the recent droughts clearly show, continuing with such populist measures can have disastrous consequences. Today, in large areas of many states, farmers, rural masses, and city dwellers alike are under enormous distress—with wells in rural areas going dry, temperature shooting up, heat waves taking many precious lives, and regular piped water supply in many pockets in cities and towns being replaced by occasional tankers, frequent power cuts, etc. This is a good recipe for things to go out of control, affecting the social fabric. Even under this situation, there is very little appreciation of the fact that this is the price we pay for short-sighted decisions of our politicians and bureaucrats.

During the drought of 2015, which badly hit the state of Maharashtra, especially the Marathwada region, drinking water for the city of Latur had to be transported through rail wagons from Western Ghats region. Marathwada region has no buffer storage in their reservoirs. The cost of transporting a unit volume of water was far higher than the cost of desalination. In the past, such drastic measures had to be resorted to only during the worst drought of 1985–87 in Saurashtra peninsula of Gujarat in western India. As shown by a recent field study, many villages in that state, which could not obtain good-quality water from surface reservoirs either through piped water supply schemes or by tankers, had to use poor-quality groundwater for drinking, cooking, and other domestic uses (Udmale et al., 2016).

In this 21 century, the news about a town being supplied with water through rail should make any sensible individual feel ashamed of how our systems function. There are hundreds of such towns and cities in India, whose water supply situation is quite precarious. But, what is being conveniently ignored is the fact that the water for supplying through rail tankers could be obtained because we have some large reservoirs that hold water even in this hottest summer. It is a different matter that if we had built an infrastructure (like the large pipeline network in Gujarat taking water from SSP to Saurashtra and Kachchh) for distribution of water from such reservoirs to the dry areas, we would have been in a much better position to tackle the current droughts (Biswas-Tortajada, 2014; Jagadeesan and Kumar, 2015). Have we thought about a situation where there are no large reservoirs in this country, and we fully depend on the wells and small tanks? Perhaps, the CSOs (civil society organizations) are the best people to answer this question. In a country that experiences high year-to-year variation in monsoon rainfall, there is no widespread realization of the need to build enough reservoirs with multiannual (carry-over) storage facility.

On the contrary, some of the CSOs, social activists, and a few influential academicians had succeeded in convincing the past governments that we do not need to build large reservoirs for achieving water security, wells and hand pumps would do the job, and the management of water supply in rural areas can be entirely left to gram panchayats. We have simply ignored the fact that groundwater is the most unreliable source of water in hard-rock areas of our country (all the states that are experiencing droughts today) and that the governments (be it the state or the local self-government) have no control over its use and leave alone the issue of the managerial capabilities of most gram panchayats to tackle droughts.

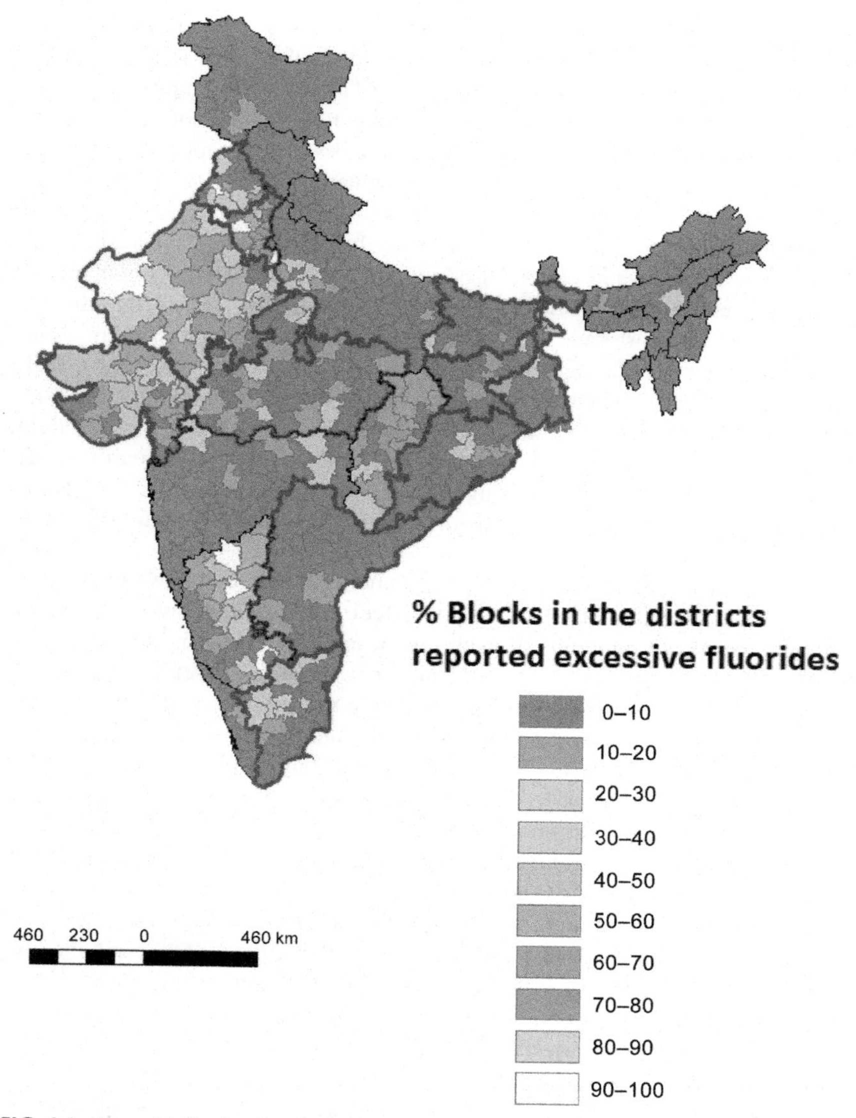

FIG. 4.1 Map of India showing districts affected by high levels of fluorides in groundwater.

Another important dimension is the quality of water, when we make a choice between groundwater and surface water. Groundwater of India is contaminated from the point of view of potability, in large geographic areas of India. The main contaminants are salinity, fluorides, nitrates, chlorides, and arsenic. Fig. 4.1 shows the districts of India affected by high fluorides in groundwater in terms of percentage of blocks in the district with high incidence of fluorides.

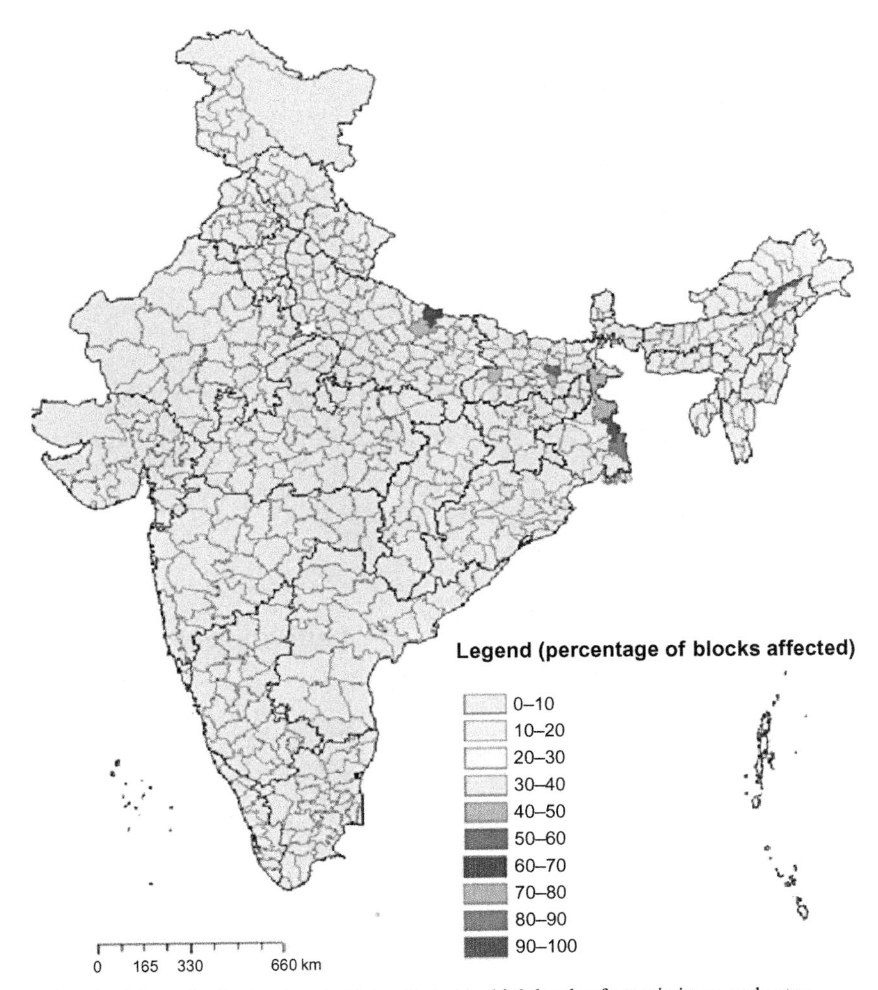

FIG. 4.2 Map of India showing districts affected by high levels of arsenic in groundwater.

Fig. 4.2 shows the districts of India affected by high level of arsenic in groundwater in terms of percentage of blocks affected. Fig. 4.3 shows the map for high nitrate levels in groundwater (source, maps prepared on the basis of GoI (2010)). Superimposing the four maps, it is evident that there are very few pockets in India that are not affected by chemical contamination of groundwater.

But the reality is that a large number of rural water supply schemes are individual village-based schemes dependent on groundwater. As the size of the geographic area affected by different types of chemical contaminants in groundwater has been increasing over the years, state governments have been trying to relax norms relating to the chemical quality of water for potable water supply

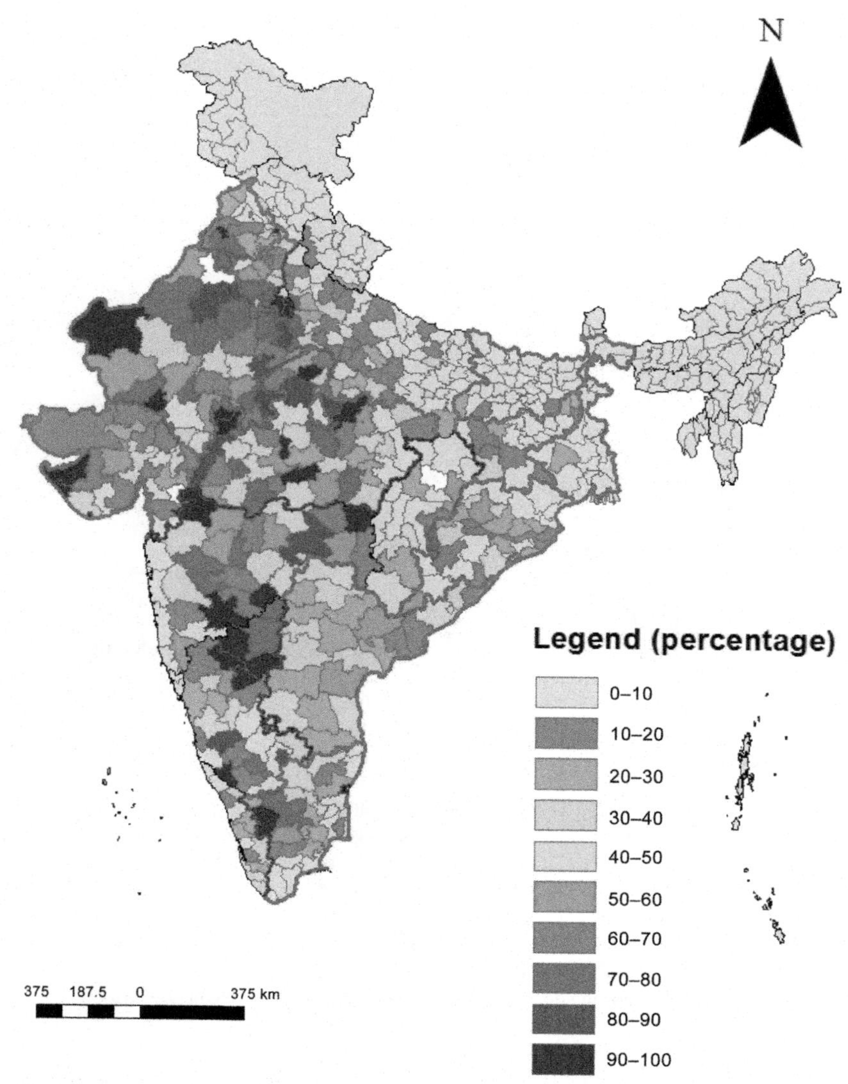

FIG. 4.3 Map of India showing districts affected by high levels of nitrates in groundwater.

so as to continue with supplying poor-quality groundwater in terms of salinity. In states like Rajasthan, which have large areas affected by high salinity in groundwater, water of TDS up to 2000 ppm is often used for rural drinking water supply.

The pathological dislike of some policy makers and CSOs toward "tap water" from large reservoir-based rural water supply schemes can come at a huge public health cost, when the communities are forced to drink untreated water of inferior quality from underground sources resulting in waterborne diseases such as fluorosis and kidney stone. In many Indian states, where problems of chemical contamination of groundwater are rampant, the village communities are buying expensive treated water for drinking and cooking needs from private water companies (Bandyopadhyay, 2016; Reddy, 2016).

4.4. HOW THE "ACADEMICS-BUREAUCRACY-POLITICIAN NEXUS" WORKS?

The unfortunate part is that when large parts of the country are reeling under droughts, the CSO statement that we have not been able to achieve "democratic decentralization of governance and management of water resources." What does this mean? This means the village communities take control of the river catchments on which the state water resources departments exercise some control today. This means the village communities take control of management of our aquifers also. The CSOs advocate that the water from the local catchments be harvested, put underground, and used within the village itself. This means leaving nothing for downstream uses. Today, many reservoirs in water-scarce regions receive much lower inflows due to too many structures built in their upper catchments. Most of this water is put underground, benefiting the large farmers who own wells. In the summer of 2016, even in the much touted *Ralegan Siddhi*, known for the watershed experiments, people have to get drinking water through tankers, as farmers have already taken out all the groundwater through their bore wells.

The idea here is not to belittle the work of CSOs and groups. In a democratic polity, the role of civil society is undisputable, and its meaningful participation is important to produce good outcomes of government decision-making. But, the leaders in the government should also use their wisdom to decide who actually represent the "civil society." Great caution needs to be exercised to make sure that a few individuals do not hijack the entire development discourse in a vast country like India and that there is wider participation. The reason is when things go terribly wrong, these privileged people don't own up. Neither can the government hold them liable for anything. The State should also make sure that the government policies do not allow private individuals to exploit natural resources at public cost (like in drought situations), and whenever that happens, this is duly compensated through the instruments of taxation and the like.

Alas, approaching complex economic problems with kid gloves, academics advise the politicians and bureaucrats that providing free power, fuel subsidies, free power connections, and free water would help eradicate poverty and bring more votes, and market-based solutions would harm the interests of the poor (Kumar, 2014). There are also those being advised that "small check dams" are viable alternatives to building large water storage infrastructure, but would not pose social and environmental problems unlike large ones; solar PV systems, considered as "clean energy alternatives" to thermal and other conventional energy sources, are also cheaper.

This is in spite of the fact that there is hardly any evidence to the effect from any part of the developing world that such solutions can replace more conventional water systems. Such an approach is being resorted to in a haste to find a place for their ideas in political and policy circles and to show quick policy impacts of their research (Kumar, 2014). Clearly, there is poverty of ideas and an "intellectual drought."

The virulent argument they make is that attempts to do away with subsidies in water and electricity sectors would be "political suicide" for ruling parties. But, the experience of West Bengal and Gujarat with farm power metering and pricing shows that this is far from the reality. For instance, the state of West Bengal is highly agrarian and is one of the poorest states in the country. The left government there could raise power tariff without any resistance from the farming community, and the power pricing policy is now continued by the subsequent governments (Mukherji and Das, 2012). A few years ago, the hugely popular government in Gujarat, led by the then chief minister, Shri Narendra Modi, had raised power tariff significantly and had also been charging for new agricultural power consumers on prorata basis. On the contrary, the governments often resort to "sops" such as free power in order to survive the revolt, from within ruling party or coalition or from other parties across the political spectrum, on issues of corruption and malgovernance. The weak governments seem to resort to such political bribing to stay in power more often than others.

As regards rainwater harvesting alternatives to large water systems, they are often embraced by governments mainly as an escape route from the hassle of overcoming the fierce opposition to large water and energy systems by local and international environmental groups, keeping in mind that such interventions are hugely popular and enjoy the support of civil society groups and activists. The most recent example is from Maharashtra, which had faced severe drought conditions during the summer of 2015. The state has promoted the following two engineering interventions to improve the water situation, under a program called *Jal Yukht Shivar*: (1) farm ponds in individual farmers' fields and (2) digging out of stream channels for increasing their effect on shallow groundwater in the form of recharge. Both the interventions had detrimental effect on the hydrology. The first one led to farmers pumping groundwater and storing in these small water storage system, instead of natural runoff getting collected in them. This increased groundwater depletion (Kale, 2017). The second one led to reduced storage of base flows in the river channels.

This "ostrich"-like approach of deferring and delaying investments on large water and power generation projects in the past has now increased the need to either go for such large systems or increase water and energy tariffs drastically to save the economy from total collapse.

4.5. USING MEDIA POWER FOR MANUFACTURING PUBLIC OPINION IN FAVOR OF WATER REFORMS

The civil society groups often succeed in convincing governments in embracing policies that are not very much in national interests, in the garb of "water sector reforms." They often do it by using media power. The latest example is the report of Mihir Shah Committee, appointed by the Ministry of Water Resources (http://wrmin.nic.in/writereaddata/Report_on_Restructuring_CWC_CGWB.pdf), government of India on restructuring CWC and CGWB. The chairman of the committee, Dr. Mihir Shah, had given several interviews to national dailies, arguing how the recommendations in the report for "reforms," if implemented, could herald a new era in India's water sector. Intriguingly, one of its major recommendations was that India should stop building large dams, and the solution to dealing with India's water crisis lies in participatory groundwater management. The committee report invited widespread criticism from the officials of Central Water Commission, for foraying into issues that were not within the scope of its work and using the platform to garner support for the antidam agenda of its members, who were mostly civil society activists.

We also analyzed the report and posted an article on LinkedIn, "There are no low hanging fruits in water management" (dated 04 October 2016), explaining how the committee has seriously faltered in their analysis of India's water sector vis-à-vis the problems it confronts and the "remedies," which was used to build a case for the kind of reforms proposed therein. Also, an article was published in *International Journal of Water Resources Development* titled "Proposing a solution to India's water crisis: 'paradigm shift' or pushing outdated concepts?" by M. Dinesh Kumar. A critical analysis of the interview of the committee's chairman with a reporter from The Hindu (dated 19 August 2016), presented below (http://www.thehindu.com/opinion/interview/%E2%80%98Time-for-a-National-Water-Commission%E2%80%99/article14578590.ece), explodes the myth that the report contains a reform agenda for the water sector in India.

The *Economic and Political Weekly* carried 13 articles, 12 of which were supposed to be critiques of the committee report, but written mostly by people who are close associates of the committee chairman (http://www.epw.in/taxonomy/term/28,686) and one by the committee chairman himself. Most of these critiques were biased and opinionated, barring two exceptions, one by Joy (2016)) and the other by Lele and Srinivasan (2016). More importantly, neither the critics nor the response of the committee chairman in his response to the critiques did make any reference to the issues raised by the author in his article. By doing this, the committee seems to have succeeded in manufacturing public opinion in favor of the recommendations of a report, which otherwise carries no substance, in terms of their ability to bring about reforms in the water sector, except its tirade against large dams and the engineering fraternity. One such interview and our rebuttal of the committee chairman's response are given in Appendix.

4.6. CONCLUSIONS

Dealing with droughts and reducing their impacts need long-term planning. Ideas like rainwater harvesting do more harm in drought-prone areas. There are no simple quick fix solutions for drought proofing of large regions. Most of the drought-prone regions have very low to medium rainfalls and also have very high to high aridity. They are also characterized by high year-to-year variation in rainfall. Because of these unique hydrologic characteristics of such regions, when droughts actually hit, ad hoc measures such as rainwater harvesting will not be effective either for saving crops from moisture stress or for ensuring drinking water supply in rural and urban areas. Overemphasis on groundwater as a source of drinking water supply also increases the exposure of communities to the scarcity resulting from drought hazards. The widespread problems of groundwater contamination in the forms of high salinity, fluoride, nitrates, and arsenic also should send a clear warning signal to the governments of the growing health risks of overdependence on groundwater.

Governments will have two options for such drought-prone regions. It can discourage water-intensive economic activities such as irrigated crop production in such regions. It can also move large urban areas or water-intensive industries out of such regions through the mechanisms of incentives and disincentives. Such measures will have huge adverse socioeconomic impacts. If these measures are not possible, it will have to implement large-scale projects to transfer water from water-abundant regions to these regions so as to minimize the socioeconomic impacts of such droughts. One reason why governments engage in the former is that they appear very cheap and quick to implement. But, what the governments need to worry about is the long-term socioeconomic impacts of droughts, in terms of poor agricultural growth, large-scale migration of people from rural areas, and low industrial outputs.

In democratic polity, civil society has an undisputable role. Its participation is essential to ensure positive developmental outcomes of government decision-making. There is a lot of stereotyping of the state as the antithesis of a deliberative space. Although it captures an essential feature of the state, it can also be deeply misleading (Evans, 2015). As noted by Evans, "effective state structures have always depended on deliberative spaces that include both key actors within the state apparatus and powerful private interlocutors. In the 21st century, deliberation has become even more crucial, because the state faces a set of tasks that require bringing in deliberation in a way that goes well beyond established traditions" (Evans, 2015, p. 51).

The government in its wisdom should choose the people who actually represent the civil society. Caution needs to be exercised to prevent hijacking of development discourse by a few influential individuals and groups who are self-serving. Unfortunately, in the past, a section of the academics has been able to influence the bureaucracy and the political class to buy into their water sector agenda through persuasive writing and constant lobbying. This happened because of the absence of effective institutional regimes to bring checks and balances for ensuring quality of evidence produced by the research that support their agenda. They have also been able to use the media to their advantage using the rhetoric of "water sector reforms." This is failure of statecraft.

REFERENCES

Bandyopadhyay, S., 2016. Sustainable access to treated drinking water in rural India. In: Kumar, M.D., James, A.J., Kabir, Y. (Eds.), Rural Water Systems for Multiple Uses and Livelihood Security. Elsevier Science, The Netherlands, pp. 203–226.

Biswas-Tortajada, A., 2014. The Gujarat state-wide water supply grid: a step towards water security. Int. J. Water Resour. Dev. 30 (1), 78–90.

Evans, P., 2015. Bringing deliberations to the development state. In: Heller, P., Rao, V. (Eds.), Deliberation and Development: Rethinking the Role of Voice and Collective Action in Unequal Societies. The World Bank, Washington, DC.

Government of India (GoI), 2010. Groundwater Quality in the Shallow Aquifers of India. Central Ground Water Board, Ministry of Water Resources. Government of India, Faridabad.

Jagadeesan, S., Kumar, M.D., 2015. The Sardar Sarovar Project: Assessing Economic and Social Impacts. Sage Publications, New Delhi.

Joy, K.J., 2016. An important step in reforming water governance. Econ. Polit. Wkly. 51 (52), 30–33.

Kale, E., 2017. Problematic uses and practices of farm ponds in Maharashtra. Econ. Polit. Wkly. 52 (3).

Kumar, M.D., 2014. Of statecraft: water, energy and food security in developing countries. In: Kumar, M.D., Bassi, N., Narayanamoorthy, A., Sivamohan, M.V.K. (Eds.), Water, Energy and Food Security Nexus: Lessons from India for Development. Routledge/Earthscan, London, United Kingdom.

Lele, S., Srinivasan, V., 2016. Focusing on the essentials. Econ. Polit. Wkly. 51 (52), 47–50.

Mukherji, A., Das, A., 2012. How Did West Bengal Bell the Proverbial Cat of Agricultural Metering? Water Policy Research Highlight. IWMI-Tata Water Policy Programme, IWMI Field Office, Anand, Gujarat, India.

Reddy, V.R., 2016. Techno-institutional models for managing water quality in rural areas: case studies from Andhra Pradesh, India. Int. J. Water Resour. Dev. https://doi.org/10.1080/07900627.2016.1218755.

Udmale, P., Ichikawa, Y., Nakamura, T., Shaowei, N., Ishidaira, H., Kazama, F., 2016. Rural drinking water issues in India's drought-prone area: a case of Maharashtra state. Environ. Res. Lett. 11 (7), 074013.

FURTHER READING

Shah, M., 2016. A 21st Century Institutional Architecture for Water Reforms in India. Ministry of Water Resources, RD&GR, Government of India.

APPENDIX

Comments on Dr. Mihir Shah's response to the interviewer's questions are in italics.

The proposed National Water Commission (NWC) subsumes the Central Water Commission (CWC) and the Central Ground Water Board (CGWB). How specifically does it improve national water management?

The CWC (set up in 1945) and CGWB (set up in 1971) were created in an era when India faced a very different set of challenges. Then, it was crucial to create irrigation capacity to ensure food self-sufficiency. But today, the challenge is different. At huge cost (around Rs. 400,000 crore), we have created 113 M ha of irrigation potential. But is this water reaching the farmers? No. As the chief minister of Maharashtra has said, the state has 40% of the country's large dams, "but 82% area of the state is rainfed. Till the time you don't give water to a farmer's fields, you can't save him from suicide. We pushed large dams not irrigation. But this has to change." Our report is trying to address this challenge.

We also highlight the fact that groundwater is the main source of water in India. This means we cannot go on endlessly drilling for groundwater through tube wells, which is what CGWB has promoted thus far. This has actually aggravated India's groundwater crisis, as water tables fall and water quality declines, with arsenic, fluoride, and even uranium entering our drinking water.

Comment: *These contentions by the committee chairman raise five important issues. First, one concerns the irrigation investments. For an investment of 4 lac crore rupee ($60 billion) for irrigating 113 M ha of land, the unit cost of irrigation comes to Rs. 35,000 per ha. If we annualize this using an average life of 25 years, it comes to a meager Rs. 3000 per ha per year. This is not a huge cost, if we consider a modest net income of Rs. 50,000 per ha of irrigated land over an entire year.*

The second issue is of changing the goal post. It is not clear why the author resorts to highlighting the issue of 82% of the area remaining rainfed in Maharashtra while looking at the national scenario. The 113 M ha of irrigation is spread all over the country and not just Maharashtra. In Punjab, 95% of the cultivated area is irrigated. The fact is that India has a gross cropped area of 180 M ha, that brings down the area irrigated to just 50%. If we want to increase the irrigation intensity, we need to invest more.

The third issue concerns the criterion for analyzing the performance of irrigation sector. If one has to analyze the performance of irrigation sector against the investments made, the aggregate area under irrigation per unit of investment is what should be considered as the performance indicator and NOT the percentage of cropped area irrigated. The attempt here seems to be to distort the facts and obscurantist. That said, CWC cannot be held accountable for the "poor" performance of state irrigation agencies.

The fourth issue is of the failure to understand the causes and the effects. It is inappropriate to blame CGWB for groundwater overexploitation, which is because of free or heavily subsidized power supplied to agriculture sector, the institutional financial subsidies for well drilling, and the lack of well-defined rights in groundwater. CGWB, which is purely a technical agency, had no role in this, whatsoever. Their mandate is scientific exploration and assessment of groundwater resources in the country.

The last issue is of fractured thinking about water resources development. The committee looks at surface water development and groundwater development in a segmented fashion. In the process, a fact, which the committee conveniently ignored, is that the groundwater overexploitation problems would have been far more severe, had we not developed surface irrigation. As per the Ministry of Water Resources, River Development and Ganga Rejuvenation (MoWR, RD, and GR) own estimates, the contribution of unlined canals to groundwater recharge is about 0.15–0.3 m^3/day/m^2 of wetted area. The committee, which is so vocal about integrated surface and groundwater management, should understand this basic fact about how integrated surface and groundwater development help sustain groundwater use in canal command areas.

That said, the chairman of the committee does not have anything to offer when confronted with the question "how the NWC would improve water management in the country?"

What are the key shifts in water management your report recommends?

One, we must take a multidisciplinary view of water. We require professionals from disciplines other than just engineering and hydrogeology. Two, we need to adopt the participatory approach to water management that has been successfully tried all over the world, as also in Madhya Pradesh, Gujarat, and Andhra Pradesh. Three, we must view groundwater and surface water in an integrated, holistic manner. CWC and CGWB cannot continue to work in their current independent, isolated fashion. The one issue that really highlights the need to unify CWC and CGWB is the drying up of India's peninsular rivers; the single most important cause of which is overextraction of groundwater.

If river rejuvenation is the key national mandate of the Ministry of Water Resources, then this cannot happen without hydrologists and hydrogeologists working together, along with social scientists, agronomists, and other stakeholders. Four, we need to focus on river basins that must form the fundamental units for the management of water. We have carefully studied the regional presence (or absence) of the CWC and CGWB and proposed a way forward whereby the NWC is present in all major river basins of India.

Comment: *Firstly, for river rejuvenation in terms of cleaning of river waters, there is no need for hydrologists and geohydrologists to work together. Please see how this has been done in many countries in Europe. We can introduce either market instruments or effective regulations, neither of which is there in our country nor the committee dealt with it in the report. On the other hand, if water withdrawal from river basins has to be cut down to reduce the environmental water stress of rivers, we need an effective institutional framework for regulating land use in catchments, water resources development, and groundwater withdrawal. The report is silent on these aspects. Simply taking river basin as the unit for planning doesn't change anything. There has to be legitimate institutions that can undertake basin-wide water resources management (at the* *state level), with powers to enforce management decisions. That constitutes the real reform in the water sector.*

Secondly, to unify CWC and CGWB into National Water Commission and make their involvement in water resource, planning of basin in different states will require constitutional amendments. A simple change in nomenclature of these central agencies will be insufficient as the state agencies have rights and control over the water resources within their territories (as per entry 17, List-II, Seventh Schedule, Art. 246 of the Indian Constitution). It will be worthwhile to examine whether anything has changed in India's watershed development program after National Rainfed Area Authority was created in 2007.

Finally, what is also not clear is why do we need a separate wing of NWC for river rejuvenation, when the whole approach is going to be integrated surface and groundwater management. The committee's thinking on the issue is simply fragmented, which will only create silos.

When the committee is so voluble about interdisciplinarity in water management, it makes a curious case by not having anyone from disciplines such as hydrology and water resources engineering.

The Central Water Commission has opposed the NWC on the grounds that several reform measures are already in place. Are you throwing the baby out with the bathwater?

Not at all; we have taken great care to ensure that all existing functions and personnel of the CWC find their appropriate place in the eight divisions of the NWC, which include irrigation reform, river rejuvenation, participatory groundwater management, urban and industrial water, water security (including droughts, floods, and climate change), and water quality.

Comment: *It is not clear, how CWC and CGWB can contribute to wings such as urban water, flood management, and water quality. There is already a Central Pollution Control Board, which is primarily concerned with water quality of all aquatic ecosystems. Further, there are two agencies, namely, the National River Conservation Directorate and the National Mission for Clean Ganga, which are responsible for improving river water quality. Similarly, Ministry of Urban Development and Ministry of Drinking Water and Sanitation are in charge of developing water supply infrastructure and providing water services in urban and rural areas, respectively. Also, there is Ganga Flood Control Commission, a subordinate office of MOWR, RD and GR, to deal with floods and its management in Ganga Basin. So, for NWC to be operational, not only CWC and CGWB but also other organizations, including many from other ministries, should also be restructured. This was simply not within the scope of the ToR given to Dr. Mihir Shah committee. Further, it is unclear what additional actions will the new agency have to perform for achieving objectives like river rejuvenation, when the entire accent is on integrated water resources management.*

This isn't the first time that you have recommended an integrated water commission.

I think this kind of fundamental change takes time to be fully understood and get actualised in policy. In actual fact, professionals involved in CWC and CGWB will get an even better chance to improve their technical capabilities and career prospects within the NWC.

Comment: *It is difficult to understand the assumptions behind this unique proposal for improving the technical capabilities of professionals in both CWC and CGWB, put forth by the committee. Normally, in any government agency, professionals are hired following stringent rules and regulations pertaining to their education, experience, and skills required for performing a particular job. Tailor-made training programs are organized by reputed government institutes, such as National Water Academy at Pune and Rajiv Gandhi National Ground Water Training and Research Institute at Raipur, to further enhance their skills. One can understand the lack of representation of other disciplines, such as agricultural sciences in these organizations, but these can be handled by employing professionals relevant to such disciplines. Also, MoWR, RD, and GR through various MoU with different countries and donors such as the World Bank is already having projects with substantial capacity building components on river basin planning, IWRM, data management,* etc. *How NWC can improve technical capabilities and career prospects of CWC and CGWB officials is not clear.*

How are we going to enhance the knowledge and capabilities of the training institutes such as the National Water Academy? The committee has made no suggestions toward this. On the contrary, they want to burden these central agencies with the task of training farmers! The NWA needs more autonomy to source experts in different disciplines from the industry and compensate them adequately. Administrative reforms are needed for that.

Water is frequently a political issue in several states. Why should states listen to an NWC?

As a committee, we took great care to get views of states on board. We have suggested that appraisal must become a demand-based exercise, done through a partnership between the central and state governments, as also institutions of national repute.

This is a key part of the reform we are proposing. We are not for a monolithic NWC. The NWC will be a knowledge institution providing solutions to water problems faced by state governments, farmers, and other stakeholders, on demand, in a truly user-friendly manner.

Comment: *MoWR, RD, and GR have been trying for quite some time to facilitate better coordination between central and state water institutions for sustainable planning, development, and management of river basins. However, water being a state subject and the absence of a national water law, which makes it obligatory on the states (sharing river basins) to come under a common platform for water management, have not led to any positive outcomes. The state-level water agencies must feel that they need the support of central institutions for informed decision-making in the field of hydrology and water resources planning. It can construct a million small WHS in a river basin falling* *in its territory; farmers can drill as many wells as they want in their farms. The central agency will only be a silent onlooker.*

Thus, the creation of NWC, without institutional reforms comprising the establishment of clearly defined water rights/water entitlement and effluent discharge norms, with laws and regulations for restricting the development and use of water resource and water pollution, will end up as a futile exercise.

On the flip side, when the committee itself has decided the water agenda for India, by writing an obituary of large dams, what scientific knowledge does it expect the NWC to generate and provide to the state governments?

Your report doesn't encourage interlinking of rivers, one of the most vocal commitments of Water Minister Uma Bharati.

Our report contains a summary of all the scholarly work available on interlinking of rivers (ILR). This work demolishes the engineering myth that water must not be allowed to flow "wastefully" into the sea. Scientists fear that the humongous ILR project could even endanger the integrity of India's monsoon cycle, which depends crucially on fresh river water flowing into the sea. However, our report is not centrally concerned with this question and is not really into the pro- versus antibig dam debate. It is much more concerned with the challenge of ensuring that the water stored in dams, present or future, actually reaches the farmers. This is a low-hanging fruit that can give us an increase of millions of hectares of irrigated area at much less than the cost of the ILR and in much less time, avoiding all interstate conflicts, land acquisition problems, and also corruption that has become a big issue in irrigation projects over the years.

Comment: *First of all, the committee did not refer to any scholarly work, worth the name on interbasin water transfers, to comment on an issue as complex as ILR. How can there be studies on the economic and ecological viability of ILR, when the very idea itself is in the conceptual stage and not a scheme with a blue print? The committee has cited the fringe literature, which are opinionated articles, from civil society, to argue that ILR is unsound and unwise. The fact of the matter is that even prefeasibility studies are not prepared for most of the river links proposed under the concept of ILR. Once that is done, we will be able to say how many of the links are actually viable based on technical, economic, and social grounds. The committee also ignored the fact that there are already many interbasin water-transfer projects under operation in our country.*

The 26.0 M ha of additional irrigation, through efficiency improvements in irrigation, as argued by the committee (see the report), is fictitious. Such a low-hanging fruit does not exist. The problem is that the committee's notion of irrigation efficiency is simply outdated and not used by water resources scientists worldwide. This is a large-scale recycling of the water supplied from reservoirs and diversion systems and lost in conveyance to the farmers' fields.

The committee's idea of transferring water from the reservoir to every farmer's field without having necessary infrastructure is simply untenable. We need to build an infrastructure to appropriate water in places where it is available in plenty and convey to regions of demand for every farmer to get irrigation water. To put things in perspective, water transfer from the relatively water-rich Narmada River to the parched areas of North Gujarat, Kachchh, and Saurashtra is already changing the rural scenario of these regions in terms of agricultural incomes and domestic water security. Similarly, water transfer from Godavari to irrigation tanks in water-scarce Krishna basin is changing the agricultural landscape of the region. Many of the metropolitan cities including Delhi, Ahmedabad, Chennai, Bengaluru, and Hyderabad are dependent on water supply from distant surface reservoirs located to meet their daily water demands. Of course, the interbasin water-transfer projects have to be scrutinized for their impact on ecology and human displacement, but to simply ride them off without proper assessment is intellectual arrogance. To ignore their role in increasing agricultural incomes and provision of water supply in water-scarce regions will have its own perils. By turning a blind eye to this reality, the attempt of the committee seems to be to call a moratorium on large water systems while not proposing anything worth the name of water reforms. So, there is surely an agenda, but it is not of water sector reforms.

Chapter 5

A Critique of Mihir Shah Committee (2016) Report on Water Reforms in India

5.1. REFORMING INDIA'S WATER SECTOR

India is facing insurmountable water problems. A country of 130 crore people, with rapid urbanization and industrialization and changing consumption patterns and life styles, the challenges facing India's water managers are immense. Managing water today is not only about developing new sources through conventional means but also about finding new sources of water and allocating the limited water among various competitive uses (Kumar, 2010) while protecting the integrity of the hydrologic system (Kumar, 2016a). Intersectoral water allocation requires greater use of sound economic principles for efficient pricing, introduction of water-use restrictions, etc. (Kumar, 2010). Water resources management requires the application of ecological sciences, ecological economics, and environmental economics. It is quite obvious that our water sector institutions have to be equipped with more technical man power, with greater competence and with people from multiple disciplines. They also call for new institutions for basin-wide water allocation and for undertaking resource management action (Kumar et al., 2012a; Saleth and Dinar, 1999).

It also needs to be appreciated that water is a state subject in India except for the interstate river basins, and most of the reforms should be to affect changes in the orientation and working of the state agencies, which plan, design, execute, and run water projects (Kumar, 2016a). The role of the central-level agencies such as Central Water Commission (CWC), Central Pollution Control Board (CPCB), and Central Ground Water Board (CGWB) is limited to hydrologic monitoring of rivers (for river discharge and sedimentation), flood forecasting, groundwater survey and assessment, and water quality monitoring of aquatic systems. In other words, their role is advisory in nature, and they have no direct stakes in the outcomes of their decisions. For instance, the CWC and CGWB can work jointly to come out with a basin management plan for an interstate river basin. But it is up to the affected parties (in this case, the riparian states), which have to agree to this plan and start executing projects as per plan. While doing so, it is quite possible that one state has to forego some of its economic interests for the benefit of

Water Policy Science and Politics. https://doi.org/10.1016/B978-0-12-814903-4.00005-1

another. As a result, in the current institutional setup, the state may show no interest in such plans, as they are not statutory in nature. Given the federal structure of the constitution, the states are more comfortable doing their own plans and executing them without their actions being regulated by any external agency.

5.2. MIHIR SHAH COMMITTEE RECOMMENDATIONS

A committee set up by the Ministry of Water Resources in December 2015, under the chairmanship of Dr. Mihir Shah, a member of the former Planning Commission of India, was to come up with specific recommendations for restructuring the CWC and CGWB. This was on the premise that these institutions are quite outdated and their work needs to be made more relevant to the changing context of India's water sector. The committee seems to have ignored some basic facts about the working of India's water administration, particularly the center-state relationships.

The committee, whose report titled "A 21st Century Institutional Architecture for India's Water Reforms" is now available on the ministry's website, made several recommendations based on its own diagnosis of the ailing sector. Some of the problems identified by the committee are the following: (1) The efficiency in public irrigation schemes is as low as 35%; (2) there is a mounting gap between potential created and potential utilized in the irrigation sector, to the tune of 26 M ha; (3) groundwater is a "golden goose," which accounts for 2/3 of India's irrigation, but its use is alarmingly unsustainable; (4) the proportion of area irrigated by canals is declining fast; and (5) there is no scope for further development of surface water in the country, particularly in view of the fact that most rivers in Peninsular India, which are highly developed, are already facing severe environmental water stress (source: Shah, 2016).

The committee's recommendations, however, are rather mere reflections of the strong ideological position of its members collectively than outcomes of any proper diagnosis of the problems in India's water sector (Kumar, 2016a). Some of them are as follows: (1) plucking the "low-hanging fruit" of 26.0 M ha of additional irrigation that can be achieved by filling the gap between irrigation potential created and potential utilized, (2) water demand management through the promotion of Water User Associations (WUAs) in irrigation commands, (3) participatory groundwater management through nationwide aquifer mapping to be completed on a war footing, (4) river rejuvenation, and (5) integrated water resources management (IWRM) that should be practiced by taking river basin as a unit for planning (see Shah, 2016).

5.3. A MORATORIUM ON LARGE DAMS AND A "PARADIGM SHIFT"

The committee suggested putting too many stipulations on future building of large dams by calling it a greatest source of human tragedy in the country in the form of human displacement and overreliance on groundwater as a future source of water for all needs. These suggestions are not based on any realistic assessment of the situation on the ground. First of all, as pointed out by Shah and Kumar (2008), the estimates of human displacement caused by building of large reservoirs (56 million people as per the estimates by Singh and Banerji, 2002), often used by antidam lobbyists are at best "guestimates" and many times higher than the realistic estimates of the possible displacement (7.845 million), arrived on the basis of submergence area and density of population in the submergence area (see Shah and Kumar, 2008).

The problem with groundwater development is that in the regions that are groundwater abundant (like in the eastern Gangetic plains), the demand for water for agriculture and other sectors is too low owing to very low landholdings, high rainfall and humidity, and several agroecological problems such as flooding. On the other hand, in the areas where the demand for water in agriculture and other sectors (urban drinking and industry) is very high and water scarcity persists, the groundwater resources are overexploited, and millions of wells are failing every year (Kumar et al., 2012b). Clearly, the committee hasn't understood the implications of this dichotomy on the feasible options for addressing India's future water scarcity problems.

The committee went on to suggest CWC and CGWB to work under a single umbrella body, named the National Water Commission (NWC), with their present chairpersons as members, along with a few other members and wings for ecology, environment, etc. and to be headed by a person with public administration background, a bureaucrat. The committee assumes that such an institution can promote IWRM planning at the basin level, for both surface water and groundwater, and take its water management agenda forward. However, the reason for the lack of integrated planning is not the lack of coordination and data sharing, but the lack of the ability to foresee how future development of the resource is going to take place in different sectors and the lack of control on that development, which still lies in the hands of the concerned state departments. The committee failed to understand this issue (Kumar, 2016a). For instance, no water agency in India can forecast how future development of groundwater resources for irrigation is going to take place as well that irrigation is almost entirely in the private domain. If the committee is concerned with institutional reform in the water sector as claimed, it should have dealt with the issue of ownership of groundwater in the report.

The committee thinks that this approach of abandoning large surface water projects and relying completely on groundwater would mark a major paradigm shift in India's water sector and would have long-lasting outcomes and impacts. This "paradigm shift" is a direct pick from the water sector component of the XII (5-year) plan document of India, which the chairman of the committee authored (see Shah, 2013). However, the fact is that most of these solutions proposed by the committee can only bring about cosmetic changes. Even if for a moment, one accepts these recommendations bundled under the "new paradigm of water management" are capable of bringing about major reforms in the sector, the larger concern remains: can the CWC and the CGWB be held responsible for the poor state of affairs of India's water sector and also made accountable for implementing the new water management paradigm for the committee to justify the restructuring proposal? The report doesn't provide any analytic base to establish the link between the poor state of affairs in the water management sector and the current structure of the two central agencies. Ideally, institutional reforms are required to affect changes in the functioning of the state water agencies that plan and develop the water resources. But the committee has not made suggestions for improving their working. On the contrary, it showered compliments on the water resources departments of Gujarat and Madhya Pradesh, which in its opinion had some remarkable work in boosting agricultural outputs through irrigation sector reforms (see Shah, 2016, p. 7 and p. 44).

5.4. FALSE PREMISE, FAULTY DIAGNOSIS

The entire report is built on false premise, faulty diagnosis, and misrepresentation of facts. Firstly, the paradigm shift being advocated is based on neither a proper analysis of the problems nor an understanding of what needs to be done to rescue the ailing water sector in India. Instead, it echoes some of the outdated concepts used for judging the performance of irrigation systems (like the irrigation efficiency concept) (see Perry, 2013, for details) and some of the stale ideas repeatedly tried for almost two and a half decades by lending agencies in India and many other developing countries with no significant positive outcomes (like creating Water User Associations) (Kumar, 2016a). Secondly, the two agencies whose work is under review (CWC and CGWB) neither are linked to the problems identified by the committee nor are parties for implementing the consequent recommendations for solving them, as it is made out to be. We will deal with them one by one and will discuss them in the subsequent paragraphs.

Instead of using the absolute area irrigated by surface sources (canals and tanks), the committee uses time-series data on (net) area irrigated by them as the percentage of total (net) irrigated area to comment on the performance of (public) irrigation and concluded that surface-irrigated area had declined over time in India (Shah, 2016, p. 25). This analysis is flawed and misleading, as the figure considered in the denominator (i.e., the total net irrigated area) has increased substantially over the years. Further, the committee conveniently ignored the fact that the reported area under "other sources" is mainly the area irrigated through lifting from public canals and that it should be considered under "surface irrigation." Also, the committee did not take cognizance of the fact that many large cities and towns in India receive water from public reservoirs, which are primarily meant for irrigation.

Thus, using concocted data on "percentage area under surface irrigation," the committee tries to build the argument that the era of dam building in India is over, as according to it the returns from continued public investments in irrigation projects have been negative during the past few decades, primarily due to incredibly low irrigation efficiencies (25%–35%) and low utilization of the newly created potential. This is the rehash of a false notion being paraded by another member of the committee for several years now (see Shah, 2011). The committee cite this as the major reason for reorienting the Central Water Commission, which in its opinion is more equipped to plan and design water resources development projects for supply management and "…current focus of the CWC on 'development' of water resources for supply-side management alone, the CWC is staffed exclusively with engineers drawn from the Central Water Engineering Group 'A' Service, and lacks almost totally a capacity in any other discipline that interfaces with water resource management" (Shah, 2016, p. 78).

5.5. HOW FAR ARE PUBLIC IRRIGATION SYSTEMS INEFFICIENT?

The efficiency in public irrigation is far higher than what is reported, if one assesses it at the basin scale (Chakravorty and Umetsu, 2003; Perry, 2007, 2013). International agencies have long been using the concept of basin water-use efficiency (see, e.g., Seckler et al., 2003). A mere 25%–35% efficiency in public irrigation schemes, as noted by the committee, means nearly 70% of the water released from the reservoirs or diversion systems would be lost. This works out to be around 280 BCM of water, annually. This "wastage" should end up in the natural sink that is the rivers and the oceans. If we acknowledge the fact that most of the large irrigation systems are located in the water-scarce states of India (Gujarat, Maharashtra, Andhra Pradesh, Tamil Nadu, Karnataka, and Telangana), most of this water should appear at the last drainage point of the rivers in this region. However, this doesn't happen. Most of the rivers have no significant amount of water draining out in normal years, as noted even by the committee. Hence, the efficiency argument is completely flawed (Kumar, 2016a; Perry, 2013). Part of the blame for putting surface irrigation schemes in poor light should also go to the engineers of the water resources department who used these outdated concepts for a long time to justify investments in canal-linking work and Command Area Development Programme.

But this is part of the larger narrative used by many "water sector experts" all over the world to come up with their package of reforms for irrigation sector (that includes promoting participatory irrigation management and creation of Water User Associations) without any analysis of the local context, arguing that it would destroy all evils of the sector, most importantly wastage of water and poor financial recovery (Molle, 2008).

5.6. IS INDIA'S EXISTING IRRIGATION POTENTIAL UNDER-UTILIZED?

The committee's remarks on the large gap between "irrigation potential created" and "potential utilized" are highly inaccurate. Such comparisons are only found in the administrative lexicons of India and do not carry much value in irrigation science. Moreover, the way estimates of "irrigation potential created" arrived at is nothing but fallacious. These figures are often unrealistic and heavily inflated, as they are based on the estimates of quantum of water available in the reservoir and a "design cropping pattern" that never happens in reality. Water inflows into reservoirs can change depending on the amount of rainfall in the catchment and many upstream developments. In the water-scarce northwest, west, and south Indian peninsula, which experience high interannual variability in rainfall, there are huge year-to-year differences in annual inflows into reservoirs. It is quite likely that the area irrigated by surface systems during drought years is much lower than that in wet years and also less than the irrigation potential created. This was illustrated in a recent analysis of irrigation sector performance in Madhya Pradesh, wherein a substantial increase in area under gravity irrigation and well irrigation was noticed in winter during wet year as compared with a drought year (Kumar, 2016b).

Also, farmers shift to water-intensive crops once irrigation water is made available, shrinking the area further (Kumar, 2016b). A modern-day water scientist with a good understanding of hydrology and socioeconomic aspects would never resort to these numbers to critique an agency's performance (Perry, 2013).

On the other hand, as regards the potential utilized, there is heavy underreporting of the actual area irrigated by canals, with no account for water lifted from canals and drains by engine owners, and the area irrigated by wells in the command that benefit from the seepage of canals and return flows from gravity irrigation (Kumar, 2016b). A large amount of water from large reservoirs earmarked for irrigation is actually diverted for drinking and domestic uses and industry, which never get reported in any of the data related to canal irrigation (Kumar, 2016a). In fact, the rate of decline in groundwater levels within canal commands account for only a small fraction of the consumptive water use for irrigation there. The rest comes indirectly from canal water through irrigation return flows.[1] The water balance illustrated above shows that we need to review the way data on canal-irrigated area are collected. Obviously, there is a need for proper accounting of water used in the basin, crucial for assessing the real performance of irrigation systems. In sum, there is no "low-hanging fruit." Every drop of water in these water-scarce basins is captured and used within the basin, though some scope exists for reducing nonbeneficial uses of water such as evaporation from barren soil or fallow.

---

1. For example, evapotranspiration requirements of rice and wheat in Moga District of Punjab are 673 and 268 mm, and the irrigation requirements (excluding effective rainfall) are 268 and 259 mm, respectively. Hence, the total irrigation requirement is 527 mm. Assuming a specific yield of 0.25 and an average decline in groundwater level of 0.5 m per annum, the net discharge from groundwater per annum is only 125 mm. Therefore, of the total ET requirement, contributions from rainfall, canal water, and wells are 414, 402, and 125 mm, respectively. However, not all canal water is directly delivered through irrigation infrastructure. Canal water recharged groundwater from water distribution and delivery infrastructure, which is being recycled though pumping from wells (Prathapar et al., 2012; Amarasinghe et al., 2010).

5.7. WHAT CAN WATER USER ASSOCIATIONS ACHIEVE?

Most scholars in the water sector today agree that the key institutional reform needed in the water sector is to affect behavioral changes for regulating the growth in demand for water in various competitive use sectors (Kumar, 2010; Mohanty and Gupta, 2012; Saleth, 1996). The foregoing analysis takes up to the point that WUAs in their present form will have too little role in achieving water demand management, except taking care of the distribution issues to some extent at the level of the tertiary canals. In the absence of water rights and water entitlements, the WUAs cannot influence the water resources department on the amount of water that the latter plans to release through the outlets. They also cannot decide on the water prices. WUAs are almost defunct in all the states, in the absence of devolution of any kind of powers to them. The state irrigation departments that are concerned with irrigation management are not willing to share any of their powers with the farmer organizations. Such delegation of powers happens only on paper.

Water demand management requires efficient pricing of water in irrigation and other sectors and rationing of or fixing volumetric entitlements in water (Kumar and Singh, 2001; Mohanty and Gupta, 2012). These are long overdue. Only such measures can bring about improvements in water-use efficiency, through optimal use of irrigation water for the crops, allocation of the available water to more efficient crops, or saving water in the existing uses and selling it to alternative uses at a high price. But these are political decisions, which need to be executed by the respective state agencies and are no way within the purview of the CWC and CGWB.

Legal reforms are needed for introducing water entitlements (Kumar, 2010; Saleth, 1996). The committee hasn't delved into these challenging issues and simply found refuge in the "success" of implementing irrigation management transfer in different parts of the world documented by international agencies such as the Food and Agriculture Organization of the United Nations and the erstwhile International Irrigation Management Institute (see, for instance, Garces-Restrepo et al., 2007; Vermillion, 1997). But as Molle (2008) points out, this is in line with the "solutions" prescribed by multilateral agencies and international NGOs for irrigation sector reform for the developing countries, in spite of no concrete evidence emerging so far to show that it is working.

5.8. IS CANAL IRRIGATED AREA IN INDIA DECLINING?

The committee resorted to a statistical jugglery when it compared surface irrigation with well irrigation. It used figures of "percentage of net area irrigated by different sources" over a long time period (1951–61 to 2005–09) to show that surface irrigation has declined and well irrigation went up consistently, while knowing clearly well that the denominator, that is, the "total net irrigated area" over this time period, had dramatically changed. Analysis of the actual gross irrigated area by different sources during 1950–51 to 2006–07 clearly shows that the gross surface-irrigated area, after steadily going up, has stagnated in the recent years and the same has happened to well irrigation as well, though well-irrigated area is much higher than the former (Kumar et al., 2009, p. 73). Stagnation in canal irrigation in the recent years is also due to the diversion of water from reservoirs to nonagricultural uses.

This comparison by the committee is very unfortunate. Water available for canal irrigation in a reservoir is determined by the runoff generated in the upper catchment. Within the bandwidth of climate variability, the runoff volume from an upstream catchment is reasonably a constant. Furthermore, except in Ganga-Brahmaputra-Meghna and Godavari river basins, there is hardly any surplus water in Indian river basins that can be developed for surface irrigation. Therefore, the area irrigated with canal water cannot increase, unless we import water into these water-scarce basins. If surface irrigation schemes are not maintained, area irrigated by canals will only decrease with the passing of time. On the other hand, well irrigation in areas underlain by alluvial and sedimentary aquifers depends on groundwater accumulated over several years. The consistent decline in groundwater levels observed across the western and northwestern parts of India affirms that most well irrigators are drawing from this storage, and therefore, this dependency is unsustainable.

Leaving such issues of misuse of statistics, such crude statistical comparisons should have been avoided by a committee that strongly considers groundwater and surface water as part of the same hydrologic system and advocates integrated surface and groundwater management. Showing its clear bias against large dams, the committee doesn't make a mention of the fact that large reservoirs (most of which were primarily built for irrigation, excluding those for hydropower) today supply water to several large cities (Mukherjee et al., 2010). Besides, in the absence of these reservoirs and command areas, groundwater development would have been less sustainable, a point argued long ago by Prof BD Dhawan in the book edited by him titled "Big Dams: Claims and Counterclaims" (see Dhawan, 1990). These two factors would have been considered by any scholar who makes an objective assessment of the sector. Not doing that shows the professional bias against public irrigation.

5.9. IS "AQUIFER MAPPING" A PANACEA FOR GROUNDWATER OVER-EXPLOITATION?

On the groundwater side, it is a well-established fact that the two major reasons for the overexploitation of aquifers are the absence of well-defined water rights in groundwater and inefficient pricing of electricity supplied in the farm sector (Kumar, 2007; Saleth, 1996). The problem is surely not due to the lack of sufficient information about the occurrence of groundwater and its flows. The farmers and the official agencies know well that the resource is fast depleting in many pockets. Participatory aquifer mapping can do nothing to halt this ongoing menace. As envisaged, it can provide microlevel details of the groundwater-bearing formations (good aquifers) within the geologic strata in different localities (Kulkarni and Shankar, 2009). Neither the committee is able to visualize how the participatory aquifer mapping gets translated into participatory groundwater management under the much touted National Aquifer Management Programme with a budgetary allocation of Rs. 3539 crore under the 12th Plan.

While we can invest some more resources in refining the current assessment methodology, we already have sufficient information to start acting in the problem areas. But those actions are going to be institutional in nature, and the state governments should have political will for that. This is quite clear from the lack of enforcement of groundwater acts despite a number of states having these on paper (such as the Maharashtra Groundwater (Development and Management) Act, 2009 (GOM, 2013). The experience of developed countries such as the United States, Australia, Mexico, and Spain in dealing with groundwater management issues and the limited success achieved by them clearly shows that the solution lies in creating robust institutions—which can clearly define water rights of individual users and enforce them or put tax on groundwater use based on volumetric withdrawal (source: based on Kemper, 2007; Rosegrant and Sinclair, 1994; NWC, 2010; Rosegrant and Binswanger, 1994; Saleth, 1996). On the other hand, programs like participatory aquifer mapping can have serious negative consequences on equity, with the rural elite using the information for their benefits—say, for instance, to buy land in areas where there is good amount of water lying underneath.

5.10. UTOPIAN IDEA OF RIVER REJUVENATION

While the committee discussed about "rejuvenation of rivers" at length, it failed to offer any practical suggestions on the ways to achieve it, except talking platitudes about integrated surface and groundwater development (Kumar, 2016a), and a special wing in the proposed National Water Commission. To get water back in the river and to maintain the base flows from aquifers, we need to cut down groundwater and surface water withdrawals in the river basins drastically. But the problem today is that most of the basins in the naturally water-scarce regions are "closed," with no water going out of the basin, uncaptured. With a large amount of arable land and high aridity, irrigation water demand is far greater than the renewable water resources in these basins (Kumar et al., 2012b), and agricultural growth is suffering in these regions due to the lack of water. Telangana, Rayalaseema, and western Rajasthan are just a few examples. Cities located in these basins are not getting adequate amount of water to supply to their rapidly growing population, as they have to compete with agricultural sector. The rivers in these regions need water to maintain the ecosystem health (Kumar, 2010). While the situation is very precarious, the committee doesn't offer any insights about the strategies for reversing the current negative water balance in these basins.

One way to achieve these multiple objectives is by improving water-use efficiency in crop production (kilogram of biomass per unit of evapotranspiration) and at the same time limiting the total crop production to the existing levels so as to affect reduction in the demand for water. This can be affected through technological interventions and institutional reforms. The report doesn't make any reference to this and the challenges involved.

Another option is interbasin water transfers from water-abundant basins, which is already happening in limited ways between basins that are characterized by sharp differences in resource endowments. For instance, water from Narmada River is diverted from the Sardar Sarovar reservoir (situated in the water-rich South Gujarat that impounds the river) through the 458 km-long Narmada Main Canal and put in several of the rivers in north Gujarat that are experiencing environmental water stress, as the canal crosses these rivers on its course (Jagadeesan and Kumar, 2015). We would require such large and sophisticated water infrastructure for interbasin water transfers to store, divert, and transfer water from water-rich basins to distant regions that are perennially water-starved. Therefore, capacity building of state- and central-level institutions has to be in disciplines such as hydrology, hydraulics, groundwater modeling, water engineering, river morphology, environmental hydrology, dam safety, ecological and environmental economics, and water law to take up the new challenges. There is no doubt that water demand management has to receive great attention in the coming years, with the cost of production of water touching astronomical heights. But to achieve this, we need legitimate, regulatory institutions that can look at issues of water allocation and water pricing at the state level, with good understanding of economics of water (Kumar, 2016a). Merely having a wing for "river rejuvenation" in the NWC does not help.

5.11. CAPACITY BUILDING

The committee in its report has suggested capacity building of CWC and CGWB officials in the field of IWRM. This suggestion should be welcomed. There is also a scope for broadening the thematic areas of the training to cover environmental, economic, and social aspects. But the real question is how such training modules from the National Water Academy, the training wing of CWC, would find takers from the state water agencies that do not show any willingness to introduce water pricing and experiment with new approaches for basin planning and institutional models for water resources management.

Further, asking the National Water Academy to build the capacities of farmers and NGOs to undertake water management activities in the field is expecting too much. The report says: "It is not enough to train water professionals (engineers) within the CWC and in the States. Water resource management has to be a participatory process with all stakeholders (including NGOs) and local communities (particularly farmers at the Panchayat level) who should be made aware of the water resource issues and who need capacity building for management of the limited water resources (e.g., participatory irrigation management, appropriate crop selection, micro-irrigation, conjunctive use, wastewater reuse/recycling, etc.). The present training activities of the NWA in this respect are practically negligible (Shah, 2016, p. 85)." This is the mandate of Water and Land Management Institute set up in different states nearly three decades ago.

5.12. 21ST CENTURY INSTITUTIONAL ARCHITECTURE?

The report has a lot of rhetoric about IWRM, but with less substance. It does not suggest any reform measures to operationalize the concept, which eventually has to happen at the state level. The IWRM is also a complex model of water governance and management. Water of varying quality is demanded by multiple sectors. Allocating this water among various stakeholders involve trade-off. The domestic and industrial sectors, which invariably have the first call on available water, largely contaminate more water than what they consume, through wastewater generation. Appropriate treatment of the wastewater generated by domestic and industrial sectors will enable the reuse of water by another sector. Environmental flows, on the other hand, often improve the quality of water, and this water too may be available for use by another sector.

By incorporating all three dimensions of water management, that is, quantity, quality, and timing, in water allocation decisions, conflict among sectors can be minimized. This would not necessarily eliminate difficult, but essential, trade-offs between agricultural and environmental sectors. These trade-offs will affect catchment's hydrologic processes and outcomes accordingly. As water scarcity looms large, the paradigm of water management should shift from freshwater for all uses and users to "timely supply of adequate amounts of water of acceptable quality" to each sector and each user. This will encourage recycling of water within catchments, and the use of alternative water supplies for agriculture within catchments.

While IWRM is a sound concept, the challenge is in operationalizing it through appropriate coordinating institutions at the river basin level. As discussed by Biswas (2008, 12), "...operationally it has not been possible to identify a water management process at a macro- or meso-scale which can be planned and implemented in such a way that it becomes inherently integrated, however this may be defined, right from its initial planning stage and then to implementation and operational phases." This is in the realm of the state water agencies that have to create such coordination mechanisms. The committee did not deal with these myriad issues. Instead, it passes judgments that these organizations have outlived their mandate and recommends restructuring of these agencies into a NWC with basin-wise headquarters, without having any vision about future water management needs and the role they have to play to make it a reality. This can only demoralize the cadre of engineers in these institutions.

Though the report is vociferous about the integrated development of surface water and groundwater at the basin level for optimal resource use and resources management, it is silent on the institutional mechanisms to achieve it. As noted by Kumar (2010), the critical issue is that various line agencies concerned with water resources development and water supply at the state level (water resources department, groundwater department, rural water supply department, and local municipalities and corporations) are not confronted with opportunity cost of overappropriating the resource or polluting the water. When an irrigation department overappropriates water from a catchment, it does not have to compensate for the damages caused by environmental water stress downstream. Further, when different sectors are competing for water, how could allocation of water across competing sectors be achieved at the basin scale, without proper water entitlements? What will be the rules for allocation? Who defines those rules and entitlements, and who enforce them? In order for the line agencies (say water resources department, water supply department, department of environment and forests, and industrial department) to conform to these allocations, legal reforms are necessary for these entitlements to have legal sanctity. The report has not delved into any of these critical issues.

If river basin organizations (RBOs) have to perform water resources management functions, including water quality management of rivers and streams, funds would be required. This can come from resource tax (from water users) and pollution tax (from the polluters), which are also part of institutional reforms. Legal reforms are needed to introduce such institutional changes (Kumar, 2010, 2014). So, creating RBOs is not an end in itself. Institutional architecture is about making these RBOs capable of performing its functions. Even for the Water User Associations, on which the report puts a lot of emphasis, to function effectively by claiming the legitimate rights of water from the irrigation bureaucracy and to decide on the price for water and recover it from the users, changes in existing laws are needed. The change in legal framework would also be determined by the sociological context. But the report did not discuss any of these. What it does is a discussion on some organizational forms, which will have no legal teeth or financial capabilities.

5.13. CONCLUSIONS

To sum up, the committee in its report has been irrational in its criticism of the Central Water Commission, which works mostly in advisory capacity, for the poor performance of water sector, for which the state water resources departments should be mainly held responsible. At the same time, the committee failed to recognize its important technical functions that are crucial for water resources assessment, project planning and designing, and project execution and monitoring. The Harvard historian David Blackbourn in his recent book, "The Conquest of Nature: Water, Landscape and the Making of Modern Germany," illustrated the way in which each succeeding generation takes the achievements of the past generation for granted and wonders how their predecessors could have been so stupid as to not have dealt with the new generation of challenges

(Blackbourn, 2007). What we forget is the fact that the state of art (of water management) is always provisional. What works today may not work tomorrow.

Finally, contrary to the committee's claim, which is reflected in the title of its report, there is no "21 century architecture" in the institutional model suggested, but only wishful thinking. This is because the committee failed to recognize the fact that institutions cannot exist in a legal vacuum and without financial viability (see Saleth and Amarasinghe, 2010). In the absence of laws, which recognize water rights of individuals and user groups, legitimize institutions such as RBOs and WUAs, and give them authority to allocate water rights/entitlements among sectors and users and levy water tax and pollution tax, these institutions will only lie on paper and won't be able to perform their functions (Kumar, 2010). Without adequate financial resources, these institutions cannot undertake any of the designated functions. In other words, the institutional architecture that this report talks about should indicate the legal and economic architectures that underpin the former and also enhance its feasibility for practical relevance and policy impact. In this sense, the report can be taken at the most as indicative rather than conclusive.

To propose 21 century architecture for water institutions, the committee should have incorporated eminent experts on institutional economics and organizational behavior, in its subcommittees. It has not done so. Instead, it has incorporated many social activists, who are known for their strong ideological positions against large water infrastructure projects (Kumar, 2016a) on "capacity building" and "institutional restructuring." Although the report has intended to be an attempt at the restructuring of water institutions, it has, unfortunately, ended just as an exercise in restructuring "water organizations," especially at the national level, though some attention is given to Water User Associations. As a result, the contents of the report failed to become a serious analytic attempt toward a practical approach to institutional restructuring of the water sector.

REFERENCES

Amarasinghe, U.A., Malik, R.P.S., Sharma, B.R., 2010. Overcoming growing water scarcity: exploring potential improvements in water productivity in India. Nat. Res. Forum 34 (3), 188–199.

Biswas, A.K., 2008. Integrated water resources management: is it working? Int. J. Water Resour. Dev. 24 (1), 5–22.

Blackbourn, D., 2007. The Conquest of Nature: Water, Landscape, and the Making of Modern Germany. W.W. Norton & Company, New York and London.

Chakravorty, U., Umetsu, C., 2003. Basinwide water management: a spatial model. J. Environ. Econ. Manag. 45 (1), 1–23.

Dhawan, B.D., 1990. Big dams: claims, counterclaims. Econ. Polit. Wkly. 25 (29), 1607–1608.

Garces-Restrepo, C., Vermillion, D., Muoz, G., 2007. Irrigation Management Transfer. Worldwide Efforts and Results. FAO Water Report 32, FAO of the United Nations, Rome.

Government of Maharashtra, 2013. Maharashtra Groundwater (Development and Management) Act, Maharashtra Act No. XXVI of 2013. Maharashtra Government Gazette, Maharashtra.

Jagadeesan, S., Kumar, M.D., 2015. The Sardar Sarovar Project: Assessing Economic and Social Impacts. SAGE Publications India.

Kemper, K.E., 2007. Instruments and institutions for groundwater management. In: Giordano, M., Villholth, K. (Eds.), Agricultural Groundwater Revolution: Opportunities and Threats to Development. CAB International, pp. 153–172.

Kulkarni, H., Shankar, P.V., 2009. Groundwater: towards an aquifer management framework. Econ. Polit. Wkly. 44 (6), 13–17.

Kumar, M.D., 2007. Groundwater Management in India: Physical, Institutional and Policy Alternatives. Sage Publications, New Delhi.

Kumar, M.D., 2010. Managing Water in River Basins: Hydrology, Economics, and Institutions. Oxford University Press, New Delhi.

Kumar, M.D., 2014. Thirsty Cities: How Indian Cities Can Meet their Water Needs. Oxford University Press, New Delhi.

Kumar, M.D., 2016a. Proposing a solution to India's water crisis: 'paradigm shift' or pushing outdated concepts? Int. J. Water Resour. Dev. https://doi.org/10.1080/07900627.2016.1253545.

Kumar, M.D., 2016b. Irrigation sector 'Turn Around' in Madhya Pradesh? Econ. Polit. Wkly. 51 (19), 67–70.

Kumar, M.D., Singh, O.P., 2001. Market instruments for demand management in the face of scarcity and overuse of water in Gujarat, Western India. Water Policy 3 (5), 387–403.

Kumar, M.D., Narayanamoorthy, A., Singh, O.P., 2009. Groundwater irrigation versus surface irrigation. Econ. Political Wkly. 44 (50), 72–73.

Kumar, M.D., Bassi, N., Venkatachalam, L., Sivamohan, M.V.K., Niranjan, V., 2012a. Capacity Building in Water Resources Sector of India. Occasional Paper 5-0112, Institute of Resources Analysis & Policy, Hyderabad.

Kumar, M.D., Sivamohan, M.V.K., Narayanamoorthy, A., 2012b. The food security challenge of the food-land-water nexus in India. Food Security 4 (4), 539–556.

Mohanty, N., Gupta, S., 2012. Water reforms through water markets: international experience and issues for India. In: Morris, S., Shekhar, R. (Eds.), India Infrastructure Report. Oxford University Press, New Delhi.

Molle, F., 2008. Nirvana concepts, storylines and policy models: insights from the water sector. Water Alternat. 1 (1), 131–156.

Mukherjee, S., Shah, Z., Kumar, M.D., 2010. Sustaining urban water supplies in India: increasing role of large reservoirs. Water Resour. Manag. 24 (10), 2035–2055.

National Water Commission, 2010. The impacts of water trading in the southern Murray–Darling Basin: An economic, social and environmental assessment. NWC, Canberra.

Perry, C.J., 2007. Efficient irrigation; inefficient communication; flawed recommendations. Irrig. Drain. 56 (4), 367–378.

Perry, C.J., 2013. Beneath the water resources crisis. Econ. Polit. Wkly. 48 (14), 59–60.

Prathapar, S., Sharma, B.R., Aggarwal, P.K., 2012. Hydro, Hydrogeological Constraints to Managed Aquifer Recharge in the Indo Gangetic Plains. Water Policy Highlight 40. IWMI-Tata Water Policy Research Programme.

Rosegrant, M.W., Sinclair, R.G., 1994. Reforming Water Allocation Policy through Markets in Tradable Water Rights: Lessons from Chile, Mexico, and California, EPTD Discussion paper 2. International Food Policy Research Institute, Washington.

Rosegrant, M.W., Binswanger, H.P., 1994. Markets in tradable water rights: potential for efficiency gains in developing country water resource allocation. World Dev. 22 (11), 1613–1625.

Saleth, R.M., 1996. Water Institutions in India: Economics, Law and Policy. Commonwealth Publishers, New Delhi.

Saleth, R.M., Amarasinghe, U.A., 2010. Promoting irrigation demand management in India: options, linkages and strategy. Water Policy 12 (6), 832–850.

Saleth, R.M., Dinar, A., 1999. Evaluating water institutions and water sector performance. World Bank technical paper no. 447. World Bank, Washington, DC.

Seckler, D., Molden, D., Sakthivadivel, R., 2003. The concept of efficiency in water resources management. In: Kijne, J.W., Barker, R., Molden, D.J. (Eds.), Water Productivity in Agriculture: Limits and Opportunities for Improvement, Comprehensive Assessment of Water Management in Agriculture. CABI Publishing in Association with International Water Management Institute, UK.

Shah, T., 2011. Past, present, and the future of canal irrigation in India. India Infrastructure Report, pp. 70–87.

Shah, M., 2013. Water: towards a paradigm shift in the twelfth plan. Econ. Polit. Wkly. 48 (3), 40–52.

Shah, M., 2016. A 21st Century Institutional Architecture for Water Reforms in India. Ministry of Water Resources, RD&GR, Govt. of India.

Shah, Z., Kumar, M.D., 2008. In the midst of the large dam controversy: objectives, criteria for assessing large water storages in the developing world. Water Resour. Manag. 22 (12), 1799–1824.

Singh, S., Banerji, P., 2002. Large Dams in India: Environmental, Social & Economic Impacts. Indian Institute of Public Administration (IIPA), New Delhi, India.

Vermillion, D.L., 1997. Impacts of Irrigation Management Transfer: A Review of the Evidence. Research Report 11, IWMI, Colombo, Sri Lanka.

FURTHER READING

Amarasinghe, U., Sharma, B.R., Aloysius, N., Scott, C., Smakhtin, V., De Fraiture, C., 2005. Spatial Variation in Water Supply and Demand across River Basins of India. Research Report 83. IWMI, Colombo, Sri Lanka.

Government of India, 2013. Guidelines on Command Area Development and Water Management Programme, Ministry of Water Resources. RD&GR, Government of India.

Kumar, M.D., van Dam, J.C., 2013. Drivers of change in agricultural water productivity and its improvement at basin scale in developing economies. Water Int. 38 (3), 312–325.

Chapter 6

Does Hard Evidence Matter in Policy Making? The Case of Climate Change and Land Use Change

6.1. INTRODUCTION

For many decades now, there is significant amount of empirical research from other parts of the world that illustrates the impact of land-use changes, which are occurring as a result of urbanization; establishment of crops after tree felling; tree plantation in areas earlier occupied by conventional field crops; and establishment of trees in grass land, on hydrology and water resources. The best researches are from Japan, Spain, Murray-Darling basin in Australia, and rainforests of Malaysia dealing with forest hydrology.

Enough empirical research is available from Brazil to show that deep-rooted trees consume more water than the crops cultivated in one or two seasons and can deplete the deep soil strata and shallow groundwater (Oliveira et al., 2005). Research has also shown that grasses are sometimes more effective in reducing runoff rates and sediment load in runoff water than deep-rooted trees (Verstraeten et al., 2002). How these effects get played out in a catchment in terms of altering the hydrology depends on the nature of trees, climate, and topography.

Yet, this scientific knowledge about the impact of land use on hydrology has not yet found much space in the policy circles in India. On the contrary, the policy making in the sectors of agricultural development, forestry, watersheds, and drinking water supplies is driven by several "misconceptions" about the impacts of increased vegetation cover, especially tree cover on streamflows and groundwater availability, increase in tree cover on sedimentation and runoff quality, and causes of streamflow reduction in rivers in the recent decades. The most important one among them, prevalent since the early 1980s, is that rainforests

Water Policy Science and Politics. https://doi.org/10.1016/B978-0-12-814903-4.00006-3

bring rains.[1] Another one is that streamflow reduction in rivers and extreme hydrologic events such as floods and droughts is caused by climate change. Another important issue is the ways researchers influence policies. Often, their claims are not supported by sufficient empirical evidence, and at times, there is mounting evidence on the contrary. Yet, persuasive writing takes their recommendation to policy.

In this chapter, we provide illustrative examples to highlight these issues. The examples cover the following: (1) hydrologic impacts of massive tree plantation, (2) basin-wide impacts of watershed development and small water harvesting, (3) impacts of intensive groundwater development on streamflows, and (4) basin-level hydrologic impacts of climate variability. We also argue that there is a need for strengthening the institutional regime that scrutinize research for their quality, before their recommendations are taken onboard by governments for policy changes. We also suggest a few ways of achieving it.

6.2. FOREST-WATER INTERACTIONS

There is no mystery about the hydrologic impact of increasing forest cover, as science of forest hydrology and catchment hydrology is quite well developed and there were a lot of scientific evidence available from the west for quite some time and now from within India through experimental stations. The science is simple: forest vegetation takes a lot of water in the form of transpiration to grow. It is well established that impact of forest catchments on catchment water yield through increase in evapotranspiration is greater as compared with grassed catchments. However, the relative impact depends on the vegetation conditions, soil types (Hamilton and King, 1983; Oliveira et al., 2005; Zhang et al., 1999), and climate (Zhang et al., 1999). The water for meeting ET demand of trees can come partly from precipitation "interception," partly from the moisture in the active root zone, partly from the unsaturated zone underlying the soil, and partly also from shallow groundwater in the catchment.

While the impact of trees on overall yield of the catchment would be negative, depending on how the increased demand is being met from the hydrologic system, the impact will be seen either on runoff or groundwater or both. If the deep soil strata (vadose zone) along with top soil contribute to evapotranspiration of

1. In the late 1970s or early 1980s, the voice of environmental movement was becoming loud in India. The southern state of Kerala already had one of the most influential environmental movements of all times, the movement against the building of a large hydropower dam in Silent Valley in the rainforests. The fierce (but nonviolent) environmental movement spearheaded by the Kerala *Sastra Sahitya Parishad* (The Federation for Science and Literature in Kerala), a membership-based organization that had many eminent scholars, scientists, and personalities from the field of literature as its active members, could force the state government to abandon this project. One of the most powerful arguments, among others, put forth by the movement as the negative impact of the dam, was that project apart from submerging vast stretches of pristine rainforests in the Silent Valley area in Palakkad, and destruction of these rainforests would seriously affect the occurrence of rains in the region.

trees, then the impact will be on both groundwater system and runoff, whereas if shallow groundwater contributes to ET, then the most significant impact will be on base flows and groundwater. The higher the leaf area index, the higher will be the transpiration (Hamilton and King, 1983; Oliveira et al., 2005). On the other hand, litter cover on the forest floor increases infiltration rate of precipitation significantly (Hamilton and King, 1983).

As is evident from the foregoing discussion, the confusion arises because of its differential impacts in different climatic settings and also because of the inability to make distinction between forests and trees. In a high rainfall, humid climate, the value of the water lost from the catchment through transpiration from the forest canopy (of trees, climbers, shrub, and grasses) cover may not be very significant when compared with the benefit of reducing the likely soil erosion that otherwise would happen in the absence of it, because of the sheer fact that the quantity of water there has much lower value in economic terms as compared with its quality. Contrary to this, in a semiarid or arid area, the value of the water that is lost from the forest cover could be far greater than the economic value of the water containing lesser amount of sediments.

Under the same hydrologic regime (in terms of rainfall, humidity, temperature, and solar radiation), the runoff from a catchment that has low density of forests would be much more than the runoff from a densely forested catchment in the same hydrologic setting (Krishnaswamy et al., 2012; Zhang et al., 1999). It is up to us to analyze the trade-off between reduced quantum of flows and improved quality of water in different contexts and decide which intervention to choose. But controversies are often created when we fail to contextualize this. One cannot suggest clearing forests, just for the sake of getting more runoff from the catchment, so long as the forest has ecosystem and economic values. At the same time, it should also act as a deterrent for those who advocate tree plantation in the rural watersheds, arguing that it would increase the streamflows and/or base flows in the catchment.

However, popular perception is that the trees that grow in the nature do not take any water, as there is no artificial application of water involved in most cases. South Africa has well documented the impact of invasive *Prosopis* species on water resources (Witt, 2010). Dense stands of *Prosopis* in the northwestern part of South Africa are estimated to use 192 MCM of water annually (Versfeld et al., 1998). Vast spreading *Prosopis* accounted for 89% of the incremental water use in the Northern Cape Province of South Africa where it was estimated to use 997 m^3/ha/year in case of dense vegetation (Versfeld et al., 1998). As regards impact on groundwater, Fourie et al. (2003) found that clearing 1 ha of *Prosopis* from the riparian zone of the Rugseer River near Kenhardt, South Africa, reduced groundwater depletion by 50 m^3/month. Many countries in Africa (particularly South Africa, Ethiopia, and Kenya) have programs for eradicating *Prosopis juliflora*, an invasive species (Witt, 2010).

A more illustrative example is available from Australia. A study of the impact of land-cover change and climate on shallow groundwater in Murray

Groundwater Basin of Australia was carried out by Leblanc et al. (2011). The study involved the use of time-series data on in situ groundwater levels. It showed that after clearance of the land of native vegetation for agricultural use, water table in the shallow aquifer in the basin rose from 1980 to 1997, causing dryland salinity in the basin. This is indicative of increased groundwater recharge due to deforestation and reduced evapotranspiration losses from the shallow water-table aquifer. The study further found that with successive multiyear droughts since 1997, the long-term trends in groundwater levels, which started with the early clearance of land by European settlers in the 1800s, got reversed temporarily. While the native tree cover provided heavy evapotranspiration of the shallow groundwater year round along with depleting the moisture in the soil profile, the agricultural crops transpired much less water, which also included the moisture in the soil profile.

Tree planting is an annual ritual in many areas, which have little green cover, without looking at its potential impact on water resources. After the first spell of rains, several lacs (100,000) of saplings of trees are planted in the wastelands and common lands in the rural areas and government land in the urban areas. If checking soil erosion is the ultimate aim, then the option of developing grass cover could as well be thought about.

The government of Telangana recently announced an ambitious scheme to plant 230 crore (2300 million) trees across the state (and 1200 million in the nonforest areas), christened as *Haritha Haram* (green garland). The National Forest Policy envisages a minimum of 33% of the total geographic area in the country be covered by forest cover for environmental stability and ecological balance. The Telangana government wanted to increase the green cover to the stipulated 33% from the current 24%. While the objectives are noble, the program is the aggrandizement of an old idea that is "social forestry." It is said that the chief minister is very keen to have fruit trees, and not the native species of trees that grow well in the different climatic regions of the state. While this might bring cheers to the ears of the environmentalists and tree protection activists, there is hardly any discussion about the impact it would have on the state's water balance, which is already precarious.

Here are some back-of-the-envelope calculations to show the enormous impact on water resources. A tree sapling would require at least 1–2L of water a day (depending on the size), and leaving aside the monsoon season (during which it would consume water from the soil profile), a sapling would consume at least 200–400L of water in the year. This water will have to be made available as the root development would take time. So, 1200 million saplings would consume 200–400 MCM of water per year. Once these saplings mature into trees, they will create a total canopy cover of 12,000 km^2 or 1.2 million ha based on the assumption that an average tree will have a canopy cover of $10\,m^2$ ($120 \times 10^7 \times 10\,m^2$).

The amount of water these trees would take from the hydrologic system once matured, based on an ET demand of around 2000mm/year, is around

24,000 MCM per annum. It is quite obvious that no power on Earth can supply this much water to those trees every year. So, they will find their own way to survive. While some water will come from the soil, the rest will come from the vadose zone and the shallow groundwater. If we assume around 200 mm of water to come from the top soil (which otherwise would be lost through barren soil evaporation), the remaining 21,600 MCM will be through runoff reduction and groundwater depletion. Can a water-scarce region afford to lose such a large quantity of water for greening? Even if only 10% of these trees finally survive, the state would lose 2400 MCM of water. This is five to six times higher than the lean season flow in river Krishna at Vijayawada in a dry year.

Most of Telangana is in the rain shadow area, with an average of 700 mm of annual rainfall, with many districts receiving less than 500 mm. Vast stretches of the state's landmass are parched and devoid of adequate irrigation facility, with less than 40% of the cultivable land under irrigation. The city of Hyderabad, which is also seeing tree plantation, faces acute water shortages. Earlier watershed approaches are propagated in the surrounding areas of the city of Hyderabad. In Maheswaram area, where the World Bank and the state government together spent a whopping sum on water conservation works like vetiver grass plantations, social forestry, and pasture development has now turned into a booming commercial area. The experiments of community forestry and social forestry have not met with expected success in the rural front. Forest areas are shrinking due to constant encroachments.

Yet, the question is not whether *Haritha Haram* would happen in reality or not. It is simply that there is no hydrologic planning in place. It is high time that the state takes a look at its water balance and how much water would be taken by the plantation, how much by the rejuvenated tanks under *Mission Kakatiya*, and how much would be left to be picked up by the medium and large reservoirs. One thing is sure. If by any chance the "green dream" of Telangana state comes true through the efforts of the present government, it would surely and certainly deplete all the water in the soil profile and in the water bodies. The only suggestion therefore is to go for tree plantation in areas that used to have good tree cover and where water is in plenty, before the state gets its act together for proper water resources planning.

6.3. GROUNDWATER-SURFACE WATER INTERACTIONS

The hydrologic science dealing with groundwater-surface water interaction is quite well developed as is evident from the 1980 publication of UNESCO that dealt with the subject (Wright, 1980). Groundwater-surface water interactions in wetlands are controlled by difference in heads of wetland water body and groundwater, the local geomorphology of the wetland and the wetland and groundwater flow geometry. The groundwater-surface water interactions can be of three types: wetlands losing water to the underlying aquifer, wetland gaining water from the underlying aquifer, and wetland gaining water from the aquifers in some locations and losing in others (Jolly et al., 2008). Llamas (1988) documented the interaction between groundwater and wetlands in Spain and showed how wetland conservation is in conflict with groundwater exploitation in the region.

GW-SW interactions in wetlands are mostly controlled by factors such as differences in head between the wetland surface water and groundwater, the local geomorphology of the wetland (in particular, the texture and chemistry of the wetland bed and banks), and the wetland and groundwater flow geometry. The GW-SW regime can be broadly classified into three types of flow regimes: (i) recharge, wetland loses surface water to the underlying aquifer; (ii) discharge, wetland gains water from the underlying aquifer; or (iii) flow through, wetland gains water from the groundwater in some locations and loses it in others. However, it is important to note that individual wetlands may temporally change from one type to another depending on how the surface-water levels in the wetland and the underlying groundwater levels change over time in response to climate, land use, and management.

Yet, there is too little appreciation of the negative impacts of increased groundwater use in upper catchments on downstream streamflows in river basins in quantitative terms. This "hydroschizophrenia" of water scientists and engineers had taken a heavy toll on management of transboundary water resources in India. Many of the water allocation decisions of tribunals set up in the past under the Interstate Water Disputes Act (1954) have only looked at only surface water, a point well-articulated in my study of Narmada river basin back in 2006 (Kumar, 2010).

A detailed empirical study in the Narmada river basin of central India showed that the main case of reduction in streamflows in the trunk river in recent years was the overexploitation of groundwater in the upper basin areas of Madhya Pradesh, which caused reduction in base flows. This phenomenon was manifested by the reduction in the rate of annual groundwater drawdown in many locations and reversal of annual groundwater level trends in many others spite of increased draft (Kumar, 2010). A recent study of Ashoka Trust for Research in Ecological Economics (ATREE) on two catchments of Cauvery river basin (Arkavati and Noyyal) also provides evidence to this effect. But the fact remains that in many of the states that encompass large parts of the upper catchments of major river basins (like Narmada, Mahanadi, Mahi, Krishna, and Godavari), groundwater use had increased remarkably in the past 3–4 decades. But the reducing flow in rivers is often attributed to climate change (Kumar, 2014a).

Poor knowledge about groundwater-surface water interactions has also taken a toll on estimation of groundwater resources in hilly and undulating terrains, leading to overestimation of recharge, as the current methodology of the Central Ground Water Board (CGWB) does not consider the base flows realistically (source, based on data provided by CGWB, 2012). This serves the interests of political leaders who represent rural areas. The latest victim of this poor resource evaluation methodology is Mahanadi basin in Chhattisgarh state. The groundwater assessment for the basin shows underutilization of groundwater resources, due to the reason that the groundwater discharge into streams is not captured in the assessment. The outcome is that the state had now decided to go for intensive use of groundwater with new agricultural power connections and subsidized electricity and solar PV systems. Once this is done, this will have serious implications for both monsoon flows and lean season flows in Mahanadi, affecting downstream uses.

6.4. BASIN-WIDE IMPACTS OF WATERSHED DEVELOPMENT AND SMALL WATER HARVESTING

Watershed development and construction of small water harvesting structures are another significant intervention altering the hydrology of several watersheds and basins, seriously affecting inflows into tanks and other minor irrigation reservoirs in their downstream areas (Batchelor et al., 2002; Glendenning and Vervoort, 2011; Kumar et al., 2008; Ray and Bijarnia, 2006). We have the proverbial case of Peter taking Paul's Water in many large and small basins in India, evident from the studies in Saurashtra and Madhya Pradesh, Rajasthan, and Karnataka. The growing interests of some of the villager leaders in activities such as pit digging and desilting of tanks can be explained not by any traditional wisdom or "popular hydrology" or faith in decentralized water management but by "political economy." To compound the problem, many research studies in the past were designed in such a way that they became part of a larger narrative that SWH is always benign.

Increase in area under rainfed crops alone can significantly reduce the "blue water flows," without any physical diversion of water actually taking place from the hydrologic system (Falkenmark, 2004; Kumar, 2010). So, we should not romanticize rainfed farming beyond a point. It is an established fact that the rainfed crops have much lower water productivity than their irrigated counterparts. Since these phenomena are very difficult to monitor and correlate with changing flow regimes of rivers, they will pose new challenges in water management in the coming decades. It is quite clear that we will require new scientific tools and institutions to deal with this menace.

Besides some of the underdeveloped basins in India, these results have implications for countries in sub-Saharan Africa, which are experiencing slow but steady increase in cropped area expansion and supported groundwater and surface irrigation (Critchley and Gowing, 2012). The degree of development of surface-water resources in these countries is currently quite low (Snelder et al., 2012). While there is a great thrust on increasing the use of groundwater for intensifying farming of small holders, there is very little knowledge about the properties of aquifers in this region, particularly the safe yield (Foster et al., 2006). Parallelly, there is growing emphasis on small water harvesting (Critchley and Gowing, 2012; Snelder et al., 2012) for supplementary irrigation for rainfed crops (Kijne et al., 2009). Therefore, catchment level studies should be undertaken to assess the optimum level of water harvesting and groundwater draft in different catchments (Snelder et al., 2012; Kumar, 2012). In the same way, future development of groundwater should be based on sound planning, supported by studies to understand the sustainable yield of aquifers and nature of interaction between groundwater and surface water.

6.5. WATER MANAGEMENT IMPLICATIONS OF CLIMATE VARIABILITY

A very important aspect in water management, which is not dealt with by climate change researchers, is climate variability. This phenomenon seems to be taking far greater toll on our water resource availability, water supply, water access, and overall water security and the structure of our economy than climate change impacts. The general tendency today is to attribute extreme hydrologic events to climate change phenomenon, without acknowledging the fact that almost all regions in India have been experiencing high variability in weather parameters such as rainfall, temperature, and humidity between years. Such variability (expressed in terms of coefficient of variation) is high in arid and semiarid regions as compared with subhumid and humid regions (Kumar et al., 2008).

The recorded streamflow at the last drainage point in Luni river basin in the hyper arid western Rajasthan is illustrative of this phenomenon. The streamflow was nil in many of the years, especially during the droughts of 1986–87, 1987–88, 1988–89, and 1989–90 (Fig. 6.1). The highest streamflow during the 40-year period was recorded in 1990–91, with a total discharge of 2011 MCM. While the mean discharge was 219.08 MCM, the discharge exceeded 1000 MCM only in 3 out of the 40 years. The estimated mean virgin flow (runoff from the catchment) in the basin was 383.71 MCM, the highest estimated virgin flow (1990–91) was 2144 MCM, and the lowest was 26.33 MCM (during 1972–73 and 1987–88). This huge variation was explained by the variation in rainfall in the basin. The weighted average of rainfall in the basin (u/s of the gauging station) varied from 148.6 mm in 1987 to 749.2 mm in 1990 (Kumar and Bassi, 2013).

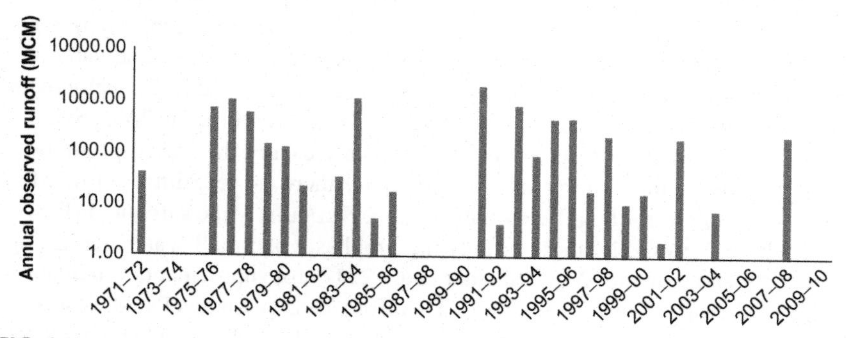

FIG. 6.1 Gauge discharge at Gandhav, Luni river basin (1970–71 to 2009–10).

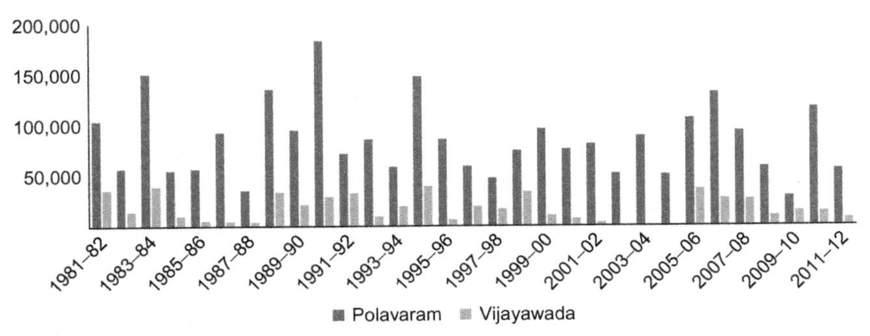

FIG. 6.2 Gauge-discharge data of Guntur and Krishna district stations in Andhra Pradesh.

As shown in Fig. 6.2, in the otherwise water-rich Godavari river basin, the annual outflows drop to 29,000 MCM in a poor rainfall year from 183,000 MCM in a wet year (1990–91). The annual outflow of Krishna river is a mere 300 MCM in a bad year as compared with 39,000 MCM in a good year (source, based on CWC as available in India WRIS). These outflows are the cumulative effect of the overflows from the existing reservoirs and barrages in these basins, and the runoff generated downstream of these systems.

So, what strategies we adopt to manage water in a basin in a dry year will have to be quite different from what we do in a wet year. Obviously, many of the popular solutions, such as rainwater harvesting and artificial groundwater recharge schemes for drought proofing, will not work in a bad year (when there is too little rainfall), whereas in a good year, they are not required at all.

Unfortunately, the impacts of climate variability on hydrology and the adaptation needs haven't received much attention of scholars from this part of the world, while there is a good amount of work in the western United States (Californian drought) and Australia. Merely from a utilitarian perspective, it is important to understand what happens to a river system during the years of extreme weather conditions if we want to predict the likely impact of climate change with lower values of predicted rainfall, on basin hydrology. A related point is that when the variation in rainfall can be as high as 1:5, causing a streamflow variation in the order of magnitude of 1:100 (as seen in the case of Luni basin), a question that needs to be posed is "what utility climate change models, which predicts a rainfall change of 10%–20% over a 50-year period, serve if they cannot capture the variability?"

The future challenge is also in designing adaptive systems for water management that can absorb the climate shocks in the form of extremely variable flows, soil moisture, and evapotranspiration. Some of the strategies that would receive attention are designing of reservoirs for multiannual storage, conjunctive management of surface water and groundwater, and designing small reservoir systems that are capable of flow regulation rather than mere water impoundment.

Urbanization resulting in increase in built-up area can increase runoff intensities and total amount of flows. But the increasing pollution load in runoff water can be a major problem, with oil, carbon, and heavy metals in it. This needs serious attention. The recharge pits constructed in urban areas for injecting storm water underground, which use simple sand-gravel filters, can do serious long-term damage to the aquifers that are already contaminated with sewage. Such "quick fix" solutions are outcomes of "bourgeois environmentalism" and campaigning by champions of voodoo science, rather than well-thought-out response of the civil society to a growing water crisis in cities (Arabindoo, 2011). The attention has to be on sound storm water management interventions such as "rain gardens," wherever they make hydrologic and economic sense. This is far better than forcing city dwellers to build RWHS through enactment of laws (as being done by Tamil Nadu government), wherein the harvested water ends up in saline formations.

As argued by Kumar (2014b), in the event of climate change, urban centers, which lose a lot of water in their distribution systems, can improve their water security more by reducing leakage up to a point where it continues to be economically viable. What matters here is the cost of leakage reduction for saving kiloliter of water against the cost of augmenting supplies by a kiloliter. In cities falling in water-scarce regions and where leakages are excessively large, leakage reduction measures would be economically viable to a large extent, as the cost of augmenting water supplies to meet the demand-supply gap in such water-scarce regions would be very high. The economics of recycling treated wastewater from cities for municipal water supply against reusing wastewater with secondary treatment for irrigation needs to be carefully analyzed.

6.6. PROPOSING MULTIPLE SOLUTIONS FOR THE SAME PROBLEM

In the early years after independence, India built institutions for fundamental and applied research. They would spend several years in scientific discovery and or developing technologies using the new scientific knowledge so generated. They would first be published in top science journals, and ordinary citizens would come to know about it several years later.

The tone and accent of the research has, however, changed over the years. Many in the new generation of researchers and academics do not believe in peer review of their work. They generally feel that their work should not be subjected to any kind of scrutiny. Part of the reason is the enormous pressure from the donors to show policy impacts of scientific research. But in order to affect immediate policy impact of their work, they would manage to get their work published in some popular magazines or journals or national dailies, thereby building peer pressure. Intriguingly, while their "solution" talks about some complex technology to recharge groundwater or to generate "cheap and clean" solar energy or an irrigation development strategy for achieving miracle agricultural growth, these articles would eventually appear in a popular social science journal, which does not deal with that particular scientific discipline. The stories would also mention how the idea was successfully implemented in a village, by a highly enterprising farmer. In the next few weeks, the newspapers would carry articles on the success of this technology. When the finance minister of the country allocates some $20 million in the budget for trying out this idea, it is touted as the great example of science-policy interface.

The articles would also claim that the government in so and so state accepted all but one of their recommendations. However, it won't take much time for the myth created around this "miracle technology" to burst. This happened in the case of miracle growth in agriculture in Gujarat and Madhya Pradesh, the impact of Jyotigram Yojna on electricity consumption and groundwater use in Gujarat, and the impact of decentralized water harvesting work in Gujarat.

When the people realize that the idea has faltered, the researchers would come up with a brilliant excuse: "government accepted only some of the recommendations we made, and again the implementation was lousy like any government scheme." The blame would also go to the finance minister for not having understood the nuances of this proposition while allocating funds for the program.

These researchers will pass occasional jibe at useful concepts like IWRM, saying that it is a concept that won't work in India and that India needs home grown solutions. While one doesn't ignore the fact that there are enormous operational challenges in implementing IWRM, these researchers do not need to give any explanations, except saying that India uses a lot of groundwater and our water economy is informal (Shah and van Koppen, 2006). However, if the government asks these researchers to come up with recommendations for revising the water institutions in India, they would merrily come up with ideas such as IWRM, arguing that "that is the only solution to problems facing the water sector today" http://wrmin.nic.in/writereaddata/Report_on_Restructuring_CWC_CGWB.pdf.[2] We can forget the tens of articles written earlier about how innovative and home grown solutions like "digging pits" in the country side would bring about water security in the country (Shah et al., 2009a), and how useless is the idea of IWRM is for a developing economy like India (see Shah and van Koppen, 2006; Shah, 2011).

Often, two to three solutions are found by the same group for a single problem in a sequence, yet claiming each time that every solution had worked (see Shah, 2000, 2010; Shah and Verma, 2008). For instance, it was claimed that the technique of "dug well recharging" was applied to solve groundwater depletion in Saurashtra with water levels rising (Shah, 2000). The story was that this became

2. The report of Mihir Shah Committee on restructuring of Central Water Commission and Central Ground Water Board is the latest example. The committee in its report had strongly argued for implementing IWRM principles at the river basin level in India, while one of its key members believes that IWRM doesn't work for India.

hugely "popular" in many other parts of India. As a matter of fact, a scheme churned out by the government to promote "dug well recharging," with financial assistance of Rs. 4000, was a disaster on many counts. The technical feasibility of recharging dug wells using runoff from agricultural fields was poor in the actual field conditions (Kumar et al., 2008). The benefits of recharging the wells were not really perceptible for the individual farmer who invests in it (Krishnan et al., 2009). The fund allocated by the government for recharge per individual well was Rs. 4000 (US $60) was too small for actually constructing the channel, filter box, and installing pipes (Krishnan et al., 2009; Kumar et al., 2008).

However, in spite of mounting evidence that disapproved their earlier claims, these researchers went ahead to invent another solution to solve the same problem in the same region, after 5–6 years, that was a decentralized approach of managing groundwater through construction of check dams by the communities with government support (see Shah, 2009). Many researchers went on to argue that decentralized water conservation work was mainly responsible for the "miracle growth in agriculture" that Gujarat clocked during 2001–02 to 2007–08 (see Gulati et al., 2009; Shah et al., 2009b).

But the researchers went on to invent another solution to solve the groundwater depletion problem in the same region that was feeder line separation for electricity. They argued that farmers were pumping water round the clock to irrigate their fields and were wasting water. It was further argued that separating the (power) feeder line for agriculture from that of domestic sector in rural areas would outsmart farmers who were stealing power yet make everyone (farmers, electricity utility, and the society at large) happy. Their reasoning was that farmers would get 8 h dedicated power supply, and they don't need to steal power as they would be able to manage with less amount of groundwater for agricultural production under the new supply regime. Interestingly, as per the claim, each time government lapped up their recommendations (Shah and Verma, 2008; Shah et al., 2008). This is probably unheard of in the history of science and technology that the rulers wasted no time in understanding the value of great scientific innovations. Nevertheless, a "permanent solution" to groundwater depletion problem was finally found through *Jyotigram Yojna.*

The myth about the miracle growth of Gujarat's agriculture was burst when it was found that the methodology adopted for estimating the agricultural GDP was flawed (see Kumar et al., 2010). Similarly, in the case of *Jyotigram Yojna*, it was found that the introduction of power supply rationing actually led to rampant theft of electricity by farmers in rural area by installing higher capacity pump sets. The power distribution companies have become more vigilant, and the theft had to be curtailed by frequent raids. A total of 2 lac agricultural power connections were raided in 2012–13 alone, and the value of the electricity stolen was 200 million rupees (DNA, 2013).

Nevertheless, the same group of researchers subsequently discovered solar power-run tube wells, with grid connectivity as a solution to create incentive for farmers to cut down groundwater use. The basic premise is that with solar power, farmers use only as much electricity as they want and sell the excess electricity produced to the state electricity board through the grid under a feed in tariff system. As per their proposal, the farmer would become an embodiment of honesty and integrity overnight. Some of these researchers were dead against and even used to laugh at the idea of metering electricity used in agrowells because of its "impracticality." But they strongly believe that "net metering" would work[3] (see Shah et al., 2016). Having said that, in the only case, the idea was tried out, the farmers did outsmart the researchers and the utility, as they figured out that pumping water using the extra energy they had from solar power and selling it is far more profitable than selling the electricity to the grid (Nair, 2016).

The researchers, however, suggested farmers using electric tube wells to surrender their power connections and go for solar PV systems so that they would have incentive to save energy and conserve groundwater. They proposed that the farmer be given a subsidy of around US $1400/kW of installed (peak) solar power, for around 15 million units of average 5 kW capacity (Shah et al., 2016), and this would cost the government exchequer US $105 billion. Again, the same farmers can get electricity at a price of 50–60 paise/kWh after subsidy, and at a price of Rs. 4–5/kWh without subsidy, against the solar power that would cost the economy Rs. 9/kWh. Obviously, no farmer would be willing to bear the full cost of it. But for this new breed of economists, only private cost matters, and economic cost is something that we shouldn't be worrying about at all.

6.7. CONCLUSIONS

To conclude, policy making requires painstaking efforts of evidence gathering. It cannot be left to the whims and fancies of a few individuals and groups who peddle falsehood in public sphere and influence politicians through persuasive writing as it is being done today. The evidence being produced by interest groups in support of their ideas should be subject to scientific scrutiny. The value of a scientific research should be assessed in terms of the benefit it brings to the society in the long run. But in a haste to show the positive impact of any research, we should not be circumventing the rules and norms set by the same society by avoiding peer reviews, debate in scientific forums, etc. The onus is on funding institutions to make sure that the protocols are strictly followed while converting research findings into policy prescriptions for the government.

For this, the institutional regimes that evaluate the research and ideas have to be more robust and respond to the growing needs of the society, particularly in sectors as important as "water." The media needs to publicize research findings that have potentially high social value but needs to exercise great caution to make sure that the arguments made by the researchers and practitioners who propagate these solutions are scientifically valid.

3. "Net metering" facility enables registering the use of electricity by the farmer from the grid and also the amount of electricity supplied by the farmers (from the solar PV system) to the grid.

The much publicized and legendary scientific debate between Albert Einstein and Neil Bohr benefited the society at large, and nothing happened to the reputation of Albert Einstein, who is still acclaimed as the greatest scientist of the modern world, though many believe that Einstein was proved wrong. Therefore, academicians and researchers should not shy away from having public debates on scientific claims and counterclaims.

While these are overarching suggestions for improving the quality of research in the field of water management in India and many other developing countries, some of the areas of specific concerns are as follows. Hydrologic impact of large-scale tree plantation needs to be carefully studies in the light of what is scientifically proven with regard to the impact of tree cover over other forms of land cover on evapotranspiration. In the wake of extreme variability in climate, the modeling studies for predicting future climate change need to be grounded to enhance their utility, by developing models that the changes in streamflows and groundwater recharge in river basins in response to historical fluctuations in hydrologic variables (such as rainfall, temperature, and relative humidity) occurring in river basins. Such models can then predict the likely future changes in hydrologic variables due to the predicted changes in rainfall, temperature, and relative humidity.

Planning for water harvesting should be supported by studies that examine the downstream impact of small water harvesting and watershed management interventions so as to analyze the trade-offs and arrive at the level of intervention that produce the optimum level of social, economic, and environmental benefits at the basin level. Likewise, planning for groundwater development should be supported by studies that examine its negative impact on streamflows and wetland so as to arrive at the optimum level of development that produce least negative ecological and economic impacts. Such studies are particularly relevant for the countries of sub-Saharan Africa that have experienced very low degree development of ground and surface-water resources so far.

REFERENCES

Arabindoo, P., 2011. Mobilising for water: hydro-politics of rainwater harvesting in Chennai. Int. J. Urban Sustain. Dev. 3 (1), 106–126.

Batchelor, C., Singh, A., Rao, M.S.R.M., Butterworth, J., 2002. Mitigating the potential unintended impacts of water harvesting. Paper Presented at the IWRA International Regional Symposium 'Water for Human Survival', 26–29 November, 2002. Hotel Taj Palace, New Delhi.

Central Ground Water Board (CGWB), 2012. Aquifer Systems of Chhattisgarh. North Central Chhattisgarh Region, Central Ground Water Board, Ministry of Water Resources, RD & GR, Government of India.

Critchley, W., Gowing, J.W. (Eds.), 2012. Water Harvesting in Sub-Saharan Africa. Earthscan/Routledge, London & New York.

Deccan News Agency, 2013. Power theft of 201 crore caught last year, May 20, 2013, Ahmedabad.

Falkenmark, M., 2004. Towards integrated catchment management: opening the paradigm locks between hydrology, ecology and policy-making. Int. J. Water Resour. Dev. 20 (3), 275–281.

Foster, S., Tuinhof, A., Garduño, H., 2006. Groundwater Development in Sub-Saharan Africa: A Strategic Overview of Key Issues and Major Needs. World Bank, GW-Mate, Washington, DC (Case Profile Collection 15).

Fourie, F.K., Mbatha, H.V., Dyk, G.V., 2003. In: The effect of vegetation (Prosopis spp.) on groundwater levels in the Rugseer River, Kenhardt, South Africa. WfW Inaugural Research Symposium, August 19–21, 2003, Kirstenbosch, Cape Town, South Africa.

Glendenning, C.J., Vervoort, R.W., 2011. Hydrological impacts of rainwater harvesting (RWH) in a case study catchment: the Arvari River, Rajasthan, India. Part 2. Catchment-scale impacts. Agric. Water Manag. 98 (4), 715–730.

Gulati, A., Shah, T., Shreedhar, G., 2009. Agriculture Performance in Gujarat Since 2000, Can Gujarat be a "Divadandi" (Lighthouse) for Other States? IWMI and IFPRI, New Delhi.

Hamilton, L.S., King, P.N., 1983. Tropical Forested Watersheds: Hydrologic and Soils Response to Major Uses or Conversions (No. 634.922 H3). Westview Press, Boulder.

Jolly, I.D., McEwan, K.L., Holland, K.L., 2008. A review of groundwater-surface water interactions in arid/semi-arid wetlands and the consequences of salinity for wetland ecology. Ecohydrology 1 (1), 43–58.

Kijne, J., Barron, J., Hoff, H., Rockström, J., Karlberg, L., Growing, J., Wani, S.P., Wichelns, D., 2009. Opportunities to Increase Water Productivity in Agriculture with Special Reference to Africa and South Asia. Stockholm Environment Institute, Stockholm.

Krishnan, S., Indu, R., Shah, T., Hittalamani, C., Patwari, B., Sharma, D., Chauhan, L., Kher, V., Raj, H., Mahida, U., Shankar, M., Sharma, K., 2009. In: Is it possible to revive dug wells in hard-rock India through recharge? Discussion from studies in ten districts of the country, in Strategic Analyses of the National River Linking Project (NRLP) of India Series 5. Proceedings of the Second National Workshop on Strategic Issues in Indian Irrigation, New Delhi, India, 8–9 April 2009. Colombo, Sri Lanka, International Water Management Institute (IWMI), pp. 197–213.

Krishnaswamy, J., Bonell, M., Venkatesh, B., Purandara, B.K., Lele, S., Kiran, M.C., Reddy, V., Badiger, S., Rakesh, K.N., 2012. The rain–runoff response of tropical humid forest ecosystems to use and reforestation in the Western Ghats of India. J. Hydrol. 472, 216–237.

Kumar, M.D., 2010. Managing Water in River Basins: Hydrology, Economics, and Institutions. Oxford University Press.

Kumar, M.D., Patel, A., Ravindranath, R., Singh, O.P., 2008. Chasing a mirage: water harvesting and artificial recharge in naturally water-scarce regions. Econ. Polit. Wkly. 43 (35), 61–71.

Kumar, M.D., Narayanamoorthy, A., Singh, O.P., Sivamohan, M.V.K., Sharma, M., Bassi, N., 2010. Gujarat's Agricultural Growth Story: Exploding Some Myths. Occasional paper 2. Institute of Resources Analysis & Policy, Hyderabad, India.

Kumar, M.D., 2012. Water management for food security and sustainable agriculture: strategic lessons for developing economies. In: Kumar, M.D., Sivamohan, M.V.K., Bassi, N. (Eds.), Water Management for Food Security and Sustainable Agriculture in Developing Economies. London and United Kingdom, Earthscan/Roultedge.

Kumar, M.D., Bassi, N., 2013. Water Accounting Study of Luni River Basin and District IWRM Planning for Pali. Technical Assistance Team of the European Union-State Partnership Programme on Water in Rajasthan, Jaipur.

Kumar, M.D., 2014a. The hydro-institutional challenge of managing water economies of federal rivers: a case study of Narmada River Basin, India. In: Garrick, D., Anderson, G., Connell, D., Pittock, J. (Eds.), Federal Rivers Managing Water in Multi Layered Political Systems. Edward Elgar.

Kumar, M.D., 2014b. Thirsty Cities: How Indian Cities can Meet their Water Needs. Oxford University Press, New Delhi.

Leblanc, M., Tweed, S., Ramillien, G., Tregoning, P., Frappart, F., Fakes, A., Cartwright, I., 2011. Groundwater change in the Murray Basin from long-term in situ monitoring and GRACE estimates. In: Treidel, H., Gurdak, J.J. (Eds.), Climate Change Effects on Groundwater Resources: A Global Synthesis of Findings and Recommendations. CRC Press, pp. 169–187.

Llamas, M.R., 1988. Conflicts between wetland conservation and groundwater exploitation: two case histories in Spain. Environ. Geol. 11 (3), 241–251.

Nair, A., 2016. Gujarat: Solar Cooperative at Dhundi Village Sells Water instead of Electricity. Indian Express. 14th August 2016.

Oliveira, R.S., Bezerra, L., Davidson, E.A., Pinto, F., Klink, C.A., Nepstad, D.C., Moreira, A., 2005. Deep root function in soil water dynamics in cerrado savannas of central Brazil. Funct. Ecol. 19 (4), 574–581.

Ray, S., Bijarnia, M., 2006. Upstream vs downstream: Groundwater management and rainwater harvesting. Econ. Political Wkly. July.

Shah, T., 2000. Mobilising social energy against environmental challenge: understanding the groundwater recharge movement in western India. Nat. Res. Forum 24 (3), 197–209.

Shah, T., 2010. Taming the Anarchy: Groundwater Governance in South Asia, Resources for Future. Routledge, Washington, DC.

Shah, T., 2011. Past, present and the future of canal irrigation in India. India Infrastructure Report 2011, Infrastructure Development Finance Corporation, Mumbai.

Shah, T., Verma, S., 2008. Co-management of electricity and groundwater: an assessment of Gujarat's Jyotirgram scheme. Econ. Polit. Wkly. 59–66.

Shah, T., van Koppen, B., 2006. Is India ripe for integrated water resources management? Fitting water policy to national development context. Econ. Political Wkly. August.

Shah, T., Bhatt, S., Shah, R.K., Talati, J., 2008. Groundwater governance through electricity supply management: assessing an innovative intervention in Gujarat, western India. Agric. Water Manag. 95 (11), 1233–1242.

Shah, T., Avinash, K., Hemant, P., 2009a. Will the impact of the 2009 drought be different from 2002? Econ. Polit. Wkly. 44 (37), 18–22.

Shah, T., Gulati, A., Shreedhar, G., Jain, R.C., 2009b. Secret of Gujarat's agrarian miracle after 2000. Econ. Polit. Wkly. 45–55.

Shah, T., Durga, N., Verma, S., Rathod, R., 2016. Solar Power as Remunerative Crop. Water Policy Research Highlight # 10. IWMI-Tata Water Policy Program, Anand, Gujarat, India.

Snelder, D., Bwalya, M., Sally, H., Malesu, M., 2012. Investing in water for agriculture in the drylands of Sub-Saharan Africa considerations for a conducive policy environment. In: Critchley, W., Gowing, J. (Eds.), Water Harvesting in Sub-Saharan Africa. Earthscan/Routledge, London and New York.

Versfeld, D.B., Le Maitre, D.C., Chapman, R.A., 1998. Alien Invading Plants and Water Resources in South Africa: A Preliminary Assessment. Report No. TT 99/98. Water Research Commission, Pretoria.

Verstraeten, G., Oost, K., Rompaey, A., Poesen, J., Govers, G., 2002. Evaluating an integrated approach to catchment management to reduce soil loss and sediment pollution through modelling. Soil Use Manag. 18 (4), 386–394.

Witt, A.B., 2010. Biofuels and invasive species from an African perspective—a review. GCB Bioenergy 2 (6), 321–329.

Wright, W.E., 1980. Surface Water and Groundwater Interaction, A Contribution to the International Hydrological Programme. Studies and Reports in Hydrology 29. International Commission on Groundwater, United Nations Educational, Scientific and Cultural Organization, Paris.

Zhang, L., Walker, G.R., Dawes, W., 1999. Predicting the Effect of Vegetation Changes on Catchment Average Water Balance. Technical Report 99/12. Cooperative Research Centre for Catchment Hydrology.

Chapter 7

Mission Kakatiya for Rejuvenating Tanks in Telangana: Making it a *Mission Possible*

7.1. INTRODUCTION

With the bifurcation of the erstwhile state of Andhra Pradesh into two states, namely, Telangana and Andhra Pradesh, dispute has emerged over the allocation of surface-water resources in the two river basins (Krishna and Godavari) of the original state of AP between the new states. Agriculture in the two semiarid and drought-prone states is heavily dependent on irrigation, and as a result, the dispute is essentially over the sharing of water appropriated by the interstate dams, namely, Srisailam and Nagarjuna Sagar, which were earlier supplying water to the geographic areas of both the states, and a new claim made by Telangana from the 80 TMC water of Godavari river allocated for the newly proposed Polavaram reservoir in Andhra Pradesh by the Godavari water disputes tribunal (Deccan Chronicle, 2017). As compared with Andhra Pradesh, Telangana is a relatively water scarce in lieu of lower rainfall, higher aridity, and poorer groundwater potential.

The overall public sentiment in Telangana is that the geographic area of the state has been historically neglected, and similar feeling prevails over the allocation of water from the large reservoirs and electricity produced by the hydropower schemes made at the time of state bifurcation. Fully knowing the public sentiment, the government of Telangana had launched many schemes in key sectors of development, including irrigation, drinking water supply, forestry, and environment. Telangana region is historically known for the Kakatiya tanks, known after the rulers of the region who built many of these structures for drought proofing of rural areas. Mission Kakatiya for rejuvenating the irrigation tanks in the plateau region of the state is one of the flagship schemes of the government of Telangana, with the state government claiming that it would increase the extent of utilization of its current allocation from the two river basins, namely, Krishna and Godavari, and help the region become drought proofed.

Southern India is known for tank irrigation and is reported to have nearly 120,000 such structures, accounting for 60% of the irrigation tanks in the country (Vaidyanathan, 2001). Over the decades, these tanks have undergone

Water Policy Science and Politics. https://doi.org/10.1016/B978-0-12-814903-4.00007-5

significant degradation, manifested by decline in storage of water and reduction in area irrigated and decline in other functions performed by them. The strategies and approach followed in the past for rejuvenating South Indian tanks were not based on any diagnostic study of the factors responsible for their degradation and found reasons for their poor performance in poor physical condition and collapse of traditional institutions that managed them (Kumar and Vedantam, 2016). In any case, they were largely ineffective, with the main reason attributed to the dismal performance being the piecemeal approach to rehabilitation of the physical structure and mechanistic approach followed in creation of tank management institutions (Asian Development Bank, 2006; Centre for Water Resources, 2000; Sakthivadivel et al., 2004).

Too much importance was attached to the presence of user groups in tank commands in determining the performance of tanks. This is evident from the observations in an ADB report that banked on an earlier work by Sakthivadivel et al. (2004). "A study under the IWMI-Tata Water Policy Program in 2004–05 examined the various livelihood options in tank irrigation under different scenarios and gender-related issues in tank rehabilitation in 40 rehabilitated tanks under different models in three states—Tamil Nadu, Karnataka, and Pondicherry. The study concluded that in tank rehabilitation works, augmenting tank water and increasing tank storage have greater impact on the livelihood options of the landless and marginal farmers. The involvement of self-help groups in tank rehabilitation and provision of funding for income generating activity has a marked effect on their livelihood. The tanks are likely to be more sustainable when all the villagers become members of a tank users' group" (ADB, 2006, p. viii). Interestingly, neither Sakthivadivel et al. (2004) nor ADB (2006) does suggest any concrete measures to augment water inflows into the tanks.

There are no studies that show that the rehabilitation program as envisaged under Mission Kakatiya would help improve the dependability of tanks as irrigation sources and enhance the performance in terms of irrigated area. In this chapter, we identify the hydrologic and socioeconomic opportunities and constraints for improving the performance of tanks in Telangana in the current context based on learnings from recent research on the hydrologic and socioeconomic processes causing degradation of tanks. It examines the validity of the claims about the impacts of the interventions under *Mission Kakatiya* and suggests the key protocols that need to be adopted for selecting tanks for rehabilitation in order to achieve the intended outcomes.

7.2. TANKS IN TELANGANA

Agriculture is the primary source of income for 78% of the population of the newly carved state of Telangana but currently produces only 30% of the total income of the state (Pingle, 2011). Nearly 85% of the cultivated area is rainfed, and tank irrigation still remains as one of the major sources for irrigation

(Deccan Chronicle, 2015). Marginal and small holdings constitute 86% of total agriculture holdings in the state, making agriculture a subsistence source of livelihood for the majority of the population (Directorate of Economics and Statistics, 2015). The state of Telangana has 47,907 tanks with an irrigation potential of 2,263,498 acres spread over 10 districts. Between 1956 and 2001, there has been a reduction in tank-irrigated area in the range of 4.5 lac acres (Nag, 2011). Currently, the state statistics claim that only 37% of the potential area is served by the tanks in the state.

In terms of proportion of the net irrigated area in the state, tanks constitute only 8.4%, and its contribution has been fluctuating between years depending on the rainfall (see Fig. 7.1). Fig. 7.1 also shows clearly that groundwater contribution to the net irrigated area in the state is major, at 86%. The total net irrigated area (from all sources) in the state has been highest during 2013–14 at 2.289 M ha and lowest during 2009–10 at 1.493 M ha. The gross irrigated area stood at 2.53 M ha during 2014–15 (Source: Directorate of Economics and Statistics-2015, Govt. of Telangana).

Tanks are man-made wetlands as per wetland classification. Out of the total of 8262 man-made wetlands in the state of Telangana (with wetland area more than 2.25 ha and covering a total wetland area of 492,059 ha) as per the National Wetland Atlas, 5714 are tanks/ponds constituting a total wetland area of 95,212 ha. The rest are man-made reservoirs, water logged areas, and rivers and streams. The district-wise number of tanks, their wetland area, and the water-spread area pre- and post-monsoon are given in Table 7.1.

This is against a figure of 47,907 tanks considered by the government of Telangana for rehabilitation. The difference is due to the difference in size norm. This means that nearly 88% of the tanks have wetland area less than 2.50 ha. What is most interesting is the fact that water-spread area of these wetlands shrinks drastically after the winter, touching the lowest point during peak summer. Table 7.1 shows that summer water-spread area (36,748 ha) is less than

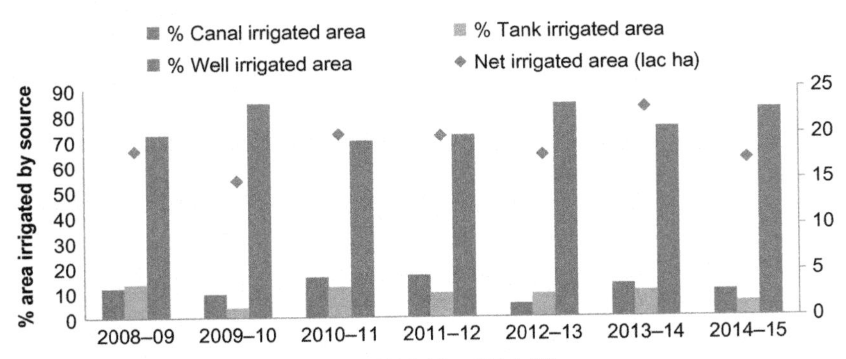

FIG. 7.1 Irrigation pattern in Telangana (2008–09 to 2014–15).

TABLE 7.1 Number of Tanks and their Wetland Area in Different Districts of Telangana

Sr. No.	Name of District	Total Number of Tanks/Ponds	Total Wetland Area (ha)	Average Wetland Area Per Tank/Pond (ha)	Total Water-Spread Area (ha)	
					Postmonsoon	Premonsoon
1	Adilabad	590	7383	12.51	5090	1947
2	Nizamabad	915	19,152	12.51	13,581	7629
3	Karimnagar	692	10,535	20.93	8802	3590
4	Medak	1066	20,116	15.22	16,250	8011
5	Hyderabad	12	107	18.87	71	73
6	Ranga Reddy	281	3287	8.92	2728	2002
7	Mahbubnagar	340	5424	11.70	3509	2510
8	Nalgonda	601	11,702	15.95	8573	5857
9	Warangal	659	9462	19.47	8433	2367
10	Khammam	558	8044	14.36	5547	2762
11	Total	5714	95,212	150.44	72,584	36,748

Source: Authors' own estimates based on National Wetland Atlas, Andhra Pradesh, prepared by the Indian Space Research Organization, Ahmedabad.

half of postmonsoon, that is, November 2010 (72,584 ha). This has major implications for the total water availability of these tanks and the various functions that these tanks can perform in different seasons. Comparison of district-wise data shows that Medak has the largest number of tanks, followed by Nizamabad and Nalgonda. But, in terms of average size, tanks/ponds in Karimnagar are the largest, with an area of 20.93 ha.

Another interesting observation is that the "ratio of the area irrigated by the tank and the wetland area," which reflects the physical characteristic of tanks and also has a significant bearing on the way in which tank degradation in terms of reduction in irrigated area has occurred in the past (Kumar and Vedantam, 2016),[1] varies widely between districts. For the analysis, we have considered the area irrigated by the tanks in 1970–71, assuming that the real deterioration in tank performance started only after this year. The wetland area (as estimated through remote sensing imageries) of the tank was taken from the wetland atlas prepared by the Indian Space Research Organization, Ahmedabad. Since the estimates of wetland area do not consider the tanks with wetland area less than 2.5 ha, it might induce some errors in the estimation of irrigated area-wetland area ratio. The ratio varies from 2.62 in the case of Medak to 14.18 in the case of Mahbubnagar. The district of Hyderabad was also not considered in the analysis as data on tank irrigation in the district were incomplete.

7.3. THE MISSION KAKATIYA

In the end of 2014, the government of the newly carved state of Telangana launched an ambitious project, titled *Mission Kakatiya*. It is about rejuvenating the 47,000 tanks and lakes spread over nine districts of the state by the year 2020, to bring them back to the past glory, the glory they had when such structures were first built during the rule of the Kakatiya dynasty, which reigned over the region. A notable trend during the dynastic period was the construction of reservoirs for irrigation (now known as "tanks") in the uplands, and around 5000 of them were built by warrior families subordinate to the *Kakatiya* rulers. This dramatically altered the possibilities for development in the sparsely populated dry areas.

The Mission envisaged enhancing the agriculture-based income for small and marginal farmers by accelerating the development of minor irrigation infrastructure, strengthening community-based irrigation management, and adopting a comprehensive program for restoration of tanks (Government of Telengana, 2015). The government has prioritized the restoration of minor irrigation tanks (see Table 7.2 for details) to restore and enhance their effective storage capacity

1. As shown by Kumar and Vedantam (2016) through regression, when the tank-irrigated area/wetland area ratio reduced, the rate of degradation of the tanks in terms of decline in irrigated area in 2004–05 in relation to the original irrigated area of 1970–71 increased.

TABLE 7.2 Distribution of Tanks in Telangana Districts and Number of Tanks Considered for Phase I of Mission Kakatiya

	District	Total Tanks	Tanks in Phase I
1	Karimnagar	5939	1188
2	Adilabad	3951	790
3	Warangal	5839	1168
4	Khammam	4517	903
5	Nizamabad	3251	650
6	Medak	7941	1588
7	Ranga Reddy	2851	570
8	Mahbubnagar	7480	1496
9	Nalgonda	4762	952
	Total	**46,531**	**9306**

Source: Government of Telangana.

to 255 TMC (7225 MCM), so as to fully utilize Telangana's allocation of 255 TMC of water from Godavari and Krishna. The restoration works sanctioned are desilting of tank beds, repair of sluices, repairing of feeder channels, etc. and are to be completed in a time span of 5 years.

7.4. TANK REHABILITATION STRATEGY: REPETITION OF THE PAST

The initiatives to rejuvenate the tanks in the region are not new. They were many attempts in the past to rejuvenate the tanks in the erstwhile undivided AP state (Kumar and Vedantam, 2016). However, this is the first attempt by any state government to rehabilitate such a large number of water bodies in one go, using funds from its own budget. The earlier works at rehabilitation of tanks were all done with bilateral funding (either from the World Bank or the ADB or the European Union). These works largely involved civil works such as strengthening of the embankments, construction/repair of waste weir/sluice, lining of canals, cleaning of supply channels, and jungle clearance. Then, there was this garnishing with water users' associations (Sakthivadivel et al., 2004). The underlying assumption was that the tanks were degraded because the tank management institutions that existed in the past collapsed (with the demolition of the *Zamindari* system and introduction of Ryotwari system and a few other factors) and that once WUAs (of ayacut farmers) are created, things would all fall in place. A related assumption was that the WUAs would desilt the tanks periodically (to restore their capacity), clear the supply channels and maintain the water distribution channels, and equitably distribute the water, and the performance of the tanks would henceforth be better, with larger inflows from their catchments and larger impoundment of water. This almost became an axiom in the development and policy circles.

If this was so easy as it is made out to be, it is equally intriguing that no scholar really bothered to find out how come in the past no village community really came forward to persuade the government to "rehabilitate" a system, which was claimed to be offering such great benefits to the poor people but was brought to disuse by external factors, through some of the civil works mentioned above. The real issue is that the tanks and tank management institutions of South India were so glorified that few questioned the validity of two underlying theories in tank management programs—first relates to what civil works can do to alter the tank hydrology, and the second relates to the impact of institutions on tank hydrology and physical performance in the current scheme of things. All these approaches inherently consider the village communities as hapless spectators to the assault on their tanks by external agents, who encroaches tank supply channels, tank beds, and catchments and not as party to this. But, this is far from the reality. As noted by Esha Shah (2008), the tanks in South India stood testimony to the increasingly extractive statecraft involving coerced labor, highly oppressive caste systems, and expropriation of surplus by elites and were symbols of enormous money and muscle power enjoyed by feudal landlords and warlords, respectively (Shah, 2008).

Nevertheless, the outcome of these interventions was that these tanks hardly performed any better than in the past few decades. The major problem was the inadequate inflow from their catchments. But, there was hardly any systematic and scholarly attempt to understand "where the water was disappearing" or in other words what was causing reduction in inflows into the tanks from their catchments, with the exception of the one sponsored by DFID in Karnataka and AP (see Batchelor et al., 2002). Their study showed how intensive watershed work and increased groundwater draft in the catchments reduced the tank inflows. Ignoring all these factors, the blame eventually went to the lackadaisical attitude of the agency toward building water user associations though there is no denial of the fact that this was also done mechanistically.

Coming back to *Mission Kakatiya*, a budget allocation of INR 2016 crore and INR 2083 crore was made for the years 2014–15 and 2015–16, respectively. It appears that even after several years of experience with tank rehabilitation, we seem to be repeating those historical mistakes of following a pure civil engineering approach to tank rehabilitation with no attention being paid to hydrology and ecology. The focus is on earthwork, waste weir construction, canal lining, etc., and more of these structural interventions essentially mean more funds for such projects. Though it is a noble idea to make water available to distressed populations in the state, who had invested in unsuccessful bore wells in this hard rock region, the approach seems to miss out on the fact that mere desilting

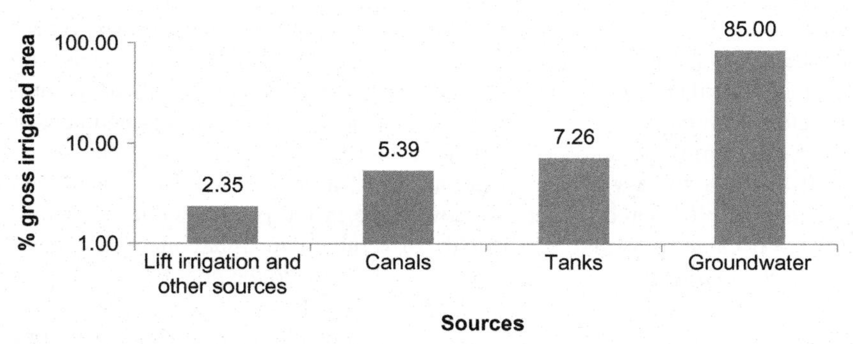

FIG. 7.2 Gross irrigated area by sources (2012–13), Telangana. *(Source: Authors' own analysis using Season and Crop Report, Andhra Pradesh, 2012–13.)*

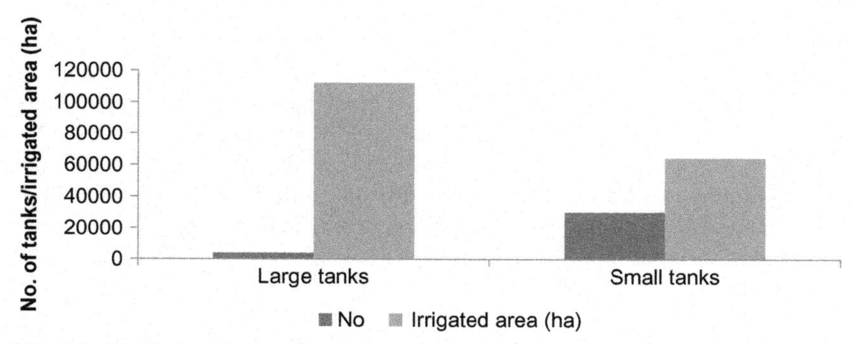

FIG. 7.3 Number and irrigated area by tanks in Telangana (2012–13). *(Source: Authors' own analysis using Season and Crop Report, Andhra Pradesh, 2012–13.)*

or deepening of tanks may not lead to overall increase in water availability. As a matter of fact, none of these interventions can alter the hydrology of the tank catchments. Unfortunately, such projects pass through the scrutiny of economists and planners, with a myriad of exaggerated benefits such as direct irrigation, groundwater recharge, nutrient rich silt as fertilizers, and fish production. One only wonders how the estimates of irrigation potential are arrived at when one doesn't really know how much inflows these tanks would receive.

In reality, almost 85% of the gross irrigated area in Telangana is from wells, and only about 7% is from tanks (Fig. 7.2). Further, 90% of the tanks in Telangana are small tanks with a command area of less than 100 acres, and together, they irrigate only one-third of the total tank-irrigated area. A total of 3864 large tanks accounts for 67% of the tank-irrigated area (Fig. 7.3). If it is so, one needs to understand the economic rationale behind picking up all the tanks for rehabilitation. Thus, it needs to be seen the difference these renovated tanks can make after a huge public expenditure. Already, there are reports on large-scale corruption involving contractors and some officials from irrigation department and poor implementation of the scheme. In some case, contractors are charged with not having adequate resources to undertake desilting and in many other cases deepening works.

Earlier research by Institute for Resource Analysis and Policy (IRAP) in undivided AP, with detailed field surveys in Kurnool, Nizamabad, and Vizianagaram, had shown that there is excessively high degree of degradation of tanks in the past four decades or so. Further, two important processes are altering the hydrology of these tanks. First is intensive use of groundwater in the catchment (through bore wells), which reduces the base flows (or groundwater outflow into the streams) that contribute to the tank inflows. Secondly, the increased cultivation facilitated by access to wells for irrigation in the catchment led to runoff from the catchment getting captured by the farm bunds and used in situ. In many areas like in northern Karnataka, there is a lot of plantation of water-guzzling trees such as eucalyptus in the catchment, which leave no water downstream (these trees act like pumps, sucking the water from the deep strata and grow very fast). This is indicated by the negative correlation between (1) density of wells in the catchment and rate of reduction in tank performance and (2) cropping intensity and rate of reduction in tank performance. The tanks whose catchments did not experience cropping intensification and increase in irrigation wells over time continue to perform well (Kumar and Vedantam, 2016).

7.5. INTENDED AND ACTUAL IMPACTS OF TANK RESTORATION

The government of Telangana ideates the following gains due to the expansion of irrigated area to cover the gap ayacut: (1) technology impacts through the adoption of resource conservation-cum-production technologies when the project is fully implemented; (2) diversification to cover irrigated area under high-value and low water-intensive crops such as chillies, maize, and vegetables; (3) development of fisheries; (3) improvement of livestock; (4) reduction in waterlogged area; (5) increase in groundwater levels and water quality there by getting the lands beyond command area under bore well irrigation; and (6) power savings due to the reduced need for well irrigation that is currently used to supplement the insufficient tank water. Some of these projections of future benefits involve unrealistic assumptions. For instance, how does one expect waterlogging in a region where groundwater resources are mined? How does one expect that farmers would grow high-value crops, with improved water availability, when it is clearly shown that paddy is the most preferred crop in the gravity irrigation systems in Telangana? With dewatered aquifers, it is quite likely that as a result of desilting of tank beds, the percolation of water would increase. But, this would be at the expense of direct irrigation from tanks. One can obviously see some gaps in the way the project is conceptualized.

With just 27 months into the program, it is too early to measure the impacts. Nevertheless, a few academic studies are available that show the impacts of this much publicized scheme on certain aspects. A study by the University of

Michigan on the impacts of Mission Kakatiya in two villages of Adilabad and Karimnagar districts found that the use of silt removed from the tank bed in the agricultural fields led to a dramatic increase in crop production of up to 500%. The study also noted a high variation in the production of cotton from 2 quintals to 15 quintals for fields where silt was not applied to the ones where it was applied, respectively (The Hindu, 2016a). A study by ICRISAT concluded that the silt recovered from the tanks helped in improving the moisture retention capacity of farms. Due to an increase in yield of 1000 kg/ha for cotton, savings on fertilizers and pesticides in the range of INR 2500 to INR 3750 per hectare were observed[2] (The Hindu, 2016b). As one can clearly see, these are surely not the major intended impacts of the project, and there are no studies so far looking at the hydrologic impacts, especially on groundwater regime, impacts on irrigated area, etc.

For that matter, studies looking at the local-level hydrologic and socioeconomic impacts of tanks will not be sufficient to conclude whether rehabilitation has actually helped or not. This is particularly true for Telangana and many regions in the south Indian peninsula. A major part of Telangana state falls in Krishna basin, and the rest falls in Godavari basin. Krishna is a "closed basin," except in very wet years (Biggs et al., 2007). Though Godavari is an open basin (Kumar et al., 2011) with substantial amount of water flowing draining into the Bay of Bengal, most of the uncaptured flows appearing in the stretches of the river in Telangana region are from Maharashtra and Chhattisgarh and not from the drainage areas in Telangana. This is because the rainfall in the basin parts that fall in the state is low to medium and aridity is high, and there are many local water storages including tanks and reservoirs. Hence, increased capture of water from tank catchments through storage capacity enhancement would be possible only in very wet years, and again, this would lead to the reduction in outflows into the tributaries of the two rivers. Such interventions can therefore affect the inflows into the downstream reservoirs in these basins and the benefits accrued therein.

7.6. HOW TO GO ABOUT DOING TANK REHABILITATION?

The findings of IRAP study on the impact of groundwater intensive use on tank hydrology have serious implications for the way tank rehabilitation program should be conceptualized. Given the fact that only about 5% of the tanks are rehabilitated so far, what is needed at present is a systematic assessment of the catchment hydrology of remaining cascades, rather than doing rehabilitation

2. The study of tanks where works were completed showed that the addition of tank silt by 50–375 tractor loads per hectare improved available water content by 0.002–0.032 g in the soil. An increase in clay was noticed from 20% to 40% in the root zone. A decrease in coarse and fine sand was also noticed, while there was no change in pH, EC, and organic carbon.

"lock, stock, and barrel," in an effort to make it a "mass movement." There is a need to pick up only those tanks that have enough water generated in their catchments. Unfortunately, there are no quick ways to assess the runoff generation potential of these tank catchments. The streams draining into these tanks are not gauged. But, going by the previous discussion, it is quite obvious that tanks that are characterized by intensive cultivation in their catchment with a high density of irrigation well (in both the catchment and the command) should be entirely excluded.

For the rest, runoff has to be estimated using some standard methodologies for each tank cascade system. The runoff coefficient would depend on the catchment land cover, the soil conditions, and the antecedent soil moisture. The usual practice in the minor irrigation departments is to use the "rational formula," which uses the catchment area, a "runoff coefficient," and the average rainfall of the catchment, or the strange formula, which assumes a runoff coefficient based on whether a catchment is good or degraded. Such methods produce highly erroneous results. Therefore, internationally accepted scientific methods need to be used to estimate catchment runoff, which take into account the three factors mentioned above. Again, given the high year-to-year variation in the rainfall in these regions, we need to estimate the runoff for typical rainfall years (very wet year, normal year, and very dry year), during which the pattern and the magnitude of rainfall change significantly. Once assessment is done, the tank capacity enhancement should not be undertaken to capture the runoff that occurs in a very wet year, as is usually done. This is because under such a scenario, there would be no outflows from the tank even in the wettest year. The project planners need to recognize the fact that the outflows from these tanks ultimately end up in a river that either drain into a tributary of Krishna or Godavari, depending on which basins they are located.

The question then comes as to what to do with the tanks that are heavily degraded. Desilting would help, as it would produce good nutrient rich soils for farmers. But, this is just a one-time activity, and it takes many years for good quality silt to get deposited in the tank bed. The initial enthusiasm of the farmers (who take their tractors to collect the silt from the tank bed) would fade very quickly after the first monsoon when they do not see much water in their tanks. For such tanks, there is no point in doing heavy earthwork for capacity enhancement, bund stabilization, waste weir construction, etc., all of which involve huge capital investments.

Contrary to this grave reality, the false narrative that is being paraded by some vested interests is that the monsoon water from local tank catchments just runs off uncaptured and that it can be stored in the tanks if their capacity is enhanced. In some years, water flows down the tank cascades. But, we need to recognize the fact that it is not going directly into the ocean. It enters the rivers downstream, and there are many large reservoirs built in Telangana and Andhra Pradesh states to capture that water on those rivers.

Unless the desilted tanks in the region get water from exogenous sources, there is no way the region as a whole will witness an increase in tank-irrigated area. The simple reason is that the total water withdrawal in the region today exceeds the renewable water generated within the region, except for the water in Godavari basin (Kumar et al., 2011). The groundwater depletion and rampant well failures in many parts of Telangana are a manifestation of the precarious water balance of the region (Sishodia et al., 2016).[3] By performing desilting and deepening in some cases, the government may end up redistributing the water in the basins of the state with resultant adverse impact in the downstream areas. Hence, instead of taking a popular approach, effort should be on improving the overall water balance of the region. Further, this should be complemented with increasing area under microirrigation systems, to manage irrigation water demand. But, the potential for adopting microirrigation systems, especially drips, would depend a lot on the cropping system in the command area.

7.7. RECOMMENDATIONS FOR FUTURE

It is reported that the total budget for *Mission Kakatiya* is around Rs. 125 billion for 5 years. There is no doubt that even if 0.5% of this money is spent in doing a scientific assessment of the hydrology of the catchments, a lot of the precious money can be saved. By doing this, it would be possible to know which local catchments have surplus water (and with what probability) that can be stored by increasing the capacity of the cascade tanks and which of the tanks would require imported water. Since microlevel studies to ascertain the feasibility of rehabilitation of tanks involve significant costs, it is necessary to use simple and quantitative criteria for short-listing tanks for conducting detailed investigation, to finally decide on the nature of rehabilitation.

As Kumar and Vedantam (2016) had indicated, tanks that have relatively higher "irrigation-wetland area ratio should be given priority as they are likely to experience lesser degree of deterioration." One of the hydrologic explanations for this is that such tanks would have higher losses from percolation as a result of increased groundwater draft. In areas where fishery is a major economic activity preferred by the tank communities, then, the minimum water required in the tank for fish production should be considered instead of water requirement for winter crop production, for assessing the suitability for rehabilitation work.

Tanks with low density of wells in the command area and catchment area and low intensity of land use in the catchment need to be given priority while choosing tanks for rehabilitation. While it is difficult to arrive at a quantitative criterion for selecting tanks for rehabilitation based on number and density of wells in the catchment, a well density of one well per hectare of command is a clear sign of diminishing importance of tanks for the farmers' livelihoods.

3. Sishodia et al. (2016) analyzed groundwater level data for 64 wells for the period 1990–2012 and showed statistically significant decline in water levels in 36 of them. Similar trends were observed when observation well data were analyzed for the period from 1990 to 2005 and 1997–2012 for another 159 and 99 observation wells, respectively.

On the other hand, a well density of 0.5 and above, that is, 1 per 2 ha of land, is a clear indication of high intensity of land use in the catchment area, with multiple cropping. This can reduce the inflows into the tanks drastically by capturing the runoff in situ and reducing the groundwater outflows into surface streams (Kumar and Vedantam, 2016).

The presence of large number of wells in the tank command may also indicate declining importance of tanks in the livelihoods of farmers in the command. The argument that farmers abandon tanks because of wells is also not true. Nevertheless, there can't be a standard norm on well density and catchment land use, for selecting or rejecting tanks for rehabilitation. The effect of well density and catchment land-use intensity on tank inflows would also depend on the rainfall of the region and the physiographic features of the catchment. In high rainfall regions with steep catchment slopes, the norm could be relaxed, whereas in semiarid and arid regions with low-to-medium rainfall, the norm will have to be more stringent (Kumar and Vedantam, 2016).

Benefits would be accrued from well-conceived and well-implemented projects. When the local people find real benefits coming from such projects—better irrigation, fish production, and water for livestock—they would participate (Kumar et al., 2016). But, before taking up various interventions for tank restoration, there should be proper conceptualization to have greater clarity on what benefits are to be derived from them and how. For this, a comprehensive understanding of the macrohydrology of the region, where such rehabilitation projects are to be undertaken, should be developed to assess how much of water flows out of the region. However, many tank rehabilitation programs are designed on the basis of the vague concept that augmenting the storage capacity of the tanks and improving their structural features would lead to increased water availability and creating user group organizations would ensure their enhanced performance. The narrative is that a lot of monsoon runoff from the catchment flows out of the tank uncaptured due to insufficient storage capacity (see Pingle, 2011) and management of available water from the tanks is poor due to weak tank management institutions (Sakthivadivel et al., 2004).

REFERENCES

Asian Development Bank, 2006. Rehabilitation and Management of Tanks in India: A Study of Select States. Asian Development Bank, Philippines.

Batchelor, C., Singh, A., Rao, R.M., Butterworth, J., 2002. In: Mitigating the potential unintended impacts of water harvesting. Paper Presented at the IWRA International Regional Symposium "Water for Human Survival." Hotel Taj Palace, New Delhi, India, November 26–29, 2002.

Biggs, T.W., Gaur, A., Scott, C.A., Prasad, T., Rao, P.G., Gumma, M.K., Acharya, S., Turral, H., 2007. Closure of the Krishna Basin: Irrigation, Streamflow Depletion and Macroscale Hydrology. International Water Management Institute, Colombo, Sri Lanka. Research Report 111.

Centre for Water Resources, 2000. Monitoring and Evaluation: Phase II and Phase II-Extension, Tank Modernisation Project with EEC Assistance. Anna University, Chennai. Final Report Volumes I and II.

Deccan Chronicle, 2015. Telangana State to Restore 46,000 Water Tanks. www.deccanchronicle.com/150122/nation-current-affairs/article/telangana-state-restore-46000-water-tanks.

Deccan Chronicle, 2017. Telangana, AP Water Experts to Argue their Cases on Sharing of River Water. Nation, Current Affairs, Deccan Chronicle.

Directorate of Economics and Statistics, 2015. Statistical Yearbook 2015. Government of Telangana, Hyderabad.

Government of Telengana, 2015. Mission Kakatiya. Government of Telangana, Hyderabad.

Kumar, M.D., Vedantam, N., 2016. Groundwater use and decline in tank irrigation? Analysis from erstwhile Andhra Pradesh. In: Kumar, M.D., James, A.J., Kabir, Y. (Eds.), Rural Water Systems for Multiple Uses and Livelihood Security. Elsevier, Singapore, pp. 145–182.

Kumar, M.D., Sivamohan, M.V.K., Vedantam, N., Bassi, N., 2011. Groundwater Management in Andhra Pradesh: Time to Address Real Issues. Occasional Paper No. 4, Institute for Resource Analysis and Policy, Hyderabad, India.

Kumar, M.D., Bassi, N., Kishan, K.S., Chattopadhyay, S., Ganguly, A., 2016. Rejuvenating tanks in Telangana. Econ. Polit. Wkly. 51 (34), 30–34.

Nag, K., 2011. Battleground Telengana: Chronicle of an Agitation. Harper Collins Publishers India, New Delhi.

Pingle, G., 2011. Irrigation in Telangana: the rise and fall of tanks. Econ. Polit. Wkly. 46 (26–27), 123–130.

Sakthivadivel, R., Gomathinayagam, P., Shah, T., 2004. Rejuvenating irrigation tanks through local institutions. Econ. Polit. Wkly. 39 (31), 3521–3526.

Shah, E., 2008. Telling otherwise a historical anthropology of tank irrigation technology in South India. Technol. Cult. 49 (3), 652–674.

Sishodia, R., Shukla, S., Graham, W.D., Wani, S.P., Garg, K.K., 2016. Bi-decadal groundwater level trends in a semi-arid south Indian region: declines, causes and management. J. Hydrol. Region. Stud. 8 (2016), 43–58.

The Hindu, 2016a. Mission Kakatiya Already Showing Positive Results. The Hindu.

The Hindu, 2016b. Mission Kakatiya Starting to Bear Fruit. The Hindu.

Vaidyanathan, A. (Ed.), 2001. Tanks of South India. Centre for Science and Environment, New Delhi.

FURTHER READING

EDU, 2016. IIT Hyderabad, BITS Pilani And NABARD Sign Agreement With Telangana Irrigation. Dept. EDU.

Chapter 8

What Has Worked for Irrigation Miracle in Madhya Pradesh: Infrastructure or Reforms?

8.1. IRRIGATION AND AGRICULTURAL GROWTH

Since Independence, India's agriculture has grown at a modest long-term annual growth rate of 2.65% (Source: author's own estimates based on the data of agricultural GDP in real terms for the period from 1954–55 to 2012–13). However, significant fluctuations in growth rate were noticed between years. By and large, these fluctuations have been attributed to the monsoon fluctuations, directly impacting on kharif crop production that is dependent on rainfall. While it is quite well established that irrigation had played a key role in agricultural growth in the country by stabilizing kharif production by insulating it against moisture stress and intensifying cropping (winter and summer crops) through the provision of water during those season, too little effort has really gone into understanding how the success and failure of monsoon influence irrigation from the existing schemes, thereby affecting agricultural outputs. The moot point is that even with no additional investment in irrigation infrastructure and their management, the performance of irrigation sector can change significantly across the years, with storage of water in reservoirs and aquifers fluctuating widely in accordance with the changes in precipitation in regions concerned.

In the recent past, there were attempts by researchers to attribute the good performance of agriculture sector in some states to the investments in decentralized water management systems consisting of small water harvesting systems and microirrigation systems (Gulati et al., 2009; Iyer, 2008; Shah et al., 2009 in the case of Gujarat) and innovative management of large irrigation systems, followed by the state (Shah et al., 2016 in the case of Madhya Pradesh). In the same manner, it was argued that increasing investments in public irrigation was not leading to expansion in irrigated area from that sector (Shah, 2009; Mukherji et al., 2009) and this "poor performance of public irrigation" (mainly surface irrigation) was attributed to poor and outdated management models adopted by the state irrigation bureaucracy. For instance, Mukherji et al. (2009) argue that "the large-scale, centrally managed irrigation schemes of the past were not designed to be demand-driven or provide the reliable, flexible and equitable

Water Policy Science and Politics. https://doi.org/10.1016/B978-0-12-814903-4.00008-7

year-round water service that modern farming methods require. Beset with problems of inappropriate design, poor maintenance, salinity, and waterlogging, many large-scale schemes are currently in decline across Asia" (Mukherji et al., 2009, p. 3). Such sweeping statements are not rare in irrigation literature.

Efforts by many national governments to rehabilitate them are ongoing, but the results are, at best, mixed. The externalities such as changes in land use in the catchment, larger hydrologic changes, upstream water diversion, and changes in water allocation decisions from public reservoirs, which can impact severely on the amount of water that can be released for irrigation (Kumar et al., 2012), were never factored in.

8.2. THE CONTEXT

Some researchers in the recent past have unleashed an assault on India's irrigation bureaucracy by saying that the current fate of public (surface) irrigation investments in the country is like a bottomless pit, wherein increase in investment for surface irrigation systems only leads to reduced area under irrigation (Iyer, 2008; Shah, 2009; Shah et al., 2016). Quoting a Reserve Bank of India (RBI) study, Shah et al. (2016) argued that in spite of well over 2.0 lac crore of investment in public irrigation and flood control during 1991–2007, the actual area irrigated by government canals decreased by 3.8 M ha. The authors look at it as colossal waste of money. While incorrectly considering an increase in "canal-irrigated area" as the only benefit from public irrigation and flood control schemes, they attribute this wastage to "deficiencies in planning, implementation, and management" and a case of poor governance, again taking the help of RBI researchers. They have prescribed a "silver bullet" to eradicate the problem that the irrigation planners in the country are grappling with, by drawing on the experience of Madhya Pradesh in "turning around" the sector, manifested by a miracle growth in irrigated area and agricultural production.

Madhya Pradesh, which is one of the largest states in India with an economy largely agrarian, has attracted national attention in the recent years for a miracle growth in agriculture, and a huge jump in the procurement of wheat, second only to Punjab, which is known as the granary of India. As per the official data, the state clocked a compounded annual growth rate of nearly 13.9% over the 5-year period 2010–15. This is far higher than the agricultural growth rate in the state during the previous 5 years, which was less than 5% (Ninan, 2017).

Ninan (2017) had tried to explain the miracle growth in agricultural output[1] and expressed concern over the possibility of sustaining such high growth rates. Gulati et al. (2017) attributed the high agricultural growth rate in the state

1. The factors identified by Ninan (2017) for the miracle growth rate in agricultural output are the low agricultural output base of that state and low average yield levels of cereal that are far lower than the yield recorded by agriculturally prosperous states such as Punjab, rapid expansion of irrigation with resultant jump in yield levels, improved electricity supply in rural areas, increase in acreage, and better access to market because of improved road connectivity.

mainly to the following five factors: (i) irrigation expansion through tube wells and canals, (ii) increased power supplies to agriculture, (iii) assured and remunerative price for wheat, (iv) expansion of all-weather roads, and (v) suitable incentives and signals for the private sector for increasing the level of investments to reap the benefits of trunk infrastructure and improved services. Shah et al. (2016) delved into the reason for the "miracle expansion in irrigated area" and credited this phenomenon to irrigation sector reform initiated by the state and pitched for other states to follow the MP reform model for improving irrigation sector performance.

This chapter first examines whether a simplistic analysis of irrigation performance based on the data on irrigated area really leads to correct diagnosis of India's public irrigation sector problem or not and then takes a critical look at the miracle performance shown by MP's public irrigation sector during the recent years and the factors that are driving this impressive performance. While doing this, it highlights the increasing tendency among a few researchers to ascribe the positive changes in irrigation and agriculture sector performances to factors that are actually trivial, unmeasurable, and subjective, who do not wish to consider the important hydrologic and socioeconomic variable that affects the irrigation performance of large water systems.

8.3. IRRIGATION SECTOR PROBLEMS: WRONG DIAGNOSIS

First of all, to illustrate the issue of "governance deficit" in public irrigation, Shah et al. (2016) resort to the numbers on the growing gap between irrigation potential created and irrigation potential utilized. Traditionally, this has been the criterion used by water resource ministries and economists to assess the performance of irrigation sector. Sadly, there are serious flaws in this performance yardstick and should have been done away with a long time ago. This is because the "irrigation potential created" is based on the estimation of effective storage of water in the reservoir available for irrigation and the amount of water required to irrigate a unit area of land for the cropping pattern assumed at the time of planning, neither of which are sacrosanct. For instance, if at the time of planning a cropping pattern with low-water-consuming crops is considered, the potential created would be very high. If the cropping pattern after the implementation of the scheme changes from the design toward water-intensive crops, the water requirement per unit area would increase. This would bring down the figures of potential utilization in area terms.

If the volume of inflows into the reservoir falls below the total effective storage of water, the figures of irrigation potential utilized would further reduce. Analysis of data on reservoir (live) storage in the month of September for a total of 81 major reservoirs available for 16 major states of India for the years 2009 and 2010 shows that the difference in live storage is in the order of magnitude of 24,766 MCM, and Fig. 8.1 shows the storage as on 10 September 2010 was

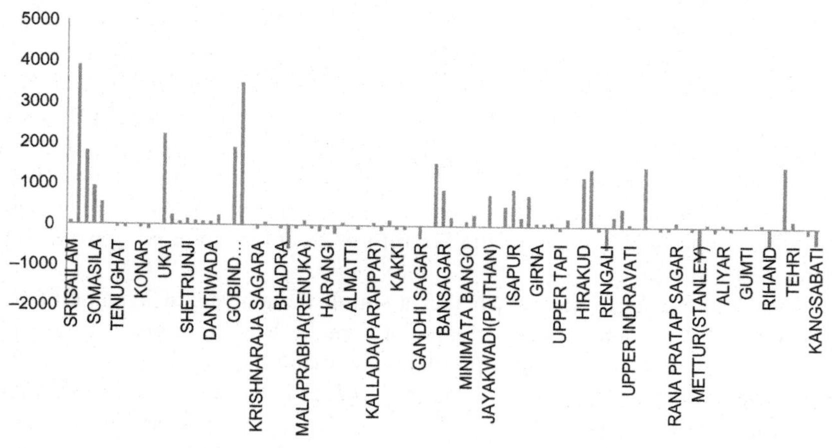

FIG. 8.1 Differences in live storage of reservoirs (2009 and 2010) in MCM.

higher than that of the same date in 2009 in most reservoirs. The cumulative live storage in September 2010 from the 81 reservoirs was 24,766 MCM higher than that in September 2009. It is to be noted that 2009 was a year of drought in many states, namely, Gujarat, Maharashtra, Rajasthan, Chhattisgarh, and Jharkhand. Such a huge difference in storage can create a difference of nearly 2.40–4.80 M ha of irrigation. Now, if we consider all the major, medium, and minor irrigation reservoirs in the country, the difference drought that can induce on the actual irrigation potential can be far higher.

The "potential utilized" can also be much lower than the "potential created" even without any reduction in inflows or effective reservoir storage, as water from large reservoirs often gets relocated for nonirrigation uses such as municipal, industrial, and rural drinking water supply. However, the blame for increasing gap between the potential created and potential utilized goes to the irrigation bureaucracy (Kumar et al., 2012).

Sometimes, the low utilization can also be due to the canals not being ready for the distribution of water in the command area. There are a few reservoirs in India where water remains behind the dam for several years, without being diverted for irrigation due to the incomplete canal network. Problem in land acquisition for construction of canals is one major reason for the long gestation. Once the scheme is planned and approved and execution of the project is started, the land value in the command area shoots up, and the farmers seldom show the willingness to give their land to the department at the price decided by the government. This is an implementation issue and also falls under governance and needs serious policy measures, to enable smooth acquisition of land. Inadequate fund allocation for completing the project, which again is linked to rising cost of the project due to time overrun, is also a problem.

Hence, out of the four factors responsible for the "gap between potential created and utilized," only one concerns the planning or design deficit, that is, overestimation of catchment yield. The other concerns implementation problem. The rest two problems cannot be attributed to either planning or implementation or management.

But, a careful analysis would show that even highlighting the problem of overestimation of catchment yield would be misplaced in the current scenario, as it would downplay another major threat facing many irrigation projects in the country today. The stark reality is that hundreds of thousands of small water harvesting structures built under various government schemes, either Integrated Watershed Management Project (IWMP) or the NREGS, impound the water in the upper catchments of the sanctioned or completed schemes and reduce the inflows into the reservoirs. The state water resources departments (erstwhile irrigation departments) are never consulted, while such schemes are planned and implemented by the rural development department under whose aegis the MNREGS and IWMP works are carried out. So, there is surely a governance problem, but it is in the way our water resource sector is being administered with many ministries having a stake in it. Further, as highlighted by Kumar et al. (2012), area irrigated by direct water lifting by farmers from canals gets reported as lift irrigated area, an issue that falls in the realm of irrigation statistics. For instance, in the newly formed state of Telangana alone, there were 13875 surface lift irrigation scheme spread of the 10 districts, as far back as 2006–07 (Source: Directorate of Economics and Statistics, Govt. of Telangana).

Given this scenario, a more sensible approach to evaluate the performance of public irrigation systems would be to examine the economic value of the outputs/benefits realized from the uses of water against the volume of water stored in various storage reservoirs plus diverted from various diversion schemes (weirs and barrages). The factors considered in the numerator should include the value of the incremental outputs generated from the use of water from wells in the command area, which benefit from canal seepage and return flows from irrigated field, and the economic value of the benefits from the use of water supplied from the reservoir for nonirrigation uses, along with the economic value of the incremental outputs generated from the use of water supplied by canals, whereas the volume of water impounded or diverted should be considered in the denominator. By doing the latter, we would be able to factor in the influence of (a) rainfall on the storage of water in the reservoirs and therefore the irrigation and other benefits and (b) new storage dams/diversion structures built on the total irrigation potential created and the benefits arising out of the same.

Not considering the indirect outputs from well irrigation, etc. on the one hand and overestimating the water use on the other mean undervaluation of the economic benefits from public irrigation, as shown by basin-wide simulation of surface irrigation systems involving hydroeconomic analysis (Chakravorty and Umetsu, 2003). However, building new storage infrastructure to effectively increase water storage from monsoon flows is also an indication of the overall sector performance, provided that there is unutilized runoff in the basin that goes into the natural sink.

8.4. WHAT IS THE MADHYA PRADESH EXPERIENCE WITH IRRIGATION "TURNAROUND"?

Shah and others in their article (Shah et al., 2016) took the case of Madhya Pradesh to build the argument that the performance of canal irrigation sector can be enhanced through minor investments that would increase the utilization of water from the irrigation potential created, thereby reducing the gap and tweaking the administration of irrigation bureaucracy by the political leadership (Shah et al., 2009). In the authors' opinion, "...the UPA governments in Maharashtra and Andhra Pradesh (AP) have used massive irrigation investments to create rent-seeking opportunities...." (Shah et al., 2016, p. 19). But "... Narendra Modi and Shivraj Singh Chouhan pursued agricultural growth through irrigation development as a political strategy for capturing agrarian vote-banks rather than rent seeking" (Shah et al., 2016, p. 20). In the subsequent paragraphs, authors become eloquent. Quoting an article of the lead author published in 2009, the paper says "as Gujarat's chief minister, Modi was the original architect of this strategy that delivered to Gujarat's agricultural growth rate of 9% per year during 2000–08 (Shah et al., 2009)."

After Shah et al. (2016), "Modi did this inter alia by: (a) expediting the construction of the Sardar Sarovar Project (SSP); (b) implementing a highly successful programme of supporting local communities for water harvesting and groundwater recharge; (c) improving quality of farm power ration through the Jyotigram scheme of feeder separation; (d) issuing over 5 lac new electricity connections to Scheduled Castes/Scheduled Tribes and small farmers (Shah et al., 2016)." Intriguingly, in the 2009 paper (Shah et al., 2009), the lead author criticized the investments in SSP by the same BJP government, saying that the project was not delivering, and gave full credit to the rain water harvesting schemes implemented by that government and the reliable and high-quality farm power supply for the so-called 9.6% agricultural growth. It is not clear what led to the change in the lead author's view of public irrigation scheme in the ensuing 7 years.

In fact, not much has changed in Gujarat with respect to completion of the SSP and the area irrigated by it since 2009. A socioeconomic impact assessment of SSP by Jagadeesan and Kumar based on an intensive and extensive field study carried out in eight districts of Gujarat and carried out during 2010–12, which evaluated the project benefits, found that not only the direct economic benefits of irrigation in terms of incremental income from crop and dairy production, increased agricultural wage employment, and drinking water supply were substantial, but also the indirect benefits (positive externalities of the project) to well irrigators, farm laborers, and drinking water supply schemes based on wells were huge (see Jagadeesan and Kumar, 2015).

The impressive growth in agriculture achieved by Madhya Pradesh, according to Shah et al. (2016), is all due to the canal reforms and improvement in farm power supply and not due to the capital investment for building irrigation infrastructure and creating new irrigation potential. The authors seem to suggest that the Rs. 36,689 crore investments made for public irrigation alone during the first tenure of the current BJP government in MP were very "marginal" but could produce an additional irrigation of 0.80 M ha in 2006 to 2.50 M ha in 2012–13. It is important to mention here that the total planned investment for SSP, which is to irrigate 1.8 M ha in Gujarat, along with producing hydropower (1410 MW) and supplying water to domestic sector and industries in that state, is estimated to be around Rs. 39,000 crore. Therefore, it should be inferred that the investment made by MP government in public (canal) irrigation was very heavy and not something that is made for repair and maintenance.

The canal reforms, which according to the authors resulted in large expansion in surface irrigation to the tune of 1.7 M ha, included the following: restoring canal management protocol; last mile investment; reducing deferred maintenance; animating the irrigation bureaucracy, including the sacking of a corrupt irrigation secretary and putting an efficient officer in place to run the bureaucracy; and vitalizing farmer organizations. Here, one can only express disbelief over the new-found confidence of the lead author, who has been a critic of promoting WUAs in irrigation commands under the existing format as a futile exercise because of the huge transaction costs involved (see Shah, 2009), in the abilities of farmer organizations in MP. For instance, Shah (2009) argued that "...despite lack of evidence of large-scale success, PIM/IMT will continue to be peddled as blanket solutions for improving system performance."

Madhya Pradesh's achievements in bringing about reforms in the irrigation sector were not worth beyond a mere mention. As Bassi and Kumar (2011) notes, "the administrative, governance and institutional reforms for promoting farmer involvement in irrigation management were not adequate, and there was a dominating presence of department officials in the working of the user associations, undermining meaningful participation of farmers in irrigation management" (Bassi and Kumar, 2011).

The model being promoted by Shah et al. (2016) for revitalizing irrigation bureaucracy, as existing in MP, is nothing short of a "fairy tale." The authors are eloquent: "in a masterful innovation, the engineer-in-chief would randomly call any of the 4000 odd mobile numbers of tail end farmers to enquire if water reached her/his field." If at all what the authors say is correct, this is surely not part of irrigation reform but autocracy. But, such anecdotes are used extensively by both the authors of the article and the report of the Mihir Shah committee (on restructuring of CWC and CGWB) as part of a larger narrative to push a new irrigation reform model in India, which is devoid of any sector reform intervention.

A key factor, which the authors highlight as contributing to irrigated area expansion, is the lining and desilting of canals. As pointed out by Chris Perry in a paper (Perry, 2007) and later on in a commentary on a water management paradigm proposed for India by Mihir Shah (Perry, 2013), while this strategy is quite convincing as an irrigation water management strategy for many who do not understand hydrologic sciences, the point is that most of the water, which is lost in seepage from unlined canals in areas with groundwater lying at a shallow depth, is actually recycled back into the system by the wells in the command area. Any intervention to stop the seepage through lining would help expand canal-irrigated area but would render the many wells that come up in the command area useless. This concept of canal lining to improve irrigation efficiency is rather outdated and is not useful for areas with extensive well irrigation. In fact, as noted by Kumar (2007), reducing yield of the wells due to well interference is a major problem in the hard-rock areas of MP. Therefore, such seepage from canals is extremely valuable for sustaining yield of wells in these hard-rock areas. Therefore, even the argument about the increase in electricity connections to the farms increasing the well-irrigated area is untenable (Kumar, 2007).

The authors' observation vis-à-vis revitalization of farmer organizations is rather misleading: "under a new law made in 1999, some 2000 WUAs were formed but mostly lay defunct. WUAs had little role when poorly-managed main system failed to deliver water to many parts of the command for years. Now that the O&M of the main system improved, water began reaching the tail-ends and defunct WUAs sprung to life." Here, the researchers found a new engineering solution (of lining the canals) to the large social problem of inequitable distribution of water across the hydraulic system, due to the power structure in the villages. Past research however shows that if the farmer institutions are strong, they could invest in maintenance of the canals and also ensure equitable distribution of water, through proper enforcement of rules and norms. One wonders what these 2000 odd farmer organizations were doing when the water was not flowing in the main system for years.

8.5. IRRIGATED AREA EXPANSION: CANAL SYSTEM IMPROVEMENT OR GOOD RAINS?

Shah et al. (2016) conveniently use the data of season-wise irrigation (kharif and winter) in 2 years, that is, 2002–03 and 2012–13 to conclude that all is well with irrigation bureaucracy and that whatever expansion in irrigated area is occurring because of the software called "reforms" and not due to the hardware called "water infrastructure" comprising reservoirs and canal systems. But a careful examination of the data presented by the authors for canals and wells separately would reveal the secret. In 2002–03, the percentage area under well irrigation during kharif was 28.2%, but this went down to 16.2% in 2012–13. If the authors' argument about the improved power supply in the state of MP is correct, then, ideally, more area should have been brought under supplementary irrigation. But, this did not happen. This glaring inconsistency is because of the higher rainfall received in most parts of MP during the year 2012–13. Comparison of weighted average annual rainfall of the two regions of MP, that is, eastern MP and western MP for the years 2002 and 2012

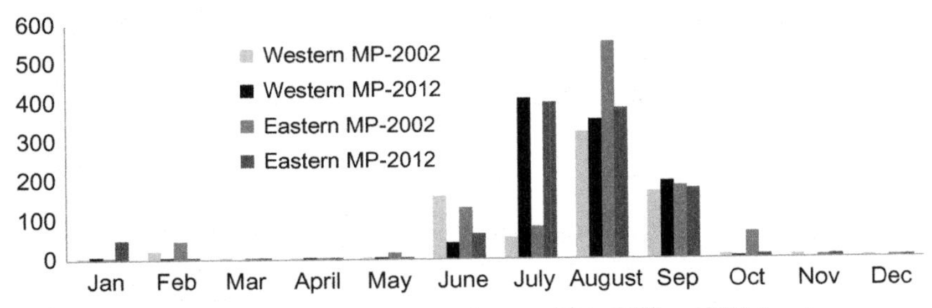

FIG. 8.2 Average annual rainfall in western and eastern MP—2002 and 2012 (mm).

(Source: https://data.gov.in/keywords/annual-rainfall), shows that the rainfall in the latter during 2012 was 34.2% higher than that of 2002 (Fig. 8.2).

The high rainfall also resulted in increased storage in surface reservoirs, including those that were built after 2002–03, which ultimately resulted in slightly larger areas (especially those areas that are characterized by lower mean annual rainfall) receiving canal irrigation.

Whereas during winter of 2012–13 (when there is no rain in any part of the state), both canal-irrigated area and well-irrigated area (as percentage of total cultivated area) were higher. This is a clear indication of the fact that the natural recharge from precipitation and reservoir inflows during this year was much higher than that of 2002–03. The fact that percentage area irrigated during the kharif of 2012–13 was much lower (24) than that of 2002–03 reinforces the point about the effect of differential rainfall. Ideally, the question that the researchers should have asked is "if there is so much water in the canals, how come only 24% of the cultivated land during kharif was under irrigation?" The answer lies in the fact that there was much less demand for irrigation water, due to higher rainfall in the western parts. A careful examination of chart (Fig. 5, p. 21) presented by the authors in their paper itself would show that all that is happening is because of greater investments in irrigation infrastructure. Fig. 5 (Shah et al., 2016, p. 21) shows that maximum increase in surface-irrigated area occurred in Narmada river basin. It is a common knowledge that out of the 28 large reservoir-based irrigation and hydropower projects planned in MP part of the Narmada basin, seven (Tawa, Barna, Kolar, Sukta, Matiari, Man, and Jobat) with a total design command area of 3.95 lac ha are already completed, and another five (with a design command of 7.10 lac ha) are ongoing. But, the article is conspicuous for the absence of any reference to this aspect. In fact, the area irrigated by public canals went up from 1.9 to 3.7 lac ha in the Narmada basin, followed by Chambal basin, with area irrigated going up from 1.5 to 2.9 lac ha.

Yet, apart from merely discussing how the irrigated area has increased over time, the authors did not discuss how the utilization of irrigation potential created in percentage terms has changed over the years in MP, particularly since 2003. This would have exposed the fallacy of the claim made by the authors in their paper. The moot point is that a lot of additional potential had been created in the irrigation sector during the period through the completion of many large reservoir-based irrigation projects.

8.6. CONCLUSIONS

It is a truism that in semiarid and arid regions, irrigation brings about boost in agricultural production and economic growth. Building large storages is critical to achieving this (Biswas and Tortajada, 2001; Dhawan, 1990; Kumar, 2009), especially when groundwater resources are very limited. The current political dispensation in Madhya Pradesh fully understood that for achieving water security for irrigation and other sectors of economy, building new infrastructure for storing surface water is of paramount importance. The factor that worked in their advantage is that many river basin subbasins (such as Narmada, Tapi, and Chambal) of Madhya Pradesh have a lot of untapped water resources, unlike many water-scarce regions of India where surface water resources are already fully exploited using large and small reservoirs.

During the past 13–14 years, several major irrigation projects were taken up by the government of MP, and many were completed, unlike what the previous governments of that state did by focusing merely on decentralized water harvesting and watershed development. With this, there has been a substantial increase in area irrigated by public schemes. The above-normal (weighted average) rainfall during 2011, 2012, and 2013 (Source: https://data.gov.in/keywords/annual-rainfall) also created favorable environment. With an increase in gravity irrigation, one would expect an increase in well-irrigated area as well, as canals and surface irrigation induce recharge of the shallow aquifers, improving well yields. Unlike the claim by Gulati et al. (2017) and Shah et al. (2016), simply improving power supply to agriculture cannot help increase well irrigation. There should be sufficient water underground within economically accessible limits for the farmers to pump out. But one of the narratives used by the researchers to push their agricultural growth model is that farmers were not able to use the water in their wells due to the lack of power supply.

Irrigation is a serious enterprise involving application of knowledge from disciplines as varied as hydrology, hydraulics (irrigation engineering), agricultural sciences, economics, sociology, public administration, and management, and there are no quick fix solutions. Therefore, major changes in the sector performance are unlikely to happen overnight. India needs serious reforms in the public irrigation sector involving systemic changes that can bring about transparency and accountability in the working of the state irrigation bureaucracy and create incentive for greater performance to achieve sustainability, equity, and social justice. That should not just be limited to cosmetic changes like promoting participatory irrigation management through water user associations (Mollinga and Bolding, 2004) and should include water right reforms (Madhav, 2007; Chapter 5, this book) and water pricing.

Unfortunately, there is an increasing tendency among a few researchers to ascribe the positive changes in irrigation and agriculture sector performances to factors that are actually trivial, unmeasurable, and subjective, without considering the important hydrologic and socioeconomic variable that affects the irrigation performance of large water systems, as evident from the analysis of agricultural growth trends in Gujarat (Kumar et al., 2010) and Madhya Pradesh (Kumar, 2016).

In the case of Madhya Pradesh, Shah et al. (2016) fail to comment on how the utilization of potential created (or what proportion of the potential created is being utilized) has changed in the state over time. The fact is that there has been significant addition to the existing irrigation potential with the completion of many large and medium reservoir projects in Madhya Pradesh during the past 13–14 years, mostly for the purpose of irrigation. By not commenting on this important development, the authors actually try and obfuscate the significant initiatives of the governments to improve the agricultural situation in that state that involve building of large-scale irrigation infrastructure in an effort to highlight the "sector reform" that never happened.

Bhattarai and Narayanamoorthy (2003) had shown on the basis of analysis of data from 1970 to 1994 for 14 major states that investments in irrigation expansion and rural literacy rate are the two most important factors responsible for agricultural growth and rural poverty reduction in India. Extending irrigation access to a large number of farmers and investing in human capital development are crucial to increasing agricultural productivity and reducing poverty in India. It is time that we do away with doing agricultural growth projections based on short-term data on agricultural GDP and irrigated area change, which obfuscate the data on change in level of investment in irrigation and show more of the effect of factors such as climate variability.

REFERENCES

Bassi, N., Kumar, M.D., 2011. Can sector reforms improve efficiency? Insight from irrigation management transfer in central India. Int. J. Water Resour. Dev. 27 (4), 709–721.

Bhattarai, M., Narayanamoorthy, A., 2003. Impact of irrigation on rural poverty in India: an aggregate panel data analysis. Water Policy 5 (5-6), 443–458.

Biswas, A.K., Tortajada, C., 2001. Development and large Dams: a global perspective. Int. J. Water Resour. Dev. 17 (1), 9–21.

Chakravorty, U., Umetsu, C., 2003. Basinwide water management: a spatial model. J. Environ. Econ. Manag. 45 (1), 1–23.

Dhawan, B.D. (Ed.), 1990. Big Dams: Claims, Counterclaims. Commonwealth Publishers, New Delhi.

Gulati, A., Rajkhowa, P., Sharma, P., 2017. Making Rapid Strides-Agriculture in Madhya Pradesh: Sources, Drivers, and Policy Lessons. Indian Council for Research on International Economic Relations, New Delhi. Working Paper No. 339.

Gulati, A., Shah, T., Sreedhar, G., 2009. Agriculture Performance in Gujarat Since 2000: Can it be a Divadandi (Lighthouse) for Other States? International Water Management Institute/International Food Policy Research Institute, Gujarat, India/New Delhi, India.

Iyer, S.S.A., 2008. Wasting $50b in major irrigation. Swaminomics. Times of India.

Jagadeesan, S., Kumar, M.D., 2015. The Sardar Sarovar Project: Assessing Economic and Social Impacts. Sage Publications, New Delhi.

Kumar, M.D., 2007. Groundwater Management in India: Physical, Institutional and Policy Alternatives. Sage Publications, New Delhi.

Kumar, M.D., 2009. Water Management in India: What Works, What Doesn't. Gyan Books, New Delhi.

Kumar, M.D., 2016. Irrigation sector "Turnaround" in Madhya Pradesh? Econ. Polit. Wkly. 51 (19), 67–70.

Kumar, M.D., Narayanamoorthy, A., Sivamohan, M.V.K., 2012. Key issues in Indian irrigation. In: Kumar, M.D., Sivamohan, M.V.K., Bassi, N. (Eds.), Water Management, Food Security and Sustainable Agriculture in Developing Economies. Routledge/Earthscan, London, pp. 17–37.

Kumar, M.D., Singh, O.P., Narayanamoorthy, A., Sivamohan, M.V.K., Sharma, M.K., Bassi, N., 2010. Gujarat's Agricultural Growth Story: Exploding Some Myths. Institute for Resource Analysis and Policy, Hyderabad. Occasional Paper No. 2.

Madhav, R., 2007. Irrigation Reforms in Andhra Pradesh: Whither the Trajectory of Legal Changes. International Environmental Law Research Centre, Geneva, Switzerland. IELRC Working Paper 2007-04.

Mollinga, P.P., Bolding, A. (Eds.), 2004. The Politics of Irrigation Reforms: Contested Policy Formulation in Asia, Africa and Latin America. Ashgate Publishing Ltd., UK.

Mukherji, A., Facon, T., Burke, J., de Fraiture, C., Faurès, J.-M., Füleki, B., Giordano, M., Molden, D., Shah, T., 2009. Revitalizing Asia's Irrigation: To Sustainably Meet Tomorrow's Food Needs. International Water Management Institute/Food and Agriculture Organization of the United Nations, Colombo, Sri Lanka/Rome, Italy.

Ninan, T.N., 2017. MP's Miracle Agriculture. Business Standard.

Perry, C.J., 2007. Efficient irrigation; inefficient communication; flawed recommendations. Irrig. Drain. 56 (4), 367–378.

Perry, C.J., 2013. Beneath the water resources crisis. Econ. Polit. Wkly. 48 (14), 59–60.

Shah, T., 2009. Past, Present and the Future of Canal Irrigation in India. In: India Infrastructure Report 2011. Infrastructure Development Finance Company, Mumbai, pp. 69–89.

Shah, T., Gulati, A., Padhiary, H., Sreedhar, G., Jain, R.C., 2009. Secret of Gujarat's agrarian miracle after 2000. Econ. Polit. Wkly. 44 (52), 45–55.

Shah, T., Mishra, G., Kela, P., Chinnasamy, P., 2016. Har khet ko pani? Madhya Pradesh's irrigation reform as a model. Econ. Polit. Wkly. 51 (6), 19–24.

FURTHER READING

Directorate of Economics and Statistics, 2015. Statistical Yearbook 2015. Government of Telangana, Hyderabad.

Iyer, S.S.A., 2009. Agriculture: Secret of Modi's Success. Swaminomics. Times of India.

Chapter 9

Thanking "Rainwater Harvesting" and Blaming the Rain God

9.1. INTRODUCTION

"Running with the hare and hunting with the hound" goes an old English proverb. We cannot take both the sides in a conflict. But this seems to be the attitude of some civil society activists in the water sector, when it comes to the method of achieving "water security for all." Big water systems involving large dams and pipelines are essential for water security in arid and semiarid tropics (Shah and Kumar, 2008; Kumar, 2009). But the activists hate them on the ground that they involve displacement of indigenous human population, forest submergence, destruction of wildlife habitats, and ecological degradation of rivers. But they are eloquent about the virtues of small-scale rainwater harvesting systems, in spite of the fact that there is hardly any evidence to the effect that they ensure water security in such regions (Kumar and Pandit, 2016). Their idea of monsoon rains is something that falls in small quantities so as to make it easy to harvest and store in small vessels to tide over the water crisis, which goes quite contrary to the well-established facts about Indian monsoon.

Monsoon rainfall in India is highly erratic, and the quantum of rainfall in a particular area varies from year to year, and the same interannual variability is higher in regions of low rainfall and vice versa (Sharma, 2012; Pisharoty, 1990). Also, the low to medium-rainfall regions are arid to semiarid and high to very high, and excessively high-rainfall regions are subhumid to humid (Pisharoty, 1990). The implications of these characteristics of Indian rainfall for the effectiveness of water management choices we make, be it for irrigation or for drinking water supply or for flood control, have hardly been appreciated by these civil society activists, who advocate local and decentralized water systems (Kumar and Pandit, 2016). Extreme variability in rainfall and its consequent impact on aridity means that the same area can experience flooding and severe dryness. However, the perspectives of civil society groups on flood management are even poorer.

In this chapter, we analyze, in the backdrop of Chennai floods, how the antagonistic view of some activists toward large water infrastructure projects, their obsession with "rainwater harvesting" as an approach for ensuring water security, and their increasing influence on the civil society at large (see Agarwal and

Water Policy Science and Politics. https://doi.org/10.1016/B978-0-12-814903-4.00009-9

Narain, 1997; Mishra, 1996; http://www.cpreec.org/pubbook-traditional.htm) weaken the institutional capability of water sector agencies to evolve long-term strategies to deal with floods and droughts and achieve water security. While doing this, the chapter also examines the effectiveness of rooftop rainwater harvesting, a solution advocated by these civil society groups for solving water crisis in Chennai and many other cities, especially the quantity and quality aspects of water collected through rooftops, and the technical feasibility and economic viability of roof water harvesting.

9.2. THE CHENNAI FLOODS

Chennai is one of the metropolitan cities located on the coast in the south Indian peninsula, with a population of nearly 6 million people. The mean annual rainfall in Chennai is nearly 1300 mm. To this, the contribution from southwest monsoon is 598 mm (52.9%), that from southwest monsoon is 422 mm (37.3%), and the rest 9.2% is from summer and winter rains. Analysis of rainfall records for 41 years (1965–2005) at four locations, namely, Cholavaram, Chembarambakkam, Redhills and Poondi, shows that year-to-year variation in rainfall is significant, with the lowest of 624 mm recorded in 1999 and the highest of 2206 mm in 2005 (Source: based on http://shodhganga.inflibnet.ac.in/bitstream/10603/27024/8/08_chapter%203.pdf). The graphic representation of the annual rainfall values for the said time period is given in Fig. 9.1. The mean annual rainfall is 1275 mm, and the coefficient of variation in rainfall, which expresses the interannual variability, is 28.3%.

High-intensity rainfalls are very characteristic of hot and semiarid region of low to medium rainfall (Kumar et al., 2006). When such rainfall occurs for longer duration, they cause flooding rivers, their floodplains, and streets and often marooning an entire city landscape. Improper land-use planning of cities, which increase the extent of impervious land, aggravates this problem by increasing the runoff intensity. Storm water drainage systems, designed for inadequate capacity, also increases the chances of urban flooding.

This is what happened in Chennai during December 2015, when it rained 374 mm in just 24 h breaking the hundred year record. During the month of

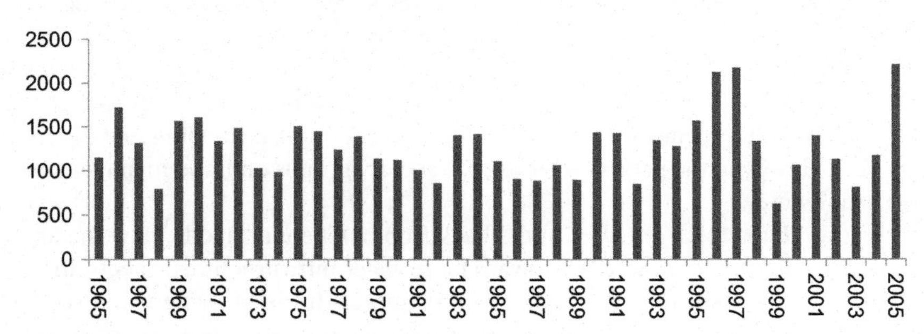

FIG. 9.1 Annual rainfall in Chennai during 1965–2005 (mm).

November, Chennai recorded a whopping 1218.6 mm of rain—three times its monthly normal rainfall of 407.4 mm. The local tanks, ponds, and lakes overflowed owing to insufficient capacity, and the shallow coastal aquifer was completed saturated. The storm water drainage system had a design capacity of 50 mm/h, while the intensity of runoff was 80 mm/h.

Urban flooding is a complex phenomenon caused by several location-specific conditions, and therefore is not limited to Chennai city. Historical analysis of urban floods in four mega cities of India, namely, Chennai, Mumbai, Kolkata, and Delhi showed that the problem is caused by a combination of factors, including heavy rainfall occurring over short durations causing high-intensity storm runoff, and flooding of the rivers and lakes, which form the natural sinks for urban drainage water (De et al., 2013).

The heavy monsoon showers of 2017 have exposed the problems of inadequate drainage infrastructure that several Indian cities are crippled with. Most of them are devoid of separate drainage and sewerage infrastructure. Urban flood (drainage congestion) has started affecting even cities in the water-scarce regions, with unprecedented jump in population density, growing wastewater outflows, and dwindling capacity of the natural sinks for urban storm water due to rampant encroachment. The monsoon floods of 2017 affected the city of Mumbai, Chandigarh, and Jaipur, and the problem was primarily drainage congestion as the storm water drainage system, which also carries the sewage, lacked the capacity to evacuate the storm water runoff caused by high-intensity storm. In the case of Mumbai, the high tides in Arabian Sea, which coincided with the high-intensity precipitation across the city, magnified the drainage problem. While the government tried to put the blame on climate change causing "extremely high-intensity rains," such phenomena are not rare in the hot tropics. What has changed is the runoff intensity and the sewage outflows. While the former has increased due to increasing proportion of urban land under impervious built-up area, the sewage generation per unit area has increased due to increase in urban population density.

Coming back to Chennai city floods, the city is historically known for acute water shortage. While the population of the city grew rapidly during the past three decades, the existing water supply sources such as lakes and tanks, and bore wells became inadequate. The reservoirs at Poondi, Cholavaram, Redhills, Chembarambakkam, Veeranam Lake, Rettai Eri and Porur Lake in Chennai, and Kandaleru Reservoir in Andhra Pradesh are the main sources of water supply for Chennai city and the adjacent urban areas. Korataliyar river is the main source of water for Poondi, Cholavaram, Redhills, and Chembarambakkam reservoirs. A complex system of reservoirs, anicuts, underground conduits, tanks, and canals enables supply of water to Chennai city. The Poondi reservoir plays an important role in Chennai water supply since this reservoir receives the Krishna water from Kandaleru reservoir through the Telugu Ganga canal, whose capacity is 28.34 m^3/s. The inflow into Poondi reservoir consists of the inflow from Nandi and Nagari rivers, diversion from Kesavaram anicut, release from

Kandaleru reservoir, and its own catchment area (Source: http://shodhganga.inflibnet.ac.in/bitstream/10603/27024/8/08_chapter%203.pdf).

When the annual rainfall is far below normal values (say by more than 30% or magnitude of annual rainfall less than 900 mm), the water situation in Chennai city dramatically changes, with disproportionately higher reduction in the runoff resulting from increase in aridity (Kumar et al., 2006; Prinz, 2002), seriously affecting the inflows into the lakes. The tanks and reservoirs and the local aquifers also dry up during such dry years, while the demand for water increases exponentially. This is when Chennai city experiences acute water shortages. When excessively high rainfall occurs, there is too much runoff generated in the catchments, and the storage space in the local systems becomes too inadequate, and such phenomena can cause urban floods. The demand for water from the public water utility for various water-consuming sectors drops considerably during such wet years. This reality of arid zone hydrology is hardly understood or appreciated by proponents of urban rainwater harvesting.

During 1978–2007, the water supply capacity of Chennai Metro Water Supply and Sewerage Board had increased from 240 million liter per day (MLD) to 645 MLD, to cope with the rise in demand from the city as the population increased from 3 million to nearly 5.375 million (Source: http://shodhganga.inflibnet.ac.in/bitstream/10603/27024/8/08_chapter%203.pdf). The city water managers were quite convinced that Chennai would not be able to meet its municipal water needs unless new sources of water were found. Desalination was one of the alternatives, while NGOs and civil society activists pushed for rooftop rainwater harvesting. For the past few years, a good portion of Chennai city's water supply comes from desalination, an outcome of a decision taken by Chennai metro water in the aftermath of the severe droughts that hit this coastal city in 1999. The water activists were not so happy with this initiative, which they call "mindless desalination." They instead wanted to promote only rainwater harvesting. What they failed to understand is that local rainwater harvesting as a solution to water scarcity would fail when the rain in the locality fails (Kumar, 2004; Kumar et al., 2006).

9.3. OBSESSION WITH SMALL RAINWATER HARVESTING SYSTEMS

The civil society's increasing obsession with the concept of rainwater harvesting is evident from their publicity campaigns. One such campaign for Chennai rainwater harvesting efforts says this: "In contrast (please read as "mindless desalination"), the benefits of RWH are there for all to see." Following the implementation of RWH systems, the city in 2005, saw a record annual rainfall of 250 cm. When surveys were carried out, they showed that the water table had gone up by 20 ft, a phenomenal increase! The results were particularly remarkable in the temple tanks, which had at one point become so dry that children used them as cricket fields! Sekhar says, "we surveyed nearly 39 temple tanks in January '06. These tanks which had earlier been empty were now half full.

The secret lies in the fact that surrounding houses had adopted RWH, and the water table all around improved significantly (Source: Rain Centre, Chennai)."

As one can infer from the above paragraph, when the water table in Chennai went up, the credit was given to rainwater harvesting efforts, and not the very good monsoon that resulted in improved natural recharge to the aquifers. When the tanks started overflowing due to excessive inflows from the catchment, again, it was attributed to the rainwater harvesting in the surrounding areas, and not the runoff generated in the catchment! It was reported that around 2.7 lac rainwater harvesting structures were built in Chennai city and its suburban areas of the city. Today, no one knows exactly about the status of these structures. According to some conservative estimates, at least 110 million dollars was spent for constructing these 270,000 structures. Asking questions about the fate of these structures built by the communities is of course forbidden.

But such was the impact of civil society campaigning for urban rainwater harvesting for mega cities on the academics and policy that an academic research that examined various options for mitigating urban water stress in Chennai city explored rainwater harvesting by urban dwelling as one of the options, along with supply augmentation through desalination and efficiency improvements (see Srinivasan et al., 2010). The simulation involved several favorable assumptions concerning the cost of RWHS and suitability of the aquifers for storing recharge water.[1] Interestingly, the cost assumed for rainwater harvesting system (for shallow aquifer recharge) was less than one-tenth of what such a system would actually cost. Further, the aquifer underlying Chennai has alluvium and hard rocks, and in many pockets, groundwater is saline due to seawater intrusion. Therefore, the conditions are not very favorable for rainwater harvesting and recharge. Probably, as a result of these, a combination of RWHS and efficiency improvement was found to be the least cost option for the city.

But when the city was flooded, the rains had to take part of the blame, along with city managers. No civil society group came forward to point out the (water) infrastructure deficiency as the root cause of the problem, which in this case is the lack of adequate storm water drainage systems. Mitigation of water scarcity was the central theme that civil society organizations in the cities were involved in. The city managers were blamed for the lack of preparations for meeting with crisis situations, though there were occasional voices raised about nondesilting of tanks around Chennai that reduced the flood absorption capacity of the region. The fact of the matter is that the storm water drainage system of Chennai was designed for a rainfall intensity of 39.44 mm/h, which corresponds to a flood frequency of once in just 2 years, with a duration of hours, as per the intensity-duration-frequency curve (IDF) prepared for Chennai (Narasimhan, 2016). Ideally, the design intensity should have been around 78 mm/h, which can take care of a flood of 25-year return period.

1. One of the several assumptions made by Srinivasan et al. (2010) in the economic simulation was the very low cost for rainwater harvesting system (US $60 per bore well as the retrofitting cost), and another assumption was that the shallow alluvial aquifer underlying the city is free from salinity.

No sound theory in hydrology or hydraulics or development economics can probably explain this interesting social phenomenon, and the answer has to be found only in political science. It was after an aggressive and often vitriolic campaign by the civil society groups, which were proponents of rainwater harvesting, that Chennai became member of an exclusive club of cities around the world to introduce rainwater harvesting in dwellings through the enforcement of an act. In fact, such was the popularity of the concept of "rainwater harvesting" that the chief minister of Tamil Nadu, (late) Smt. J. Jayalalitha made it part of her manifesto in 2003 before the state assembly elections. During 2003–04, it was enforced by an authoritative state under the leadership of Smt. Jayalalitha and promoted enthusiastically by environmentalists to raise awareness about the city's much-destroyed hydrologic ecosystem (Arabindoo, 2011).

However, the fact is that Chennai city and its neighboring districts, located on the eastern coast, are also subject to cyclonic storms causing high-intensity rains. This can cause serious drainage congestion. To complicate the matter, there are two rivers passing through the city, namely, Adyar and Cooum, and numerous tanks and lakes of varying sizes located on its western fringes with reducing carrying capacity due to siltation. Inundation of these water bodies can during such storms can magnify the flooding problems in the city (De et al., 2013). The storm water drainage system of the city should have adequate capacity to evacuate the runoff from high-intensity storms generated in the city and the incoming floodwaters carried by the water bodies. But this issue has not received any attention from either civic body or the civil society organizations that crave for decentralized solutions for tackling urban water stress.

9.4. INEFFECTIVENESS OF SMALL RAINWATER HARVESTING SYSTEMS IN URBAN AREAS

During 2015, when the northeast monsoon lashed over the city of Chennai and its neighboring districts, the storm water drainage system simply did not work, the large areas of the city were under water for many days (and continue to be so), and the life of its citizens went out of gear. The storage tanks constructed by some residences in Chennai might have collected water. But drinking water shortage was being felt all over. What might have happened to the recharge wells, which were to inject water underground, would be anyone's guess. Ideally, water harvesting and recharging works when there is sufficient space in the aquifer for storing the inflow of water. Also, the system that diverts the water into the aquifer (recharge shaft) should have adequate carrying capacity, higher than the rate at which runoff is generated.

In the case of simple rooftop rainwater collection tank, there should be immediate demand for the water collected so that the need for storing it for future use, a factor that increases the storage volume requirement of the tank, reduces. When the rainfall is distributed over long time periods (say 6–7 months in a year), it is possible that there is a demand for water from the tank in the household during times of inflow and water can thus be drawn from the tank. Because of this, the effective water storage provided by the tank will be greater than its storage capacity (Kumar, 2009). Against this, when the duration of rainy season becomes short and rainfall is concentrated in a few rainy days, it is unlikely that there is immediate demand for that water from the tank and water will have to be stored. The reason is that all the local storage systems such as tanks and lakes and wells will normally have sufficient water when there is high rainfall. Thus, the effective storage volume and the tank storage capacity remain the same.

This is where both small local water harvesting systems and aquifer recharge schemes fail. Let us take the case of rooftop rainwater harvesting tanks. When there is a large inflow of water from the natural catchments during a high-intensity storm during the rainy season, there is neither space to store that water nor demand for it in the natural form in the same locality or catchment for immediate use. So, extra storage space will have to be created in the sumps, which would involve additional investments. As regards aquifer recharge, a recharge system, which is capable of evacuating the runoff and injecting it into the aquifer (after filtration) in a short time span, would cost far higher than the one that is capable of evacuating runoffs of low intensity. More importantly, when the rainfall is very high, the aquifers get replenished naturally and have no space to accommodate the incoming flows. Whereas in drought years, there will be too little water that can be harvested from the roof catchments against the heightened demand for the water from the household for water from such systems.

Decision to invest in rainwater harvesting tanks should be driven by economic considerations of cost of rainwater collection tanks against the cost of public water supply per unit volume of water supplied (Kumar, 2004; Pickering and Marsden Jacob Associated, 2007). Earlier simulation studies show that rooftop rainwater harvesting systems offer poor hydrologic opportunities in low- to medium-rainfall regions and in situations with low "per capita roof area." They are also economically unviable when compared with conventional water supply systems in most situations, and the viability of the system reduces further with reduction in magnitude of annual rainfall, which is also associated with reduction in number of rainy days (Kumar, 2004). Further, the rooftop rainwater harvesting systems were found to be hydrologically feasible and economically viable in hilly and mountainous regions having very high rainfall—above 2000 mm (Kumar, 2009; Institute for Resource Analysis and Policy, 2010). As Table 9.1 (Source: IRAP (2010)) shows, the unit cost of rooftop rainwater harvesting system reduces with increase in rainfall and roof catchment area. This is because the storage capacity required for harvesting unit volume of water from the roof catchment decreases with increase in rainfall.[2]

2. This is by virtue of the fact that in areas of very high to excessively high rainfall, precipitation occurs in several days and the monsoon season is spread over long duration, unlike low- to medium-rainfall regions where it occurs in few rainy days and the monsoon season is of very short duration. Due to this reason, there can be simultaneous withdrawal of water from the tank during the times of inflow (Kumar, 2009).

TABLE 9.1 Unit Cost of Production of Water Through RWHS in High Rainfall Areas

Roof Catchment Area (m^2)	Storage Capacity Required for Different Rainfall Regimes			Cost Per Cubic Metre of Water Harvested (Rs/m^3)		
	2000 mm	**2500 mm**	**3000 mm**	**2000 mm**	**2500 mm**	**3000 mm**
40	42	54	52.5	70.22	69.2	56.1
50	51	65.75	75	65.38	67.4	61.3
60	60	77.5	97.5	61.31	63.4	66.4
80	75	97.5	123.75	57.48	59.8	57.5
100	90	117.5	150	55.18	55.0	52.9
120	90	120	157.5	45.98	44.6	46.3
150	105	141.25	187.5	42.91	39.9	41.8

Source: Institute for Resource Analysis and Policy, 2010. Tool Kit for Integrated Urban Water Management Vol. 2: Technical Report. Final Report Submitted to Arghyam. Institute for Resource Analysis and Policy, Hyderabad, India.

9.5. QUALITY OF WATER FROM ROOF TOPS

Too little attention is being paid to the (microbial and chemical) quality of water collected using rooftop water harvesting tanks and recharged through wells. Obviously, no one would dare drinking the water in the lakes, tanks, and ponds that were overflowing. Similarly, no one would like to drink water collected from rooftops. Going by the rhyme of the ancient mariner, Samuel Taylor Coleridge, "water, water, everywhere, and all the boards did shrink. Water, water, everywhere, nor any drop to drink." It is also not about collecting a few kiloliters of rainwater in tanks. In what form it comes, the quality of water and the perception of the communities about the quality of water are important. Evidence gathered from rural areas during the interviews of the adopters of rural water tanks promoted by NGOs suggests that they do not use the water collected in rainwater tanks for drinking and cooking, and use only for washing clothes and utensils.

The water collected from rooftops in cities is known to contain carbon and that from urban catchment contain heavy metals, oil, and grease. As noted by Chang et al. (2004), during rainwater collection from rooftops, chemical compounds on the rooftops can leach into acidic rainwater, airborne pollutants can be captured by the incident rains, rainwater can absorb the organic substances, and high temperature of the rooftop can cause organic decomposition of the materials collected on the roof catchment.

Field studies world over has shown incidence of chemical (Magyar et al., 2007; Melidis et al., 2007; Polkowska et al., 2002; Simmons et al., 2001) and microbial contamination (Evans et al., 2006; Simmons et al., 2001) of the rainwater collected from rooftops that make it unfit, though in some cases, the level of concentration of chemical contaminants was below permissible levels as per the potability standards for the countries concerned (http://www.rainwaterresources.com/rooftop-runoff-source-contamination-review). Consuming the chemically and biologically contaminated water can pose serious health risks, as experience in rural Auckland suggested (Simmons et al., 2001). Simple slow sand filters currently promoted for treating the water from roof catchments is inadequate to remove these chemical pollutants. Small-scale units at the household level for carrying out sophisticated treatment of the raw water for potability are simply unviable.

9.6. DEALING WITH DROUGHTS AND FLOODS: UNDERSTANDING THE SOCIO-ECONOMIC AND INSTITUTIONAL CONTEXT

"Rainwater harvesting in urban areas" as an idea sounds very appealing, and no one should be over critical about it. International agencies have been promoting rooftop rainwater harvesting for urban areas of developing countries where either formal water supply systems do not exist or the poor people, especially those living in squatters and slums, are not served by municipal water connections. Often, examples from developed countries such as Japan, Germany, Thailand, and Australia are being cited to justify such system for developing countries of Latin America, Africa, and Asia (see, for instance, UN-HABITAT, n.d.). But such recommendations are not based on proper analysis of the socioeconomic and institutional setting in which they work and the situation that exist in the cities of these developing countries. Unlike the developed countries where the water from roof catchments is used by households for low-end uses such as car washing and gardening to save the cost of water, to protect the environment and to comply with the existing laws (Blackburn et al., 2010; White, 2010 in Australian context), people from poor countries need water for basic survival including drinking, cooking, and personal hygiene needs. Therefore, quality of the water from rooftop catchments and its dependability cannot be compromised.

In that respect, what kind of treatment would be required for the water that comes from the roof catchments—contaminated with carbon from vehicular emissions for potability? What is the cost of treatment systems that would be required? Will simple filtration work for removing the contaminants in the water collected from rooftops? What should be the size of the tank that can collect all that water, which can come in a few high-intensity storms? How much it would cost in comparison with the cost of building public water supply systems? If this fast gushing water has to be injected underground, what should be the criteria for designing that recharge system? Obviously, the design of the system is going to involve complex considerations (runoff intensity, infiltration rate of the soil, aquifer storage space, etc.), and not simple ones that a local mason can work out and execute. But no proper thinking has gone into these issues. More importantly, the unpleasant truth is most of the people living in slums of Indian cities who really need that water do not have good roof, and for then, rooftop rainwater harvesting does not make any sense.

In hot and arid tropics, when there is so much uncertainty in the climate, particularly the magnitude of annual precipitation, which determine the probability of occurrence of droughts and floods, if we want water security, with reliable supplies of high-quality water in adequate quantities, the systems that provide that supply will be complex and also expensive. With a huge population living in a small area that continues to rise, it will be naive to think of a cheap and simple system (like a 5001 syntax tank) that provides 24 × 7 water supplies. The research that examines the viability of rainwater harvesting for urban areas of developing countries in semiarid tropics is conspicuous for its silence on the cost-effectiveness (see Shadeed and Lange, 2010).

The rainfall does not occur in equal amounts every year, in the same month and same day, according to our daily demands. As a matter of fact, for Indian monsoon, everything changes from year to year—its magnitude, intensity, duration, and pattern (Pisharoty, 1990; Chapter 10, this book). But bourgeois environmentalists try to produce such twisted view of the reality as a part of a larger narrative to prove a point that we can manage our water needs without having large reservoirs (capable of storing high seasonal flows) and distribution systems (Arabindoo, 2011). The concerns about quality of water, particularly for drinking purpose, have been not been addressed.

As regards control of urban floods, to capture floodwaters in cities and dispose it off, we need well-designed drainage systems that can quickly evacuate the runoff from high-intensity storms. Inadequate capacity of the storm water drainage system appears to be a major reason for the floods in Chennai. In cities such as Chennai, large rainwater collection tanks, which can store the runoff from roofs, might also provide flood cushioning to some extent. Simultaneously to reduce the runoff from other built-up areas, rain gardens will be useful (Dussaillant et al., 2005; Kumar, 2014). If this water is to be stored for future use, we would require a much more sophisticated system—pavements and built-up areas free from oil, carbon, and garbage—detention systems for sedimentation, and then systems for filtration and disinfection and all integrated with a centralized water distribution system so that heavy investments for treatment of the storm water can be avoided (Kumar, 2014). For technical and economic efficiency, such systems will have to be large and centralized.

If the rainwater collection tank is to be built for flood cushioning, what matters is the cost of the tanks that can smoothen out the flood peaks against the additional investment required for providing a higher-capacity storm water drainage system that is capable of disposing off the highest-intensity storm. Such economic analysis would involve complex hydroeconomic simulation of urban floods using historical data. If the purpose is the latter (flood control + rainwater collection for domestic water supply), then the cost of rainwater collection tanks plus treatment system will have to be compared with the additional cost of providing a higher-capacity storm water drainage system plus the extra cost of supplying the water through the public system or the tankers to meet the summer water needs of the dwellings. However, such analyses to facilitate an informed debate on urban water management are missing.

9.7. CONCLUSIONS AND WAY FORWARD

The civil society organizations that promote traditional water harnessing systems such as ponds, tanks, *tankas*, and *nadis* as alternatives for modern water systems involving large reservoirs, diversion systems, and pipeline networks in India have a lopsided view of the water problems and solutions and therefore do not seem to recognize the limitation of these traditional systems. They are against centralized approaches to planning and management of large infrastructure for water supply, irrigation, and flood control, fearing state control of water resources (based on Mishra, 1996, and several remarks of Rajendra Singh, the leader of Tarun Bharat Singh, which works in western Rajasthan, as reported by Gupta, 2011). This particular ideology motivates them to lobby against large dams and technically complex water supply schemes (Kumar and Pandit, 2016).

Notably, none of them have ever taken up the issue of building urban drainage probably because of the reason that addressing them would require sophisticated engineering infrastructure.

For a better and informed debate on water management, the governments need to generate more knowledge about the science of managing water through investment in water research institutions that are autonomous and aggressively engage in the application of the same science to solve the growing problems of water scarcity, droughts, and floods facing its citizens. As on today, there are only three autonomous institutions that are directly under the Ministry of Water Resources, River Development and Ganga Rejuvenation of Government of India. Of these, only one (National Institute of Hydrology) undertakes research in the field of water management. Here again, the research is on hydrologic issues. There is a need for undertaking interdisciplinary and multidisciplinary research in the field of water. For this, either new institutions need to be created or the mandate of the existing institutions needs to be broadened.

REFERENCES

Agarwal, A., Narain, S., 1997. Dying Wisdom: Rise, Fall and Potential of India's Traditional Water Harvesting Systems. Centre for Science and Environment, New Delhi.

Arabindoo, P., 2011. Mobilising for water: hydro-politics of rainwater harvesting in Chennai. Int. J. Urban Sustain. Dev. 3 (1), 106–126.

Blackburn, N., Morison, P., Brown, R., 2010. In: A review of factors indicating likelihood of and motivations for household rainwater tank adoption. Stormwater 2010, National Conference of the Stormwater Industry Association. Sydney, Australia, pp. 1–10.

Chang, M., McBroom, M.W., Beasley, R.S., 2004. Roofing as a source of nonpoint water pollution. J. Environ. Manag. 73 (4), 307–315.

De, U.S., Singh, D.P., Rase, D.M., 2013. Urban flooding in recent decades in four mega cities of India. J. Indian Geophys. Union 17 (2), 153–165.

Dussaillant, A.R., Cuevas, A., Potte, K.W., 2005. Raingardens for stormwater infiltration and focused groundwater recharge: simulations for different world climates. Water Sci. Technol. Water Supply 5 (3–4), 173–179.

Evans, C.A., Coombes, P.J., Dunstan, R.H., 2006. Wind, rain and bacteria: the effect of weather on the microbial composition of roof-harvested rainwater. Water Res. 40 (1), 37–44.

Gupta, S., 2011. Demystifying "Tradition": the politics of rainwater harvesting in rural Rajasthan, India. Water Alternat. 4 (3), 347–364.

Institute for Resource Analysis and Policy, 2010. Tool Kit for Integrated Urban Water Management Vol. 2: Technical Report. Final Report Submitted to Arghyam, Institute for Resource Analysis and Policy, Hyderabad, India.

Kumar, M.D., 2004. Roof water harvesting for domestic water security: who gains and who loses? Water Int. 29 (1), 43–53.

Kumar, M.D., 2009. Water Management in India: What Works, What Doesn't. Gyan Books, New Delhi.

Kumar, M.D., 2014. Thirsty Cities: How Indian Cities Can Meet Their Water Needs. Oxford University Press, New Delhi.

Kumar, M.D., Ghosh, S., Patel, A., Singh, O.P., Ravindranath, R., 2006. Rainwater harvesting in India: some critical issues for basin planning and research. Land Use Water Resour. Res. 6 (1), 1–17.

Kumar, M.D., Pandit, C., 2016. India's water management debate: is the "Civil Society" making it everlasting? Int. J. Water Resour. Dev. https://doi.org/10.1080/07900627.2016.1204536.

Magyar, M.I., Mitchell, V.G., Ladson, A.R., Diaper, C., 2007. An investigation of rainwater tanks quality and sediment dynamics. Water Sci. Technol. 56 (9), 21–28.

Melidis, P., Akratos, C.S., Tsihrintzis, V.A., Trikilidou, E., 2007. Characterization of rain and roof drainage water quality in Xanthi, Greece. Environ. Monit. Assess. 127 (1–3), 15–27.

Mishra, A., 1996. Rajasthan ki Rajat Boondein. Environment Division, Gandhi Peace Foundation, New Delhi.

Narasimhan, B., 2016. Storm Water Drainage of Chennai: Lacunae, Assets and Ways Forward. IIT Madras, Chennai, India.

Pickering, P., Marsden Jacob Associates, 2007. The Economics of Rain Water Tanks and Alternative Water Supply Options: A Report Prepared for Nature Conservation Council of NSW, Australian Conservation Foundation and Environment Victoria. Marsden Jacob Associates, Australia.

Pisharoty, P.R., 1990. Characteristics of Indian Rainfall. Monograph, Physical Research Laboratories, Ahmedabad, India.

Polkowska, Z., Gorecki, T., Namiesnik, J., 2002. Quality of roof runoff waters from an urban region (Gdansk, Poland). Chemosphere 49 (10), 1275–1283.

Prinz, D., 2002. The role of water harvesting in alleviating water scarcity in arid areas. In: Key Note Lecture at the International Conference on Water Resources Management in Arid Regions. Kuwait Institute for Scientific Research, Kuwait.

Shadeed, S., Lange, J., 2010. Rainwater harvesting to alleviate water scarcity in dry conditions: a case study in Faria catchment, Palestine. Water Sci. Eng. 3 (2), 132–143.

Shah, Z., Kumar, M.D., 2008. In the midst of the large dam controversy: objectives, criteria for assessing large water storages in the developing world. Water Resour. Manag. 22 (12), 1799–1824.

Sharma, B.R., 2012. Unlocking Value Out of India's Rainfed Farming Areas. IWMI-Tata Water Policy Program, Anand, Gujarat, India. Water Policy Research Highlight No. 10.

Simmons, G., Hope, V., Lewis, G., Whitmore, J., Gao, W., 2001. Contamination of potable roof-collected rainwater in Auckland, New Zealand. Water Res. 35 (6), 1518–1524.

Srinivasan, V., Gorelick, S.M., Goulder, L., 2010. Sustainable urban water supply in south India: desalination, efficiency improvement, or rainwater harvesting? Water Resour. Res. 46 (10), 1–15.

UN-HABITAT, Rainwater Harvesting and Utilisation. Water, Sanitation and Infrastructure Branch, United Nations Human Settlements Programme, Nairobi, Kenya. Blue Drop Series Book 2: Beneficiaries and Capacity Builders.

White, I., 2010. Rainwater harvesting: theorising and modelling issues that influence household adoption. Water Sci. Technol. 62 (2), 370–377.

Chapter 10

Adapting to Climate Variability and Reducing Carbon Emissions: Strategies That Work for India

10.1. INTRODUCTION

Monsoon weather systems are a result of the land-sea temperature differences caused by solar radiation (Huffman et al., 1997). It is quite a well-established fact that there are many uncertainties associated with climate-change predictions for countries like India, which experience monsoon weather pattern (Loo et al., 2015).[1] The predictions of climate trends for India by various climate models differ widely vis-à-vis not only the magnitude of change in rainfall and temperature but also the nature of change, that is, whether positive or negative change. The degree of uncertainty about the predicted variable is likely to be even higher when these model outputs are used for further predictions of hydrology, that is, runoff and groundwater recharge, of river basins using hydrologic models, and crop yields using biophysical models. While these issues related to climate predictions are widely discussed in various forums, a critical issue that is left out is the implications of climate variability for the utility of model predictions even if we assume that the models are robust and perform well within the bandwidth of permissible errors.

Yet the government of India, the provincial governments, and civil society organizations are preoccupied with serious thinking on complying with international protocols to reduce carbon emissions and measures to reduce the impacts of climate change on the society. What is really worrisome is that whether it is in the use of climate models for predicting future temperature and rainfall, or in the use of biophysical models for predicting future crop yields, or in the design of strategies for mitigating emissions, or in the design of interventions for making regions "climate proof," there is too little application of common sense and practical wisdom. Central to this concern is the fact that there is too little

1. The small-scale regional circulations are more vulnerable to variations in monsoon rainfall (Rajeevan et al., 2008). Therefore, a general measurement of strength of monsoon systems is not enough to represent the temporal and spatial distributions in monsoon rains. Many studies have tried to link the year-to-year variability in monsoon with the El Niño-southern oscillation (ENSO) (Kripalani and Kulkarni, 1997; Ranatunge et al., 2003).

Water Policy Science and Politics. https://doi.org/10.1016/B978-0-12-814903-4.00010-5

attention being paid to the implications of climate variability (interseasonal and interannual) on hydrology and water flows in river basins, energy demand in agriculture, and biomass/crop production.

The result is that significant investments are being made in water, agriculture, and energy sectors, without having any comprehensive and integrated understanding of the likely impact of climate change on these sectors. In a haste to convince the international community, many economic decisions are taken by the central and state governments that are seriously flawed, affecting the power sector economy, financial institutions, and India's manufacturing sector adversely (Iyer, 2015, 2017). This chapter discusses some of the fundamental flaws in climate research in the Indian context and analyzes the physical efficacy and economic viability of some of the climate mitigation and adaptation strategies. It also looks at some of the alternatives ways of reducing carbon emissions that would help the economy and save precious natural resources and the politics behind the stepmotherly attitude toward such alternatives.

10.2. VARIABILITY: INDIA'S CLIMATE REALITY

We ordinary Indians seem to know where the rainfall would increase and where it would decrease and what exactly would be the rainfall by 2050, 2080, and so on. But there is something that really worries people who are concerned with climate research in India. It is the tendency among the "climate crusaders" to run roughshod on those who question climate-change theory and confront them with the more visible phenomenon of "climate variability," and how this reality on the ground, which ideally should pose a challenge to the climate modelers, is systematically ignored by the latter. The disturbing fact is that their work forms the only ground on which most of the suggested actions to stop global warming and arrest climate change are made. The subcontinent is known for climatic variability for several millennia.

An analysis of rainfall data for 90 years and data on annual potential evaporation by the renowned climatologist Prof PR Pisharoty (late) who extensively worked on Indian monsoon had shown the lower the amount of mean annual rainfall in a region, the higher the year-to-year variation and vice versa. The analysis also showed that the lower the rainfall, the higher the aridity, and the greater the erratic nature of the rainfall with precipitation occurring in fewer rainy days (Pisharoty, 1990).

10.3. PREDICTING HYDROLOGICAL CHANGES DUE TO CLIMATE IN AN ERA OF HIGH VARIABILITY

When questions are raised about how the climate variability is factored in in their "model predictions," the standard answer is "Oh, the variability would increase; and we would have greater frequency of extreme weather conditions, with more floods and droughts." But there is no analysis to support this part of the prediction. The fact is that the values of the "predicted variable" of the climate models (such as the change in temperature, rainfall, etc.) are much lesser than the interannual variability we experience in these climate parameters.

They all come with "doomsday prophecies" vis-à-vis the impact of climate change. They would essentially cover such findings as follows: The river would have lesser streamflows (if the river is already known to produce low runoff), or the river would have more frequent floods (if it is known to be a flood-prone river); the paddy and wheat outputs would reduce; and sorghum will have lesser yield in 2080. Unfortunately, these are all based on aggregate figures.

When we know that there is so much interannual variability in rainfall and consequently streamflows, what on earth compels us to say that the values of these variable would increase by "X" percentage, after a time period of say 50 years or 80 years? When we know that the rainfall in an area can range anywhere between 200 and 1000 mm in an area like Anantapur, what sense does it make to say that the (mean) annual rainfall would reduce by 8% or 12% after 50 years? When we know that the annual discharge of Godavari river is found to vary from 29,000 to 183,000 MCM (based on 31-year data from 1981–82 to 2012) (source: CWC), what purpose does it serve if the climate-change-hydrology model for the basin just predicts that the average runoff in Godavari would increase by 10%?

From a purely utilitarian perspective, what one would like to know is the "reference levels" to which these increase and reduction are likely to occur and how it would look like in dry and wet years. Future projections based on mean annual rainfall or normal rainfall will not make much sense. Is it that we don't even know these basic facts about surface hydrology in arid regions, before we attempt such modeling?

10.4. PREDICTING CROP YIELDS UNDER CHANGING CLIMATE

The model predictions for analyzing the impact of climate change on crop yields also suffer from the same inadequacy as in the case of hydrologic impacts. The models compute the yield impacts so accurately for a timescale of say 50 or 80 years. One would realize the absurdity of such predictions, if we consider the fact that the crop yield is determined by a complex function, with solar radiation; micro climate, namely, temperature, humidity, and wind speed; and ecology, with soil nutrient and moisture availability playing a key role (Thornton et al., 2014). But most crop models seem to function in a linear fashion, and the complex interactions between climate parameters at the "local scale" are conveniently ignored.

Internationally, scholars have highlighted the importance of studying the impact of climate variability on biological systems (Thornton et al., 2014; Rowhani et al., 2011), and recent studies had tried to examine the impact of rainfall patterns and temperature stress on crop yields. Rainfall variability is a principal cause of interannual yield variability both at the plot level and aggregate level (Thornton et al., 2014). Hlavinka et al. (2009) found a statistically significant correlation between a monthly drought index and district-level yields in the Czech Republic for several winter- and spring-sown crops, each one of which has a different sensitivity to drought. Both intra- and interseasonal changes in temperature and precipitation have been shown to influence cereal yields in Tanzania (Rowhani et al., 2011). The increases in rainfall variability will have substantial impacts on primary productivity and on the ecosystem provisioning services provided by forests and agroforestry systems. Despite the uncertainty surrounding the precise changes, climate variability needs to be taken into account (Thornton et al., 2014). Rowhani et al. (2011) noted that the impacts of climate change to the middle of this century on crop yields in parts of East Africa may be underestimated by between 4% and 27%, depending on the crop, if only changes in climatic means are taken into account and climate variability is ignored. Thornton et al. (2014) noted that changes in temperature and rainfall quantum and patterns would combine to bring about shifts in the onset and length of growing seasons in the future. However, similar initiatives are missing in India.

While temperature rise (beyond a threshold) can have a negative effect on yield for many crops, what is ignored is that increase in temperature, which has an effect on rainfall (either positive or negative), can alter the local temperature for a few days when the rain occurs. Instead of running "sophisticated" crop models to predict the impact of climate change on yields, we just need to see how the yield of a particular crop fluctuates between years, characterized by sharp difference in rainfall, average temperature, humidity, etc., within the same agroclimatic zone, on the same week or month or day. Also, when the temperature and humidity can fluctuate widely on the same day or between two consecutive days and such fluctuations are even higher than the difference between highest and lowest daily values experienced in a year, the stresses they can induce on crops need to be investigated.

The daily weather data for Aurangabad (Maharashtra) illustrate these points.

Fig. 10.1 shows the daily values of relative humidity (morning and evening) in Aurangabad over a period of 2 years, that is, 2009 and 2010. Fig. 10.1 illustrates the following points: (1) The highest difference encountered in the relative humidity values between morning (RH-AM) and evening (RH-PM) of any day over the entire year during 2009 and 2010 (74% and 68%, respectively) is higher than the difference in relative humidity values for both morning and evening between the most humid day and the least humid day of the year (63% and 74% for 2009 and 63% and 63% for 2010); (2) relative humidity is excessively high in the range of 80%–90% during the rainy season; (3) the RH values for both morning and evening for the same day of the month can vary significantly between years (Table 10.1).

Fig. 10.2 shows the daily maximum and minimum temperature values for two consecutive years (2009 and 2010) recorded in Aurangabad. It illustrates

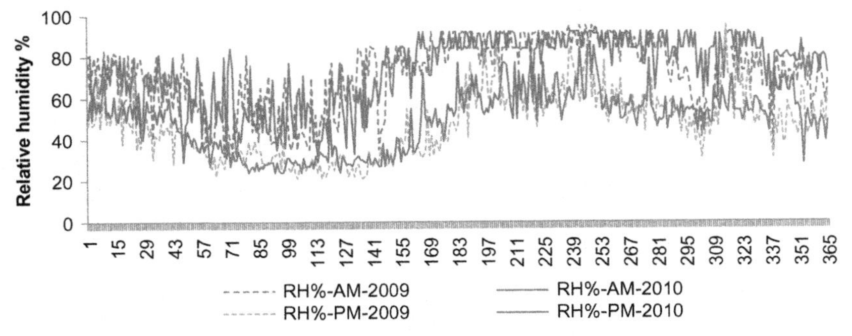

FIG. 10.1 Relative humidity and wind speed—Aurangabad (2009 and 2010).

TABLE 10.1 Characteristics of Relative Humidity (Location: Aurangabad, Maharashtra, and India)

Year of Monitoring	Relative Humidity		
	Difference Between Max of Daily RH-AM and Min of Daily RH-AM	Difference Between Max of Daily RH-PM and Min of RH-PM	Highest Difference Between Daily RH-AM and Daily RH-PM Over the Year
2009	63	74	74
2010	63	63	68

Source: Author's own estimates based on data presented in Fig. 10.1.

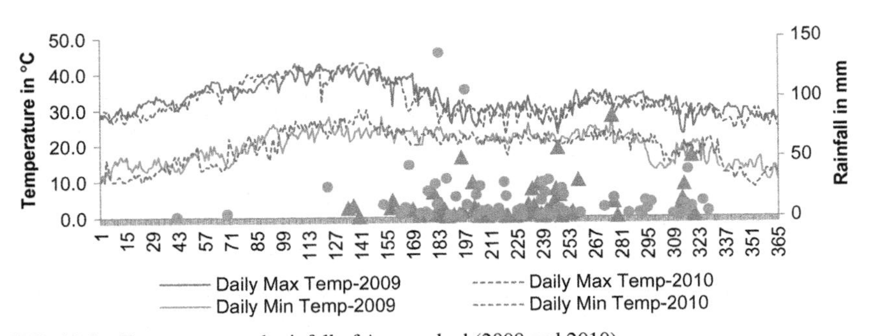

FIG. 10.2 Temperature and rainfall of Aurangabad (2009 and 2010).

TABLE 10.2 Characteristics of Temperature (Location: Aurangabad and Maharashtra)

Year	Temperature (°C)		
	Difference Between Max of Daily Max and Min of Daily Max	Difference Between Max of Daily Min and Min of Daily Minimum	Max Difference Between Daily Max and Daily Minimum
2009	19.8	19.0	32.8
2010	20.2	22.0	35.0

Source: Author's own estimates based on data presented in Fig. 10.2.

the following points: (1) There is wide variation in the temperature between the hottest day and the coldest day of the year; (2) the maximum difference between the daily maximum and the daily minimum temperature recorded in the entire year (32.8°C for 2009 and 35°C for 2010) is higher than both the difference between the highest of the daily maximum or minimum temperature and the lowest of the daily maximum or minimum temperature (19.8°C for max and 19°C for min 2009, respectively); (3) the decline in daily maximum temperature owing to occurrence of rainfall over a period of time is higher than the decline in minimum temperature over the same time period; and (4) finally, there can be significant variations in temperature (daily maximum or daily minimum) on the same day of the month, between 2 years, and this difference can be in the order of 4–5 degrees (Table 10.2).

It will be statistically possible to estimate the effect of changes in mean weekly or monthly temperature and humidity between 2 years on crop yields, provided we have proper controls on other parameters affecting crop yield such as variety, seed quality, irrigation requirements, and soil-nutrient regime.

Our universities should invest in doing such practical, problem-solving-, and policy-oriented research than doing the mundane studies using the "run-of-the-mill" models whose ability to simulate the real conditions is questionable. Such studies can very well throw useful insights into how under various climate-change scenarios the crop yields would change.

India has 15 agroclimatic and 17 agroecological zones. It is not possible in the wildest of imagination that in a country with so many agroclimatic and agroecological zones and with a myriad of crops grown from apples in the high mountains of the Himalayas to wheat and paddy in the semiarid and subhumid alluvial plains to coconut and banana in the hot and humid coastal areas (like in no other part of the world), the impact of temperature change on other climatic parameters would be uniform and the impacts of those changes on yields would be negative for all crops. Intuitively, while climate in some regions becomes less favorable for a particular crop, in some other regions, it must become more favorable for that crop.

10.5. STRATEGIES FOR REDUCING CARBON EMISSIONS: ARE WE GETTING OUR PRIORITIES RIGHT?

As regards strategies for reducing the carbon emissions, India has already made a commitment to have 175 GW of renewable energy. Nearly 57% (100 GW) of this is solar. This can happen if we have overflowing coffers. But we should also do something about the 4+ million inefficient diesel pumps in the Ganges basin pumping out groundwater and chocking the air. Concurrently, we should have plans to electrify these regions, which are heavily dependent on a fossil-fuel-like diesel. Solar pumps are never a great option for most farmers of this basin, owing to the high capital cost of the system, very small operational holdings, and poor economic conditions of the marginal farmers who dominate the region.

There are around 16 million electric motors in India for pumping groundwater, using roughly 120 billion units of electricity every year. If we assume that the entire electricity comes from fossil-fuel-based power plants, the carbon emission would be around 31.2 million ton per annum (Kumar et al., 2013).[2] If we can achieve 20% reduction in electricity use in farming (which is very modest and quite possible through farm water management and pump-efficiency improvements), the carbon emission reduction can be huge (6.24 million ton). What we need is political will. For that, we need to start metering electricity use by farmers and charging for it.

These issues of energy inefficiency and the negative externality-free electricity induced on society through groundwater depletion were surely burning issues even before the "climate change" took the center stage in our environment-development discourse. But, there were no actions for almost 20–25 years. Do we need the "climate change" excuse to address these long pending issues, which have not taken the shape of "ongoing disaster"? Won't it be too early to assume that putting solar PV systems over 100,000 ha of land (1000 km^2) is a better option and it will have no negative impact on climate? What about the carbon footprint of producing solar panels, running into 1000 km^2 in area?" Our environmentalists do not seem to have problem with this, while they are against large hydropower projects, another renewable energy source. The politics around slow pace of hydropower development in the country will be taken up in the next section.

Though we still do not know whether climate change is a reality or not, meteorologic droughts are here to stay. We need to work to reduce their impacts on our society. We need to build infrastructure to take water from water-rich areas

2. This is based on the norm that in 1 kWh of electricity produced from fossil fuels, there is a carbon emission of 0.26 kg.

(we know that there are still a few basins that are very water-rich) to the farthest farms in the dryland areas to save the crops during such climate extremes. The widespread distress among farmers today is not because farmers are not getting remunerative price for the crop in the market, but they are not having adequate access to the hardware called "water" to save their crops from damage during droughts. In order to find water, heavy investments are made by farmers in drilling bore wells in hard-rock areas (Telangana, Andhra, and Vidarbha region of Maharashtra), with utter despair. The strong relationship between well failures and farmers' suicides can be well established if one looks at data from districts such as Anantapur. We need to have infrastructure to supply water to the dryland areas of semiarid and arid India so that the farmers could take risk to grow high-value crops.

While agriculture is still "gambling with monsoon" in India, it is more so in the semiarid hard-rock areas, where wells are a major source of irrigation water. Unfortunately, we haven't got any long-term plan to provide surface water to these areas and thought that the farmers would happily use well irrigation forever. The mammoth investments required for water transfer projects would not go to waste and on the contrary produce high returns, if we consider the societal cost of losing hundreds of thousands of precious lives every year, the social tension and stress they create, and the economic loss incurred by our farmers during droughts.

The indirect benefits of supplying surface water such as reduced energy cost of pumping groundwater and the augmented recharge to groundwater, which improve the sustainability of well irrigation, improved quality of groundwater, etc., can hardly be ignored (Jagadeesan and Kumar, 2015). Table 10.3 shows the economic benefits from energy saving in well irrigation in canal command area of Sardar Sarovar Project, resulting from reduction in depth of groundwater pumping owing to improved recharge of the shallow aquifers from irrigation return flows. The economic benefit ranged from the lowest of Rs. 2034 per hectare in Ahmedabad district to the highest of Rs. 9170 per hectare in Mehsana district, depending on the differences in average reduction in pumping depth and the volume of groundwater use per hectare.

But we need to charge for this expensive water, and our farmers would be more than willing to pay for this, if we can ensure reliable supplies. Yet there are very few initiatives for water transfer underway in India such as the recently commissioned Polavaram project, which links Godavari and Krishna rivers. It would transfer a small portion of the (average) 89,000 MCM of water that drains into the Bay of Bengal, annually.

But our environmentalists and antidam activists and some "civil society organizations" hate such ideas, which according to them means "ecological disaster." As is evident from a recent review of the book by Jagadeesan and Kumar (2015), done by a well-known activist, no matter how rigorous a methodology one follows, how big a sample one chooses, and how much analytic outputs one produces to show the direct and indirect impacts (positive) of large

TABLE 10.3 Average Groundwater Use for Crop Production in Well Command Area and Energy Saving per Hectare of Irrigation (m^3/ha)

District	Volume of Groundwater Use per Hectare After Narmada Water	Average Reduction in Pumping Depth (m)	Energy Savings in Kilowatt Hour per Cubic Meter of Groundwater Pumped (Cx0.055)	Total Saving in Energy for Groundwater Pumping by Farmers per Hectare (kWh)	Total Economic Benefit per Hectare (INR)
Ahmedabad	866.0	8.54	0.47	406.8	2034
Bharuch	1247.8	2.24	0.12	153.5	767.5
Mehsana	2542.6	13.12	0.72	1834.0	9170
Narmada	2665.1	3.76	0.21	551.4	2757
Panchmahals	1383.3	5.08	0.28	386.8	1934
Vadodara	3610.0	5.90	0.32	1170.7	5853.5

Source: Kumar, M.D., Jagadeesan, S., Sivamohan, M.V.K., 2014. Positive externalities of irrigation from the Sardar Sarovar Project for farm production and domestic water supply. Int. J. Water Resour. Dev. 301, 91–109.

water systems, they never get convinced. They can get too deep to find out the "serious flaws" in the sampling design, in terms of failure of the researchers to cover equal number of small, large, and medium farmers (Dharmadhikary, 2015). The presumption is that the large water systems help only the large farmers. This is not hydraulics but sheer ideology. But when it comes to small water harvesting, one should not ask questions about the positive impacts, as it is a "faith." There, numbers about the hydrologic impacts, economic benefits, etc. don't count.

10.6. THE SLOW PACE OF HYDROPOWER DEVELOPMENT

India's hydropower potential is estimated to be around 148,701 MW, as per the estimates of the Central Electricity Authority (CEA). Of these, what can be tapped economically is 84,044 MW. A large share of this potential is concentrated in the northeastern and northern part of the country, the former being in the Brahmaputra basin (66,065 MW) and the latter in the Ganges (20,711 MW) and Indus basins (33,832 MW). Of all the five regions, the northeastern region has the highest hydropower potential of around 58,532 MW. But only a small fraction (1252 MW) of this is currently tapped. Infrastructure to generate another 2954 MW of hydropower is under construction in that region. Overall, the total generation capacity developed in the hydropower segment is 33,013 MW, and another 13,206 is under construction.

It is a well-understood fact that for energy security, especially for managing peak demands in electricity, maintaining a good energy mix is extremely important. However, the share of hydropower in India's electricity production capacity has dwindled remarkably from 46% in 1966 to a mere 17% in 2014. The current installed capacity for power generation is dominated by thermal power with a total share of 68% (PWC, 2014). Such skewed pattern of development with overreliance on thermal power can have serious negative implications for least development, as is evident from Table 10.4. It shows that the levelized cost of energy (LCOE) is the third lowest for hydropower. But this is just one aspect. Hydropower has many advantages over other types of energy-production system. Unlike thermal and nuclear power stations, its response time to widely fluctuating demands is negligible. Therefore, the country's ability to meet peak power demands and respond quickly to wide demand fluctuations have been seriously compromised by the lack of addition of hydropower generation capacity concomitant with thermal power (PWC, 2014).

Also, hydropower is clean energy. While this advantage can be claimed by solar, wind, and tidal power, the issue is about the reliability. While solar power is available during daytime, the power demand peaks during nighttime. So even to go for large-scale solar power as a strategy to have clean energy in the future, sufficient amount of matching power needs to be generated from hydropower.

TABLE 10.4 Levelized Cost of Energy From Different Sources

Type of Electricity Generation System	Cost/MWh (US $)
Conventional coal fired	95.60
Conventional combine cycle gas fired	66.50
Integrated coal-gasification combined cycle (IGCC)	115.90
Nuclear	96.10
Wind	80.30
Offshore wind	204.10
Solar PV	130.00
Solar thermal	243.10
Hydropower	84.50

Source: US Energy Information Administration.

The issue of reliability applies to power depending on tidal waves and wind. However, these aspects are hardly matters of discussion among the civil society groups that protest against hydropower dams and advocate solar energy as an alternative.

There were several initiatives from the central government to rapidly increase the hydropower potential of the country over the past one and a half decades, starting with the National Electricity Policy of 2003. It laid the foundation for open access and trading of electricity that best favored hydropower projects that are naturally best suited to meeting peak power demands. The National Electricity and Tariff Policy of 2006 introduced differential tariff rates for peak and nonpeak power and uniform guidelines for SERC, which was aimed at bringing greater transparency in the power sector. Another important policy measure is the National Rehabilitation and Resettlement Policy of 2007, which emphasized the need for a more transparent and participative process of rehabilitation of project-affected people. The Hydropower Policy of 2008 emphasized the need for full development of the hydropower capacity and increasing private sector participation (PWC, 2014).

There are many reasons for the slow pace of development of hydropower in that region. They are (1) geotechnical problems (high seismicity and unstable geologic formations), (2) high rate of sediment load in the river flows (in the Brahmaputra), (3) the presence of pristine forests that can be submerged by the construction of storage reservoirs, and (4) threat to aquatic life especially native fish population due to reduction in river flows downstream.

While the third problem is being resolved through the construction of run-of-the-river hydel projects that avoid construction of high dams (Price Water House, 2014), the first, the second, and the fourth problem are rather unresolved. The first one has the potential to destabilize the superstructure of a dam. The high rate of sediment load can threaten the life of the reservoir through fast siltation. These problems are extremely serious in the case of Brahmaputra River and its tributaries. Brahmaputra carries one of the world's highest sediment loads, but what complicates it is the fact that there is a lot of uncertainty in the accessible estimates of the quantum of annual sediment transport. The average annual sediment load ranges from 260 to 720 million ton (Fischer et al., 2017). Because of these three problems, the environmental groups have been fiercely opposing construction of hydropower dams in the region. The opposition has also been targeted at run-of-river model of hydel projects citing problem of movement of heavy machinery causing ecological disturbances and increased vehicular traffic causing air pollution. The concerns related to ecological degradation are genuine and need to be addressed (Rao, 2014).

However, what is unfortunate is the myth surrounding hydropower dams and the toll it takes on the fate of many planned hydropower projects. The former is conveniently used by the civil society movements against large dams depending on the situation. Often, the blame for flooding of rivers is falsely put on hydropower dams located upstream without any proper scientific enquiry of the complex problem of floods, as in the case of the devastating floods in Uttaranchal (Puttaswamy, 2015; Reuters, 2014). The obvious question that should have come in the minds of people is "what would have been the extent of damage due to floods in the absence of such reservoirs that impound large amount of flows from the upper catchment." In certain other cases, droughts in the lower plains of basins are attributed to reduced flow due to hydropower dams, as in the case of high dams of Mekong built in its upper catchment in China being blamed for water scarcity in Vietnam (Biswas and Tortajada, 2017). Obviously, if the hydropower dams are responsible for the latter, that is, causing water scarcity, it cannot be held responsible for the former. The hydro dams actually regulate the flows that occur during very few months of the year by storing the monsoon flows and increase the lean season flows by slowly releasing it downstream. The simple fact is that if not anything to reduce the flows, it cannot cause increase in flows during monsoon, which are responsible for floods.

10.7. THE SOLAR FIASCO

After having seen the negative effects of ruthless building of rainwater harvesting structures in the country side, the new fashion is "solar pumps." Some researchers think that growing crops would no longer be profitable for the farmers, and instead, the farmers in India, who own wells, should now produce "solar crops" (Gulati et al., 2016; Shah et al., 2016). Shah et al. (2016) proposed that farmers should run their wells using solar energy and sell the surplus energy to the utilities and that large government subsidy should be made available for its purchase. The business model proposed by the team is as follows. The government gives a subsidy of Rs. 40,000 (US $670) for a solar PV system having a power generation capacity of 1 kW, which is sufficient to replace a connected load of 0.50 kW in the existing power distribution system. If the farmer has a 5 hp (3.75 kW) electric pump, it will be replaced by a 10 hp (7.5 kW) solar PV system, and for that, the government subsidy would be Rs. 4 lakh (US $6700).

The inbuilt assumption is that the farmer would then stop buying electricity from the utility and would instead also start supplying the surplus electricity (after using it for running his own pump) to the same utility. The condition to make this model work is that since the farmer is helping to reduce carbon footprint in agriculture, the utility should purchase this power at a price 10–12 times higher than the price at which it now offers electricity to the same farmer (50 paise per kWh). This proposal has been lapped up by the International Renewable Energy Agency (IRENA). The virtues attached are numerous: reduction in carbon emission from agriculture, reduced groundwater pumping, reduced energy subsidy, happy farmer, and happy utility.

This is a bizarre idea. With 16 million such pumps (as proposed by its proponents), the government will have to spend only INR. 6.4 lakh crore (20 times the amount spent on National Rural Employment Guarantee Act every year), that is, US $100 billion, while the total cost of production and supply of electricity in agriculture is less than US $9 billion per year. There can't be a better recipe for ruining the government finances and emptying our coffers. It is a different matter that a lion's share of this money (if at all the government decides to spend) would end up with the rich farmers, who would go for large-capacity solar PV systems. One doesn't need to apply rocket science to say this.

Further, the assumption that the farmers would be able to run his/her motor as and when needed using a solar PV system is fallacious. The availability of solar energy required to produce the electricity for running the pump is highly uncertain during rainy season, due to cloud cover. The minimum requirement for the farmer to make his energy-production system self-sufficient is to have large-capacity batteries, which are prohibitively expensive today. If not, he/she would have to depend on the power supply from the grid. The argument that the utility has to buy power from the farmers at a rate far higher than the price at which they supply power is not economic prudence.

Even if for the time being, we assume that the utility agrees to such a power purchase model, it would require "net metering" (two-way metering). The question is "What on earth makes one believe that the farmer would become honest overnight not to tamper with the meters while using power from the grid and also while selling power to the grid?" As a matter of fact, in this case, there is greater incentive to tamper with meters. In other words, if we have faith in the farmers, why can't we just install meters for every electric pump and supply reliable and high-quality power to the farmers and charge for it? This way, we can find solution to a few of the problems facing the groundwater and energy sectors.

Obviously, under both scenarios, monitoring would be required to make sure that there is no tampering of meters. Hence, it is far more sensible to start metering electricity consumption in the farm sector to improve electricity-use efficiency and cut down power consumption in agriculture sector, thereby improving the financial working of power utilities. Thereby, we can cut down carbon emissions. For this, we need the political will and not large public funds to spend on subsidies for unproved technologies like the "solar PV systems for small Indian farms." The route, which is suggested now, is expensive, indirect, and nonworkable. On the contrary, it would lead to further depletion of groundwater, huge burden on the government exchequer (for subsidizing solar pumps), and dive the power utilities to total bankruptcy. But the entire agenda seems to be privatization of electricity generation (leaving it in the hands of rich and influential farmers) and junking of the power utilities.

To sum up, there are no "wonder drugs" or magical solutions to the problems facing agriculture, groundwater, and energy sectors. If the excuse for getting rid of meters for agrowells was "transaction cost," the same can be a good reason for the electricity utilities to shy away from "net metering".[3] Also, the solar PV systems are very expensive, and the widely publicized benefits of producing "clean energy" are too meager in economic terms to offset the huge capital costs. Solar power generation could be viable at the "utility level" (with economies of scale), particularly with sophisticated operation and maintenance, and not at the level of micro farms. Bassi (2015) shows that even replacing diesel pumps by pumps powered by solar PV systems is not a viable option, even when we consider the benefits of emission reduction (Bassi, 2015).

Manufacturing of solar PV systems itself leaves a lot of carbon footprint. Also, there are serious concerns about the life of the system itself, particularly for hot and semiarid to arid climate. All the current calculations of the benefits from solar PV systems are done on the basis of the assumption that it will have an average life of 25 years. A review of field tests on the performance of solar modules reported from other parts of the world over the past 40 years shows an average degradation rate closer to 1% per annum, which is higher than the lower value of 0.5% degradation rate, necessary to maintain the 25-year warrantee period. More importantly, the effect of climate on degradation rate of solar cell is not addressed (Jordan and Kurtz, 2012).

10.8. EMISSION REDUCTION BENEFITS OF ENERGY METERING IN AGRICULTURE

Metering of electricity and water involves transaction costs, though this transaction cost can be reduced significantly through the use of mobile phones and

3. This is because in the latter case, the chances of malpractice by farmers are much greater, as farmers can tamper with the meters to show higher supply while supplying power to the utility and lower while tapping power from the grid.

information technology. Today, technologies exist and have been attempted, not only for metering but also for controlling energy consumption by farmers (Aarnoudse et al., 2016; Zekri, 2008). The prepaid electronic meters, which are typically operated through scratch cards and can work on satellite and Internet technology, are fit for remote areas to monitor energy use and control groundwater use online from a centralized station. The technology builds on services provided by the Internet and mobile (satellite) phone services, which have improved remarkably in India over the last 15 years, especially in the rural areas, with a phenomenal increase in the number of consumers (Kumar et al., 2011). No matter how large the transaction cost is, we ultimately will have to go for formalizing our energy economy and also water economy for better productivity, growth, and sustainable resource use. While we need to offer subsidies to poor farmers for electricity and water, we cannot do away with measurements. Studies clearly show that with electricity metering and pro rata pricing in the farm sector, efficiency of both water and energy use in agriculture could increase (Kumar et al., 2013; Pfeiffer and Lin, 2013). With rationing of energy supply, the productivity can further be enhanced, and aggregate consumption of electricity and water can be brought down, thereby reducing carbon footprint in agriculture (Kumar, 2013).

As per estimates provided by David and Herzog (n.d.), the cost of capturing 1 kg of CO_2 emission from thermal power generation is US $0.049 or INR 0.49 (ppp adjusted), and 1 kWh of power generation produces 0.96 kg of CO_2. Hence, the social benefit associated with preventing carbon emission by saving 1 kWh of electricity would be equal to Rs. 0.47. The total electricity consumption in agriculture sector was estimated to be 107.77 billion units per annum in 2008–09. If we assume that there would be 20% energy saving obtained from water productivity improvement in irrigated farm production alone due to efficient pricing,[4] the total social benefit would be to the tune of 709 crore rupees per annum, for a total electricity saving of 2156 crore (21.56 billion) units[5] (Kumar, 2013). Hence, let us not ruin our agricultural economy, water economy, and power utilities with pervasive subsidies, by making the cosmic blunder of trying to make some whimsical ideas work. What farmers need today is not subsidy, but assured supply of water and electricity to grow crops and good markets, and they would be more than willing to pay for them.

10.9. FINDINGS AND CONCLUSIONS

Currently, the knowledge and information about climate change and its impacts on water resources and biophysical systems are quite uncertain, due to the potentially significant effect of climate variability on both. If we want to understand

4. Energy saving can also come from improvement in pump efficiencies, occurring as a result of pro rata pricing of electricity. However, we have not considered this.

5. Here, we assumed that only 70% of the total electricity consumed in the country comes from thermal power, and the rest comes from clean energy sources such as nuclear power and hydropower.

the impacts of climate change on water, we should have good records of our weather, hydrology, and biophysical systems, to ascertain how hydrologic conditions change from location to location and from time to time with changes in weather parameters and land use and land cover. More importantly, we should analyze them properly to find out the causes and effects, before planning for climate mitigation and adaptation. This is not being done today in India.

The analysis of hydrologic changes should focus on long-term changes in streamflows in rivers that are ungauged and soil moisture levels, along with changes in weather parameters—rainfall, temperature, relative humidity, and wind speed—and land use so as to segregate the effects of climate change, climate variability, and land use. Crop models for analyzing the effect of climate change on crop yields need to capture the effect of variability in rainfall magnitude and its pattern, temperature, and other weather parameters too, and this will be possible only if we do continuous monitoring of biophysical systems along with weather parameters. If climate change becomes a reality, such database and *knowledge* would only strengthen our ability to devise strategies for adaptation. Such database showing the fluctuation in crop yields due to climate variability can also be used to test the veracity of predictions made by climate models.

At present, the priorities in the action plan on climate change seem to be misplaced. There are many measures India should adopt to reduce carbon footprint in agriculture. Energy pricing for groundwater pumping in farming sector is one of them, and this alone would also help enhance agricultural productivity and water productivity in agriculture while reducing carbon emissions. Also, the priority has to shift from solar pumps to hydropower development, which brings in the same clean energy benefits as that of the former while being very cheap.

REFERENCES

Aarnoudse, E., Qu, W., Bluemling, B., Herzfeld, T., 2016. Groundwater quota versus tiered groundwater pricing: two cases of groundwater management in north-west China. Int. J. Water Resour. Dev. 33 (6), 917–934.

Bassi, N., 2015. Irrigation and energy nexus: solar pumps are not viable. Econ. Polit. Wkly. 50 (10), 63–66.

Biswas, A.K., Tortajada, C., 2017. Myth and realities of dams and droughts. . China Daily, June 26.

David, J., Herzog, H., n.d. The Cost of Carbon Capture. Massachusetts Institute of Technology (MIT), Cambridge, MA.

Dharmadhikary, S., 2015. A weak deference. Frontline.

Fischer, S., Pietroń, J., Bring, A., Thorslund, J., Jarsjö, J., 2017. Present to future sediment transport of the Brahmaputra River: reducing uncertainty in predictions and management. Reg. Environ. Chang. 17 (2), 515–526.

Gulati, A., Stuti, M., Kacker, R., 2016. Harvesting Solar Power in India. Indian Council for Research in International Economic Relations, New Delhi. Working Paper No. 329.

Hlavinka, P., Trnka, M., Semeradova, D., Dubrovsky, M., Zalud, Z., Mozny, M., 2009. Effect of drought on yield variability of key crops in Czech Republic. Agric. For. Meteorol. 149 (3–4), 431–442.

Huffman, G.J., Adler, R.F., Arkin, P., Chang, A., Ferraro, R., Gruber, A., Janowiak, J., McNab, A., Rudolf, B., Schneider, U., 1997. The global precipitation climatology project (GPCP) combined precipitation dataset. Bull. Am. Meteorol. Soc. 78 (1), 5–20.

Iyer, S.S.A., 2015. 100,000 MW of costly solar power can sink "Make in India." Swaminomics. Times of India.

Iyer, S.S.A., 2017. Dark side of solar success: it may kill thermal power, banks. Swaminomics. Times of India.

Jagadeesan, S., Kumar, M.D., 2015. The Sardar Sarovar Project: Assessing Economic and Social Impacts. Sage Publications, New Delhi.

Jordan, D.C., Kurtz, S.R., 2012. Photovoltaic degradation rates-an analytical review. Prog. Photovolt. 21 (1), 12–29.

Kripalani, R.H., Kulkarni, A., 1997. Rainfall variability over South-east Asia-connections with Indian monsoon and ENSO extremes: new perspectives. Int. J. Climatol. 17 (11), 1155–1168.

Kumar, M.D., Scott, C.A., Singh, O.P., 2011. Inducing the shift from flat-rate or free agricultural power to metered supply: Implications for groundwater depletion and power sector viability in India. J. Hydrol. 409 (11), 382–394.

Kumar, M.D., 2013. Raising agricultural productivity, reducing groundwater use and mitigating carbon emissions: role of energy pricing in farm sector. Indian J. Agric. Econ. 68 (3), 275–291.

Kumar, M.D., Scott, C.A., Singh, O.P., 2013. Can India raise agricultural productivity while reducing groundwater and energy use? Int. J. Water Resour. Dev. 29 (4), 557–573.

Kumar, M.D., Jagadeesan, S., Sivamohan, M.V.K., 2014. Positive externalities of irrigation from the Sardar Sarovar Project for farm production and domestic water supply. Int. J. Water Resour. Dev. 30 (1), 91–109.

Loo, Y.Y., Billa, L., Singh, A., 2015. Effect of climate change on seasonal monsoon in Asia and its impact on the variability of monsoon rainfall in Southeast Asia. Geosci. Front. 6 (6), 817–823.

Pfeiffer, L., Lin, C.Y., 2013. In: The effects of energy prices on groundwater extraction in agriculture in the high plains aquifer. Selected Paper Prepared for Presentation at the 2014 Allied Social Sciences Association (ASSA) Annual Meeting. Philadelphia, PA, USA, January 3–5, 2014.

Pisharoty, P.R., 1990. Characteristics of Indian Rainfall. Monograph, Physical Research Laboratories, Ahmedabad.

Price Water House, 2014. Hydropower in India: Key Enablers for a Better Tomorrow. Price Water House Coopers Pvt Ltd., New Delhi.

Puttaswamy, H.J., 2015. Response to the opinion article entitled "Environmental over enthusiasm" by Chetan Pandit published in International Journal of Water Resources Development on 20 January, 2014. Int. J. Water Resour. Dev. 31 (4), 780–784.

Rajeevan, M., Gadgil, S., Bhate, J., 2008. Active and Break Spells of the Indian Summer Monsoon. NCC research report.

Ranatunge, E., Malmgren, B.A., Hayashi, Y., Mikami, T., Morishima, W., Yokozawa, M., Nishimori, M., 2003. Changes in the southwest monsoon mean daily rainfall intensity in Sri Lanka: relationship to the El Niño–southern oscillation. Palaeogeogr. Palaeoclimatol. Palaeoecol. 197 (1–2), 1–14.

Rao, N., 2014. Benefit sharing mechanism for hydropower project: pointers for Northeast India. In: Kumar, M.D., Bassi, N., Narayanamoorthy, A., Sivamohan, M.V.K. (Eds.), The Water, Energy and Food Security Nexus: Lessons from India for Development. Earthscan/Routledge, London, United Kingdom, pp. 57–72.

Reuters, 2014. Hydro-Power Plants Blamed for Deadly Floods in India. Reuters India.

Rowhani, P., Lobell, D.B., Linderman, M., Ramankutty, N., 2011. Climate variability and crop production in Tanzania. Agric. For. Meteorol. 151 (4), 449–460.

Shah, T., Durga, N., Verma, S., Rathod, R., 2016. Solar Power as Remunerative Crop. IWMI-Tata Water Policy Program, Gujarat, India. Water Policy Research Highlight No. 10.
Thornton, P.K., Ericksen, P.J., Herrero, M., Challinor, A.J., 2014. Climate variability and vulnerability to climate change: a review. Glob. Chang. Biol. 20 (11), 3313–3328.
Zekri, S., 2008. Using economic incentives and regulations to reduce seawater intrusion in the Batinah coastal area of Oman. Agric. Water Manag. 95 (3), 243–252.

Chapter 11

Canal Irrigation Versus Well Irrigation: Comparing the Uncomparable

11.1. INTRODUCTION

Crops need moisture, and farmers in dryland areas need irrigation water. It does not matter which source the water comes from—whether from a tank, a well, or a canal—so long as it comes in adequate quantity, quality, and reliability. Unfortunately, in the recent years, a myopic view favoring only private well irrigation in preference to canal irrigation has emerged among a few irrigation experts in India (see Shah, 2016, 2009; Mukherji et al., 2010). It is a well-known fact that major irrigation projects contributed remarkably to expanding irrigated area and stabilizing agricultural production in the country since Independence (Hazell and Wood, 2008; Kumar, 2003; Perry, 2001; Shah and Kumar, 2008).

But, in the last three decades, private well irrigation witnessed rapid growth surpassing flow irrigation in its contribution to the net irrigated area. This was because of massive rural electrification, heavy electricity subsidies, and institutional financing for wells and pumps (Kumar, 2007). However, a distorted thinking of considering one system superior to the other tends to become pervasive. This is because of a poor understanding of determinants of irrigation growth and the fundamental difference between well irrigation and surface irrigation (Kumar et al., 2012). Such a view has been pampered in the water circle for more than two decades now.

For quite some time, these groundwater enthusiasts considered that the growth in groundwater abstraction structures (wells) is a strong indication of the growing contribution of groundwater in India's irrigation landscape and all is well with groundwater development, unless there is a decline in well numbers (see DebRoy and Shah, 2003). The fascinating figures of growth in well-irrigated area in India were used to reinforce this point (Kumar, 2007). What was conveniently ignored is the fact that while the area under well irrigation was increasing steadily at the national level, in many pockets, it was already on the decline (with serious problem of mining), including areas that show increase in well numbers, and that there is no connection between numbers of wells that tap the aquifers and health of those aquifers. Such aggregate and simplistic views

Water Policy Science and Politics. https://doi.org/10.1016/B978-0-12-814903-4.00011-7

about irrigation development had come at a huge social cost in terms of declining investment for public (surface) irrigation. It is important to see how such a mechanistic approach of analyzing the irrigation sector using simple statistics misleads the sector planners.

In this chapter, we highlight how comparison between private well irrigation and large-scale public surface irrigation using simple statistics is flawed and highlight some of the unique virtues of large surface irrigation that are largely ignored by researchers who promote well irrigation as an alternative to gravity irrigation. It also highlights the equity concerns posed by intensive groundwater irrigation, which is supported by heavy public subsidies for electricity, and also how groundwater markets become highly monopolistic under the current institutional and policy regimes governing the use of groundwater, characterized by flat-rate pricing of electricity and the lack of well-defined water rights that favor large well owners.

11.2. GROUNDWATER TOO COMPLEX A RESOURCE TO MANAGE

It is well established that groundwater hydrology in India is too heterogeneous to be amenable to any simple regional comparison using well numbers. There are deep and extensive alluvial formations in India, and they extend over the entire Indo-Gangetic Plain. But, two-thirds of the geographic area is underlain by hard rocks (Kulkarni and Shankar, 2009; Kulkarni et al., 2011). While a well yield of 25 L/s is quite common in the Gangetic Plains, in the hard rock, often, it is as low as 1–2 L/s. While a tube well in the Gangetic Plains can irrigate 30–40 ha of land, a bore well in the hard-rock areas can hardly irrigate 0.4–1.0 ha. But, the ardent supporters of well irrigation ignore this conveniently while analyzing the crude statistics of groundwater structures (Kumar et al., 2013a).

It is also well known that in the same locality, geohydrologic environment changes remarkably over time. In the alluvial areas, it can change from shallow water table conditions in phreatic aquifers to deep piezometric levels in high-yielding confined aquifers. In hard rock, it can be from shallow water table in large open wells to deep water table in bore wells. Accordingly, the irrigation potential of wells also could change drastically. While the shallow open wells in the alluvium irrigate 1.2–1.6 ha of land, the irrigation potential of deep tube wells could be as high as 30–40 ha, as mentioned above. In the case of hard-rock areas, just the opposite usually happens. While the large open wells could irrigate 1.6–2.0 ha, the bore wells will hardly be able to irrigate 0.4–0.80 ha. Also, their life is very short (Kumar et al., 2013a).

Given the complexity of groundwater systems, the researchers should exercise some caution while drawing conclusions on the basis of trends emulating from the data on well numbers, particularly when the data used for comparisons for different time periods do not belong to the same states or geographic areas. Policies formulated on the basis of conclusions drawn from the analysis of insufficient and rudimentary data can have serious unintended consequences.[1] Doing simple statistical projections (of growth in well numbers) based on past trends without understanding the underlying physical processes is tardy. Many in fact assumed that the number of groundwater structures can keep increasing with time with proportional increase in area irrigated, without being constrained by the lack of availability of groundwater (particularly in the semiarid, hard-rock regions), arable land, finance, etc. The outcome of this skewed view of irrigation is that well irrigation is pitched as an alternative to gravity irrigation.

11.3. COMPARING SURFACE IRRIGATION AND GROUNDWATER IRRIGATION

The reality is that surface irrigation systems provide more dependable sources of water than groundwater-based systems in most parts. For flow irrigation, there should be a dependable source of water and a topography permitting flow by gravity to the places of demand. Ideally, the design itself ensures sufficient yield from the catchment to supply water to the command areas for an estimated duty of the design command, or in other words, the design command is adjusted to match the flows available from the catchment. Hence, the design command of the scheme is by far realistic for reliable "dependable yield" estimates, unless major changes occur in the catchment that changes the flow regimes and silt load.

But, in the case of groundwater, thousands of farmers draw water from the same aquifer. Since they all operate individually, "safe yield" of the aquifer is not reckoned while designing the well. So, the productive life of a well is not in the hands of an individual farmer who owns it, but depends on the characteristics of aquifer, wells, and total abstraction. In the entire hard-rock region, the wells go dry, or the yield reduces drastically when aquifer is "overexploited" or when monsoon fails. As a matter of fact, in spite of the explosion in well numbers, the well-irrigated area had not increased here during the past decade. Growth is almost decelerated in most parts of India since the 1990s. Most of the well irrigation in India is in the arid and semiarid regions of northern, northwestern, western, and peninsular India. Among this, per capita annual withdrawal is highest in some of the northern and northwestern states.

1. One such conclusion is that minor irrigation is all about groundwater. To make this point, comparison is made between numbers of groundwater abstraction structures against other MI (minor irrigation) structures (see Mukherji et al., 2013). The importance of groundwater in India's irrigation cannot be judged from a mere number of structures. From the point that the surface MI structures account for only 6% of the total MI structures in the country, an argument is made that we can ignore them (this is simply absurd). Many tanks irrigate hundreds of hectares of land, often 50–100 times the land irrigated by wells. The tanks account for 25% of the net irrigated area. By this standard, we should ignore all large irrigation schemes in the country, as they are only a few thousands of them, against around 19 million wells. They are comparing the uncomparable (Kumar et al. 2013b).

Well irrigation "diehards" believe that the past growth trend in groundwater-irrigated area would continue in the future too (Mukherji et al., 2013). Interestingly, intensive irrigation could sustain for many decades only in a few pockets of India such as alluvial Punjab and Haryana and UP, which have plenty of canals. These regions are already saturated in terms of irrigated area, and expansion is not possible, whereas in Rajasthan, Gujarat, Andhra Pradesh, Telangana, and Tamil Nadu, problems of overexploitation halt further growth. Most of the untapped groundwater is in eastern Gangetic Plains, devoid of sufficient arable land. Peninsular India and central India have a lot of unirrigated land. Agriculture is prosperous here, and demand for water is also high. But, well irrigation is experiencing a "leveling off" and sometimes decline due to "overexploitation" and monsoon failure. In the hard-rock areas of Narmada river basin in Madhya Pradesh, the average area irrigated by a single well has declined over a 25-year period. Such a phenomenon is occurring due to well interference. In such situations, an increase in number of wells does not result in increase in total irrigated area. Hence, it is wrong to assume that well irrigation could sustain the same pace of growth in the coming years.

The spatial mismatch in resource availability and demand can be effectively addressed only by large surface water projects and not by groundwater projects. Surface irrigation can expand in the future also with investments in large reservoirs and transfer systems that can take water from the abundant regions of the north and east to the parched, but fertile lands in the south, though their economic viability and social costs and benefits will have to be ascertained. But, the same is not true for wells, as the technical feasibility and economic viability of transferring groundwater in bulk is questionable (Kumar et al., 2012).

However, now and then arguments are made that the government investment in surface irrigation systems should be diverted for better management of aquifers. Such arguments stem from the presumption that surface irrigation systems perform badly and are based on outdated concepts, which treated the water diverted from reservoirs in excess of crop water requirement as "waste" (see Shah, 2009). It is not well informed by the knowledge of water-use hydrology of surface irrigation systems. Most of the seepage and deep percolation from flow irrigation systems replenish groundwater and are available for reuse by well owners in the canal command. This recycling process not only renders many millions of wells productive but also saves the scarce energy required to pump groundwater by lowering pumping depths.

Prof B. D. Dhawan, one of the renowned irrigation economists, looked at the economic returns from surface irrigation systems in his book wherein he examined the merits he claims and counterclaims about the benefits of big dams. He had highlighted the social benefits generated by large irrigation schemes through the positive externalities such as improving well yields, reducing incidence of well failures, and increasing the overall sustainability of well irrigation, by citing the example of Mulla command in Maharashtra (Dhawan, 1990). The point is that canal irrigation generates large social benefits by protecting groundwater ecosystems, by reducing energy cost for pumping groundwater, and by sustaining irrigated area (Dhawan, 1990; Jagadeesan and Kumar, 2015). But, irrigation planners have, by far, nearly failed to capture them in the cost-benefit calculations. Data from the government of (undivided) Andhra Pradesh showed that the command irrigated regions of the state have the lowest number of groundwater "overexploited" mandals.

Added to these are the multiple-use benefits that canal water generate such as fish production, brick making, water for domestic use, and cattle in rural areas. Fishery is possible in large reservoirs (Jagadeesan et al., 2016).[2] An alternative is to store a small share of the water released from the reservoir during monsoon and winter season in shallow ponds in the command area, which can act as reservoirs for freshwater fisheries and can be stocked with fingerlings. In areas such as Mahi command in South Gujarat and Hirakud command in Odisha and in the deltas of Krishna and Godavari, the water from canals is used to replenish local ponds for raising (Jagadeesan and Kumar, 2015).

The pathological hatred for large public irrigation doesn't stop there. It is even suggested that flows from the small canals or small water harvesting/artificial recharge structure should be used for recharging aquifers (Shah, 2009; Mukherji et al., 2009). This is fallacious as the arid and semiarid regions, where aquifers are depleting, have extremely limited surface water. Any new interventions to impound water would reduce the d/s flows, creating a situation of "Peter taking Paul's water." Such indiscriminate water harvesting is also leading to conflicts between upstream and downstream communities as reported by several scholars in the recent past.

Bringing water from water-surplus basins to peninsular India just for aquifer recharge would require large head works, huge lifts, long canals, intermediate storage systems, and intricate distribution networks. The need for vast precious land for spreading water for recharge would make it also socially unviable, while further increasing economic costs. Since the hard-rock aquifers have extremely poor storage capacity, efficient recharge would require synchronized operation of recharge systems and irrigation wells. This would call for advanced hydraulic designs and sophisticated system operation. Therefore, such an approach of using imported surface water for recharge would sound like "catching the crane using butter." The fact is that practicing environmentally sound artificial recharge is a very expensive affair.

2. The reservoir fishery is already happening in many large reservoirs. But, its productivity would be limited. Large reservoir fishery is facing problems of low productivity in India, due to rapid drawdown in water levels experienced after the monsoon, when the water level would be highest (Jhingran, 1986). In order to raise productivity levels in large reservoirs and fisheries, scientists and managers need to understand the pattern of irrigation water demand so as to adjust their stocking and harvesting programs accordingly (Saha and Paul, 2000).

Hence, the best option would be for the farmers to use this expensive canal water for applying to the crops in that season of import (mainly monsoon season) and use the recharge from natural return flows for growing crops in the next season. Opportunities for using water from "surplus basins" for recharging depleted aquifers exist at least in some areas. Examples are alluvial North Gujarat and north-central Rajasthan.

11.4. DECLINING SURFACE IRRIGATION OR MISUSE OF STATISTICS?

Often, there is misuse of statistics, when statistics of area irrigated by different sources are compared (see Chapter 5 in this book). Crude numbers of "irrigated area by source" are used to make the point that investing in surface irrigation systems is like the *Labour of Sisyphus* (see Shah, 2009). Which model of irrigation is best suited for the area in the future can be judged by the nature of topography, hydrology, and aquifer conditions. Some researchers and activists use "declining area under canal irrigation" to build a case for stopping investments in surface irrigation. But, the reasons for this declining trend can be understood if we look at the real factors that influence the irrigation performance of surface systems.

First, increased pumping of groundwater can significantly reduce streamflows in basins where groundwater outflows contribute to surface flows, thereby affecting the inflows into the reservoirs. Also, small water harvesting systems are adding to the reduction in inflows into large and small reservoirs.

Second, farmers in most irrigation commands install diesel pumps to lift water from the canals and irrigate the fields. Such instances are increasing with pump explosion in rural India. The better control over water delivery, which farmers can secure by doing this, is the reason for their preference for energy-intensive lifting to gravity flow. The pumping devices also enable the illegal diversion of water for irrigating plots that are otherwise out of command due to topographical constraints. Sadly, such areas get reported as pump-irrigated areas in official statistics.

Third, large reservoirs, primarily built for irrigation in this country, are being increasingly used for supplying water to big cities and small towns as recent studies show. A recent analysis involving 301 cities/towns in India shows that with increase in city population, the dependence on surface water resources for water supply increases, with the dependence becoming as high as 91% for larger cities (Kumar et al., 2014; Mukherji et al., 2010). Many large cities depend entirely on surface water imported from large reservoirs, built primarily for irrigation. During droughts, they become the only source of water even to supply for municipal and rural domestic needs, as wells in rural and urban areas dry up.

Fourth, farmers in canal command areas, especially at the head reaches, tend to put more area under water-intensive crops, ignoring the cropping pattern considered in the design. This is one of the reasons for shrinkage in the irrigated command area. As discussed by the author in Chapter 8 of this book, the figures of "irrigation potential created" are largely fictitious and should be done away with.

Last, but not the least, reservoirs are experiencing problems of sedimentation causing reduction in their storage capacity and life. Therefore, it is likely that with the passage of time, the area under surface irrigation declines, if nothing is done to revive the reservoirs. Hence, it is quite obvious that with cumulative investments in surface irrigation systems going up with time, there may not be proportional rise in surface-irrigated area (see discussion in Chapter 5).

11.5. UNSEEN BENEFITS OF GRAVITY IRRIGATION

An important trait of gravity irrigation from surface water sources is the quality of water. Groundwater in many parts of India suffers from chemical contamination due to geogenic factors. High levels of salinity are encountered in the groundwater. Continuous use of saline groundwater to irrigate crops causes soil salinization, thereby resulting in poor yield of crops. Scientific evidence also suggests that soil salinity also reduces the water productivity of crops in relation to ET, due to the decline in transpiration coefficient or the amount of biomass produced from unit volume of water transpired (Geerts and Raes, 2009).

Another important point is that several of the vegetables (chili, tomatoes, cucumber, and brinjal), fruits (watermelon, berry, pomegranate, and mandarin orange) and flowers (marigold), grown in the hot and arid, water-scarce regions, cannot be raised with native groundwater there, as groundwater in these regions contains considerable amount of salts. They are grown using water from wells replenished by return flows from canal-irrigated fields and seepage from large reservoirs. The farmers growing these crops are able to generate very high returns from irrigation with this freshwater, in economic terms (dollars per cubic meter of water). Return flow from canals and seepage from reservoirs reduces the salinity of the water, whereas the access to wells enhances control over water delivery. With rising market demand for high-value fruits, vegetables, and flowers, there is increasing preference among farmers for water free from salts.

Evidence available from different parts of the world suggests that there is a strong nexus between surface irrigation development and sustainability of well irrigation (Sarkar, 2012). Therefore, it is not prudent to invest in well irrigation without investment in large surface reservoirs and conveyance systems in semiarid and arid areas. It is high time for the "diehards" of well irrigation to understand that water, whether well water or canal water, has to come from the same hydrologic system. As world-renowned water resource experts Prof Asit K Biswas and Prof Cecilia Tortajada put it, "large dams have a very important role to play in human development of developing countries, and there is really no other choice."

11.6. GROUNDWATER AND EQUITY

Public irrigation systems were long criticized for being biased against small and marginal farmers. The point made by the critics was that access to water from large public systems was decided by power structure existing in the villages, with the influential rural elite with large holdings gaining effective control over the system and the water delivered by it (Mollinga and Bolding, 2004). However, most of the research had focused on how the farmers at the head end of the hydraulic system often had appropriated the resource at the cost of those whose land is located at the tail end. So far, there is hardly any empirical evidence available from field research, which suggests that the tail-end farms are all owned by poor small and marginal holders, so as to validate the hypothesis about the institutional bias of gravity irrigation system. Well irrigation was pitched against this as a more democratic resource, with nearly 50% of them being owned by small and marginal farmers (DebRoy and Shah, 2003).

The work led by the lead author of the abovementioned paper argued in another paper that India's groundwater structures are mostly owned (66.5%) by small and marginal farmers (Mukherji et al., 2013, Table 13). This argument was made as a result of a highly flawed analysis. The authors used the "wrong denominator" for knowing who controls the groundwater economy. Ideally, the authors should have used the total number of wells owned by each category of farmers against the total number of operational holders under that category. Further analysis could have been done by looking at the characteristics of wells owned by each category of farmers. Instead, they used the total number of wells owned by farmers under each category (in the denominator) against the number of wells from the 30 states and union territories to arrive at their conclusion. It is as good as saying that the percentage houses owned by low and lower middle groups are far higher than that of high-income group, and therefore, houses are constructed only by the poor people.

What the authors conveniently forget is the fact that around 80% of the farmers in India belongs to small and marginal category (see Table 11.1) and a very small fraction of them own wells and pump sets (Fig. 11.1, based on Kumar, 2007). Against this, less than 1% of the farmers are large holders, and around 69% of the large farmers own wells and pump sets. So, there exists a high inequity in access to groundwater abstraction structures. What the authors have attempted is a good example of misuse of statistics.

They stretch their argument by saying that "groundwater irrigation continues to remain as the only source of irrigation for India's poorest farmers" (Mukherji et al., 2013, p. 117). The authors do not think it is important to say anything about the degree of access small and marginal farmers enjoy in public irrigation scheme. Is it that they assume that small and marginal farmers do not have any access to water from public irrigation (canal) systems, wherein the landowners do not have to invest anything toward capital expenditure? The field research in Punjab shows that in the event of groundwater overexploitation, canal water becomes the major source of water for poor small and marginal farmers to sustain their income from irrigated production, as they lose out on investing in technology to chase the dropping water table (Sarkar, 2012).

TABLE 11.1 Calculation of Well-Irrigated Area for Different Landholding Classes

Landholding Category	Total No. of Holders	Total Holding (ha)	Average Holding (ha)	Percent of Farmers Owning Wells			Estimated Net Area Under Well Irrigation (ha) and Percentage of Total Area
				OW	STW	DTW	
Marginal	83,694,372	32,025,970	0.4	2.5	3.5	0.01	1,921,559 (6.0)
Small	23,929,627	33,100,790	1.4	10.0	12.0	0.02	7,282,174 (22.0)
Medium	20,502,460	74,481,092	2.8	13.4	13.2	0.09	19,886,452 (26.7)
Large	1,095,778	18,715,131	17.1	20.4	16.5	0.40	6,905,883 (36.9)
Total	135,597,577	158,322,983	1.16				35,996,067 (22.7)

Source: Authors' own estimates based on agricultural census (2005–06) and data from minor irrigation census on well ownership in India (Kumar, 2007).

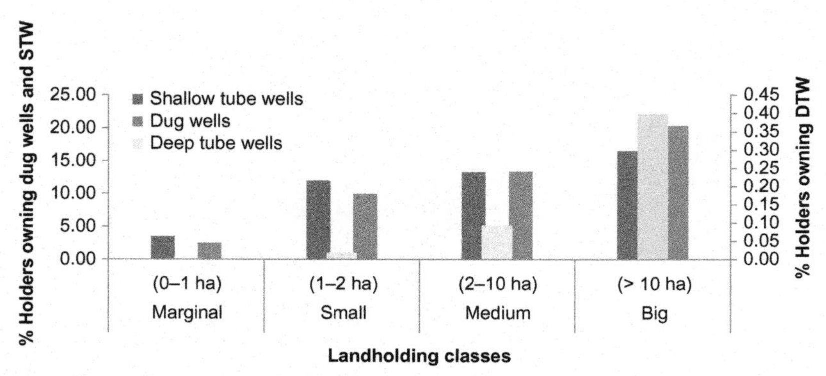

FIG. 11.1 Classwise ownership of different types of wells. *(Source: MI Census 2000–01.)*

The authors extend the argument further by saying "that small and marginal farmers own a major share of India's groundwater resources, and that the eastern Indian states are not very prosperous makes this (read it as free power connection and high diesel subsidy) a pro-poor strategy…" (p. 123, last paragraph). It does not take much effort for one to realize that the groundwater economy in India is controlled by medium and large farmers and all these benefits would end up with rich farmers who are running wells using diesel pumps now. The fallacy in the authors' argument can be understood from Table 11.1 and Fig. 11.1.

Table 11.1 shows that India had nearly 83.6 million marginal and 23.9 million small farmers, while there are 1.05 million large and 20.5 million medium farmers, as per agricultural census of 2005–06. As per our estimates, the 107.5 million small and marginal farmers together irrigate only 9.1 M ha of land (net) from wells, against 21.55 million medium and large farmers irrigating 26.8 M ha of land (net). The average well-irrigated area of small and marginal farmers is 0.09 ha, while that of a large farmer is around 6.8 ha, which is about 80 times higher than that of the former. When we consider medium farmers also, the average per farmer (for medium and large farmers) becomes 1.24 ha. In terms of proportion of area irrigated, it is hardly 14% of the total area for small and marginal farmers, against 29% for medium and large farmers. So, undoubtedly, India's groundwater economy is controlled by medium and large farmers.

A far more dangerous conclusion emerging from the analysis by the authors (Mukherji et al., 2013) is about the growing equity in access to groundwater. In fact, according to them, groundwater development was never iniquitous. The authors' contention is that a far greater percentage of the wells are owned by small and marginal farmers in India today, than in 1987. Yes, it is true that the number of wells owned by small and marginal farmers has gone up in the past 25 years or so. But, one wonders why the authors do not mention about what has happened to the number of operational holders belonging to this category. One does not require the help of rocket science to suggest that with rising cost of drilling wells (in most areas) and the high risk involved in well construction, wells are increasingly becoming a product of the rich farmers.

11.7. GROUNDWATER MARKETS AND EQUITY

With groundwater becoming very scarce in many (arid and semi) regions and the cost of accessing it becoming prohibitively high in those regions, it is increasingly coming under the control of a few resource-rich people, with unclear legal ownership rights over it. The rich well farmers who have the wherewithal to drill deep wells and install high-capacity pumps keep investing in expensive wells to chase the falling water table. They keep offering irrigation services to the neighboring farmers, who do not own wells, at prohibitive prices—be in Kolar in Karnataka and Coimbatore in TN, North Gujarat, or Madhya Pradesh. They enjoy high-monopoly power, and in fact, the price of water and the monopoly price ratio keep increasing as the number of potential water buyers against the sellers keeps increasing.

No one really knows how old the phenomenon of groundwater markets in India is. One should suspect that water-sharing arrangements between well owners and farmers without wells are as old as the practice of well irrigation itself. But, the research on informal groundwater markets is at least 40 years old. Most of the earlier research on groundwater markets viewed it as a case of "natural oligopoly" but can be converted into a powerful instrument to promote equity in access to groundwater and reduce poverty, through the right kind of energy-pricing policies that create incentive for well owners to increase the production volume and sell the surplus to his neighbors at lowered prices (Shah, 1993; Mukherjee and Biswas, 2016). Such arguments, however, had no empirical basis.

Ideally, the researchers should have asked this question: "how much would it cost if the farmer invests his own money to drill a well and pump water to water his farm as against buying water from a neighboring well owner?" A simpler approach would have been to look at the monopoly price ratio for the water-selling farmers, the ratio of the price at which water is sold in the market, and the cost of production of water from wells. Obviously, the argument that *groundwater market promotes equity in access to groundwater and therefore helps many millions of poor small and marginal farmers* will be tenable only if the cost in the former case (farmer investing in his own well and pump) is more or less the same as the price at which water is being bought or if the monopoly price of water is very low or close to one. However, such analysis was never attempted.

The fact is that the equity impact of groundwater markets (as they operate today in different parts of our country) in India is a billion dollar question, if we consider the following facts: (1) the amount of money the government spends in the form of electricity subsidy for well owners is close to 8 billion dollars (approximately 8 billion dollars per annum); (2) a large share of the subsidy benefit goes to the large and medium farmers (as studies by the World Bank in Punjab and (erstwhile) Andhra Pradesh have shown); and (3) groundwater markets cover nearly 15% of the total well-irrigated area in the country (source: estimates by Saleth, 1999).

The champions of groundwater markets stretched their argument further to build a strong case for offering electricity to this privileged group of farmers (well owners) based on flat rate on the pretext that they could produce groundwater at a very low cost and thus offer water to their neighbors at competitive prices, thereby passing on a share of the subsidy benefit to the nonwell-owning water buyers (Mukherji et al., 2009). The groundwater markets were romanticized due to the sheer scale of irrigation it could produce and the hundreds of millions of farmers it could prima facie benefit. Such an argument was palatable for many, including some state electricity utilities. Flat-rate system for electricity was introduced for agriculture by these utilities, considering the fact that metering electricity consumption in the farm sector was getting more and more difficult with explosion in well numbers in the rural areas.

Obviously, its impact on the sustainability of resource use and intragenerational equity in access to the resource was hardly analyzed empirically. Some social scientists, who were not so convinced about the arguments of "positive impacts of private water markets," did question this superficial way of looking at access equity in groundwater, which treats groundwater market as a natural oligopoly (Palmer-Jones, 1994; Saleth, 1994), and raised concerns about the negative ecological, equity, and efficiency effects of water markets. They argued for more fundamental changes in the legal and institutional framework governing the use of groundwater (Saleth, 1994; Kumar, 2000).

In fact, recent empirical research (Uttar Pradesh, Bihar, and Gujarat) clearly shows that the monopoly power of well owners, which determines the price at which water is sold by them against the cost they incur, has nothing to do with the electricity pricing policies that determine the cost of production of water. Instead, it is influenced by the market conditions (the number of buyers against the sellers and overall demand-supply situation). In a given area, irrespective of the differences in cost of production of water (between diesel well owners and electric well owners), the selling price of water was found to be more or less the same. Many electric well owners, whose marginal cost of pumping water was almost zero, sold water at as much a higher a price as the diesel well owners, who incurred very high cost of pumping water due to the high fuel cost (Kumar et al., 2013b). Hence, the argument that a large portion of the benefits from the heavily subsidized electricity supplied to well-owning farmers would eventually be passed on to poor water-buyer farmers is nothing short of a fallacy.

On the contrary, under similar market conditions, under the flat rate (pricing based on connected load), the large well owners, whose implicit cost of irrigation is very low and return from crop production is high, may enjoy monopoly power and decide the price at which water should be sold in the market. Conversely, the smallholders owning wells, whose implicit cost of irrigating own farm is high, are left without much choice but to look for buyers to whom they could sell water to earn extra income, at a price decided by the market as they will have limited bargaining power. Such prices obviously offer high profit margins for the large farmers, but not for the small farmers.

It was found that the monopoly price ratio (MPR)—the ratio of the price at which a commodity is sold and the actual cost of production—charged by electric well owners, who incur zero marginal cost and very low implicit cost of pumping water, is far higher than that charged by the diesel well owners who incur high marginal cost of pumping (Kumar et al., 2013b).[3]

Metering- and consumption-based pricing creates a level-playing field for all landholding classes, while improving efficiency of use of water and electricity, and is unlikely to cause an upward trend in the monopoly price of water and thus promotes equity in access to groundwater. If electricity is charged on pro rata basis, both small and large farmers will incur the same unit cost of pumping water, and the large farmers will not enjoy a comparative advantage over small farmers. In view of the fact that there are no fixed costs of keeping pump sets, the small landholders are not under pressure to sell water at a very low price to stay in the market. Hence, both large and small farmers would have equal incentive to invest in wells and sell water and have equal opportunities to make profits in the shallow groundwater areas. This will lower the price at which water would be sold in the market. Empirical evidence from West Bengal substantiates this view[4] (Kumar et al., 2014).

West Bengal introduced metering- and consumption-based pricing of electricity in agriculture way back in 2006. Today, interestingly, 85% of the agricultural connections in Gujarat—a state that had long been stereotyped as a difficult turf for anything like this—is now metered. Impacts of such measures on groundwater-use efficiency and equity can now be studied using a large sample. Unfortunately, it is common to find researchers manipulating data to present a distorted picture of the scenario that serve certain interests, and policy prescriptions are dished out.

The argument that the zero marginal cost of pumping water would create incentive to pump more water from the well is by and large correct, but that it would create competitive water markets is fallacious as shown by field

---

3. Studies from eastern UP, West Bengal, and south Bihar show that these markets can be highly monopolistic in the sense that the price, which well owners charge from the buyers, is much higher than the actual cost incurred by them for abstracting water, depending on supply and demand in the market (Kumar et al. 2013a, 2014). Electric well owners who typically incur very low cost for abstracting water, owing to very low cost of energy, charge as much as what a diesel pump owner, who incur much higher energy cost, charges, and therefore, the monopoly price of electric well owners was found to be very high (Kumar et al. 2013a). Over a period of three decades or so, the monopoly prices charged by these well owners have however dropped, with greater proportion of farmers owing wells and pump sets (Kishore, 2004).

4. Analysis of primary data from villages in WB (source, based on Mukherji, 2008) shows that the monopoly power enjoyed by diesel well owners (estimated in terms of monopoly price ratio, MPR = 1.90), who are confronted with positive marginal cost of pumping groundwater, was much lower than that of electric well owners under the flat-rate system of pricing electricity (MPR = 16.70).

evidence. A survey in West Bengal showed that the diesel well owners were selling water more aggressively to make profits than those who had electric submersible pumps and paying for electricity on the basis of connected load. Against an average area of 22.8 ha irrigated by diesel well owners, the total area of water buyers was 19.2 ha, that is, nearly 84% of the pumping was done to provide irrigation service to neighboring farmers. Conversely, in the case of electric well owners, against a total area of 27.0 ha irrigated by the tube well, the area irrigated by water buyers was 22.3 ha (82%) (source, based on the data provided in Mukherji, 2008).

This argument had been paraded by some researchers for the past 25 years (Mukherji et al., 2009; Mukherjee and Biswas, 2016; Shah, 1993) without producing an iota of empirical evidence to support their claim or even providing counterintuitive evidence. As a matter of fact, the empirical analysis provided by Mukherjee and Biswas (2016) for Madhya Pradesh shows that groundwater markets are monopolistic, while they argued conversely. The result is that we have a blinkered view of groundwater markets. In this blinkered view, the village water trader emerges as the savior of the poor.

Intriguingly, the profound equity impact of large irrigation systems were hardly captured by social science research in India till recently. The ongoing Sardar Sarovar Narmada project, which is irrigating nearly 0.9–1.0 M ha of land in Gujarat (around 25% of the gross irrigated area in that state), made a remarkable dent on groundwater markets, which thrived for many decades. Well owners are forced to lower the water prices, owing to overall reduction in demand for irrigation water. Gravity irrigation is rejuvenating millions of wells in the areas that are now receiving gravity irrigation. Field evidence suggests that while the area directly irrigated by "purchased water" had declined sharply, the energy consumption per unit of well-irrigated area had also declined owing to rising water table. This is seen on the electricity consumption in the agriculture sector, with a steep decline at the state level. This is a silent revolution. In Punjab, there was greater equity in the distribution of agricultural income of farmers in villages receiving canal water as compared with those in villages with tube well irrigation but facing problems of groundwater overexploitation (Sarkar, 2012).

11.8. IMPROVING EQUITY IN ACCESS TO GROUNDWATER

There are mechanisms and instruments available with the government to ensure greater equity in access to this precious resource in the coming years. It has to begin with defining groundwater rights and enforcing it, under a new law that recognizes private property rights in groundwater. To promote efficient use of the resource, water rights can be made tradable. One major challenge in instituting groundwater rights is to arrive at realistic figures of sustainable level of abstraction of groundwater in the selected management area and evolve norms for allocation of water across sectors and among users within sectors.

Since groundwater availability is going to fluctuate between years, the water rights cannot be absolute ownership rights, and instead, the right to use groundwater will have to be treated as water use rights or entitlements. The challenge will be bigger in hard-rock areas, as these areas, unlike the alluvial and sedimentary formations, do not have any groundwater stock and are dependent purely on annual replenishment from precipitation.

It might take many years, even after the government comes out with a new law on groundwater, which is now under a common pool regime, to evolve institutional mechanisms to define and enforce water rights. But a serious look at institutional reforms is needed, rather than talking platitudes like "community management," "local water governance," "crop water budgeting," etc., which carry no meaning in the current legal and institutional context.

It is argued by some professionals that aquifers are getting mined because too little is known about them, that there is a need to do their intensive mapping, and that this would bring about a new paradigm in water resources management, in the form of participatory groundwater management. The contention was that good mapping of the aquifer can tell us the community's relation with groundwater, how different communities are connected (by a single aquifer), where to put the water for effective recharge, and ultimately about groundwater governance (Kulkarni and Shankar, 2009; Kulkarni et al., 2011). However, the link between aquifer mapping and groundwater governance was never explained. Intuitively, ideas like micro-level mapping of aquifers, if implemented, would only benefit resource-rich farmers and add to the existing inequity in access to groundwater, as the information generated from such mapping exercises could be appropriated by these village elite (also see Chapter 5, this book).

Our scientists from the state groundwater development agencies could hardly ask, "what value this would add to what is already known from several decades of our work," given the overwhelming support for this idea from the civil society. The erstwhile Planning Commission of India allocated Rs. 4000 crore for "aquifer mapping," which was to improve groundwater governance in the country. It was vehemently argued that groundwater was completely left out in the water management debate all these years (Shah, 2013). But a simple analysis of the degree holders from our universities would show that India produces at least 20 times more geologists/geohydrologists than hydrologists.

Countries like Chile and Mauritius have made great progress in institutional reforms. Developed countries like Australia, Israel, the United States, and Spain have been trying it for a long time. In the Middle East, the water rights system and water trading (for irrigation) of Oman are thousands of years old. Once water-use rights/entitlements are given, the individual farmers who do not have the wherewithal to drill deep wells would find ways to exercise their rights over the resource. For instance, they can work out arrangements with existing well owners to share the water in the aquifer by paying money for the pumping services or sell their rights to the large farmers. When there is "security of tenure," it would attract great investments in infrastructure for transfer of water from field to field and even from village to village.

The government of India and the state governments concerned have to initiate reforms to ensure improved equity in groundwater access instead of leaving the hundreds of millions of poor small and marginal farmers (who do not own wells) at the mercy of the rich well owners, whom its policies currently benefits. We need efficient water markets, and not monopolistic ones. What we need to remember is the grave reality that only 4%–5% of the small and marginal farmers own wells, against a figure of 70% for large farmers. Therefore, through legal and institutional reform, the "terms of the trade" need to be changed in favor of the small and marginal farmers, who can then participate in water markets as proud sellers of their *water rights* to the needy farmers rather than as buyers.

"Aquifer mapping" in its current format as such will not produce the extra knowledge about resource availability and quality that is crucial for improving governance and management of groundwater.

If the communities have to manage the aquifers, they would need the support of higher level institutions that are legitimate. After all, it involves decisions on how much to be allocated to different uses and users and enforcement mechanisms. These institutions will have to define water rights, allocate it among uses and users, and monitor the use. They have to operate at the aquifer scale (Kumar, 2000, 2007; Mohanty and Gupta, 2012). Such institutions are absent today. The future investments should be in crafting and developing such institutions at the scale at which management efforts become effective. The existing knowledge about resources availability and quality is sufficient to initiate management actions.

11.9. CONCLUSIONS

Research on the science of irrigation is a serious business, especially when it is used for policy making, given the fact that the outcomes can affect the livelihoods of hundreds of millions of farmers. There are complex considerations involved in understanding the performance of irrigation systems. In the case of wells, it cannot be based on the simple criteria of "well numbers" and "how many of them are electrified." It is important to know where these wells are located (the aquifer characteristics), the well characteristics, the quality of power supply, the amount of land available for cultivation, and finally the amount of land irrigated and the cropping system adopted by the irrigators.

Large surface-water-based systems produce multiple impacts, and expansion in irrigated area is just one of them. Unlike groundwater, canal water is free from salts. The well irrigators in canal command in semiarid and arid region areas are able to grow fruits, vegetables, and flowers using the good-quality water in their wells that come from canal seepage and return flows and earn very high income per unit volume of irrigation water. Even the assessment of irrigated area of large water systems appears to be complex, in lieu of the fact that many use water directly from canals and many well irrigators in the command area indirectly benefit from canal seepage and return flows from irrigated fields.

In the research on irrigation and rural poverty, the questions such as "who owns the irrigation source" and "who controls the water economy" are very vital. The question that needs to be posed is "what proportion of the small and marginal farmers has access to water, against medium and large farmers." There, we find that it is abysmal in the case of wells, a primary factor responsible for the inequity in access to groundwater resources in this country. A total of 107.5 million small and marginal farmers together irrigate only 9.1M ha of land, against the 65M ha of land they own, with an average of 0.084ha per farmer, whereas 21.5 million medium and large farmers together irrigate 26.7M ha of land out of the total of 92M ha of land they own, from groundwater sources, with an average of 1.24ha per farmer. This inequity is alarming and is growing. How to address this inequity should be the concern of the researchers and policy makers (Kumar et al., 2013a).

Public gravity irrigation systems will have a big role in the future in improving access equity in groundwater irrigation as they can keep the wells running with seepage from canals and percolation from irrigated fields, destroy monopolistic water markets, and maintain the water table within the reach of the small and marginal farmers. The future investments by the governments should be in crafting and developing groundwater management institutions at the scale at which management efforts become effective. There is nothing wrong in spending a few million dollars generating additional knowledge about a resource. But, the belief that the problems of resource overexploitation can be solved by the communities merely with this knowledge can be risky. It is not the lack of sufficient knowledge of the aquifers that is posing hindrance to the evolution of groundwater management institutions, but the lack of political will and to an extent the lack of proper understanding of what is required. We need effective institutions to manage the resource, and the answer doesn't lie with the local communities alone.

REFERENCES

DebRoy, A., Shah, T., 2003. Socio-ecology of groundwater irrigation in India. In: Llamas, R., Custodio, E. (Eds.), Intensive Use of Groundwater: Challenges and Opportunities. Swets and Zetlinger Publishing Co., The Netherlands, pp. 307–335.

Dhawan, B.D., 1990. Big Dams: Claims, Counterclaims. Commonwealth Publishers, New Delhi.

Geerts, S., Raes, D., 2009. Deficit irrigation as an on-farm strategy to maximize crop water productivity in dry areas. Agric. Water Manag. 96 (9), 1275–1284.

Hazell, P., Wood, S., 2008. Drivers of change in global agriculture. Philos. Trans. R. Soc. Lond. B: Biol. Sci. 363 (1491), 495–515.

Jagadeesan, S., Kumar, M.D., 2015. The Sardar Sarovar Project: Assessing Economic and Social Impacts. Sage Publications, New Delhi.

Jagadeesan, S., Kumar, M.D., Sivamohan, M.V.K., 2016. Externalities of surface irrigation on farm wells and drinking water supplies in large water systems: the case of Sardar Sarovar project. In: Kumar, M.D., James, A.J., Kabir, Y. (Eds.), Rural Water Systems for Multiple Uses and Livelihood Security. Elsevier Publishers, Amsterdam, Netherlands, Oxford, UK and Cambridge, USA, pp. 229–252.

Jhingran, A.G., 1986. Reservoir Fisheries Management in India. Central Inland Capture Fisheries Research Institute, Barrackpore. 65 pp..

Kishore, A., 2004. Understanding agrarian impasse in Bihar. Econ. Polit. Wkly., Rev. Agric. 39 (31), 3484–3491.

Kulkarni, H., Shankar, P.V., 2009. Groundwater: towards an aquifer management framework. Econ. Polit. Wkly. 44 (6), 13–17.

Kulkarni, H., Vijayshankar, P.S., Krishnan, S., 2011. India's groundwater challenge and the way forward. Econ. Polit. Wkly. 46 (2), 37–45.

Kumar, M.D., 2000. Institutional framework for managing groundwater: a case study of community organisations in Gujarat, India. Water Policy 2 (6), 423–432.

Kumar, M.D., 2003. Food Security and Sustainable Agriculture in India: The Water Management Challenge. Working Paper # 60. International Water Management Institute: Colombo, Sri Lanka.

Kumar, M.D., Reddy, V.R., Narayanamoorthy, A., Sivamohan, M.V.K., 2013a. Analysis of India's minor irrigation: faulty analysis, wrong inferences. Econ. Polit. Wkly. 48, 76–78.

Kumar, M.D., 2007. Groundwater Management in India: Physical, Institutional and Policy Alternatives. Sage Publications, New Delhi.

Kumar, M.D., Scott, C.A., Singh, O.P., 2013b. Can India raise agricultural productivity while reducing groundwater and energy use? Int. J. Water Resour. Dev. 29 (4), 557–573.

Kumar, M.D., Sivamohan, M.V.K., Narayanamoorthy, A., 2012. The food security challenge of the food-land-water nexus in India. Food Sec. 4 (4), 539–556.

Kumar, M.D., Bassi, N., Sivamohan, M.V.K., Venkatachalam, L., 2014. Breaking the agrarian crisis in eastern India. In: Kumar, M.D., Bassi, N., Narayanamoorthy, A., Sivamohan, M.V.K. (Eds.), Water, Energy and Food Security Nexus: Lessons from India for Development. Routledge, London, pp. 143–159.

Mohanty, N., Gupta, S., 2012. Water reforms through water markets: international experience and issues for India. In: Morris, S., Shekhar, R. (Eds.), India Infrastructure Report. Oxford University Press, New Delhi.

Mollinga, P.P., Bolding, A., 2004. Research for strategic action. In: Mollinga, P.P., Bolding, A. (Eds.), The Politics of Irrigation Reform. Contested Policy Formulation and Implementation in Asia, Africa and Latin America, Ashgate, Aldershot, pp. 291–318.

Mukherjee, S., Biswas, D., 2016. An enquiry into equity impact of groundwater markets in the context of subsidised energy pricing: a case study. IIM Kozhikode Soc. Manag. Rev. 5 (1), 63–73.

Mukherji, A., 2008. In: The paradox of groundwater scarcity amidst plenty and its implications for food security and poverty alleviation in West Bengal, India: what can be done to ameliorate the crisis? 9th Annual Global Development Network Conference, 29–31 January 2008, Brisbane, Australia.

Mukherji, A., Das, B., Majumdar, N., Nayak, N.C., Sethi, R.R., Sharma, B.R., 2009. Metering of agricultural power supply in West Bengal, India: who gains and who loses? Energy Policy 37 (12), 5530–5539.

Mukherji, A., Facon, T., Burke, J., de Fraiture, C., Giordano, M., Molden, D., Shah, T., 2010. Revitalizing Asia's Irrigation: To Sustainably Meet Tomorrow's Food Needs. IWMI and FAO of United Nations, Colombo, Sri Lanka and Rome, Italy.

Mukherji, A., Rawat, S., Shah, T., 2013. Major insights from India's minor irrigation censuses: 1986–87 to 2006–07. Econ. Polit. Wkly. 48 (26–27).

Palmer-Jones, R., 1994. Groundwater markets in South Asia: a discussion of theory and evidence. In: Moench, M. (Ed.), Selling Water: Conceptual and Policy Debates over Groundwater Markets in India. VIKSAT-Pacific Institute-Natural Heritage Institute, Ahmedabad.

Perry, C.J., 2001. World commission on dams: implications for food and irrigation. Irrig. Drain. 50 (2), 101–107.

Saha, C., Paul, B.N., 2000. Flow through system for industrial aquaculture in India. Aquacult. Asia 5 (4), 24–26.

Saleth, R.M., 1994. Groundwater markets in India: a legal and institutional perspective. Indian Econ. Rev., 157–176.

Saleth, R.M., 1999. Water markets in India: economic and institutional aspects. In: Easter, K.W. (Ed.), Markets for Water: Potential and Performance. Kluwer, Dordrecht, The Netherlands.

Sarkar, A., 2012. Equity in access to irrigation water: a comparative analysis of tube-well irrigation system and conjunctive irrigation system. In: Kumar, M. (Ed.), Problems, Perspectives and Challenges of Agricultural Water Management. Intech Publishers, Rijeka, Croatia.

Shah, M., 2013. Water: towards a paradigm shift in the twelfth plan. Econ. Polit. Wkly. 48 (3), 40–52.

Shah, M., 2016. Eliminating poverty in Bihar: paradoxes, bottlenecks and solutions. Econ. Polit. Wkly. 51 (6).

Shah, T., 1993. Groundwater Markets and Irrigation Development: Political Economy and Practical Policy. Oxford University Press, New Delhi.

Shah, T., 2009. Taming the Anarchy: Groundwater Governance in South Asia. Resources for Future, Washington, DC.

Shah, Z., Kumar, M.D., 2008. In the midst of the large dam controversy: objectives, criteria for assessing large water storages in the developing world. Water Resour. Manag. 22 (12), 1799–1824.

Chapter 12

Green Revolution Versus "Dream Revolution": Agricultural Growth and Poverty Reduction in Eastern India

12.1. INTRODUCTION

Eastern India has the highest rates of poverty among all regions in India. For the years 2011–12, it ranged from the lowest of 19.98% in West Bengal to 33.74% in Bihar to the highest of 39.93% in Chhattisgarh (Planning Commission, 2014). While the governments are struggling hard to find ways to fight poverty, the thirst of some academicians, researchers, and "think tanks," especially economists, in understanding poverty in eastern India and finding "silver-bullet" solutions, seems to be unsatiated. The recent commentaries and articles from some of them, relating to agricultural growth and poverty reduction in West Bengal and Bihar, which appeared in newspapers and magazines, see the problem of stagnation in agriculture and rural poverty in that region, through the "groundwater lens" (IWMI, 2012; Mukherji, 2008; Mukherji et al., 2012; NRRA, 2011; Pant, 2004; Shah, 2016; Shah et al., 2016).

While the authors of these articles are unanimous in their view that groundwater intensive use for irrigation could bring about agricultural growth in this region, some of them recommend subsidizing power connections/energized (solar) pumps for farmers in the "groundwater-abundant" states, with a contention that this would lead to a second Green Revolution, poverty reduction, etc. (see Shah, 2016; IWMI, 2012; Mukherji et al., 2012; Shah et al., 2016). The core of the argument is that it would reduce the price at which water is traded in the market, thereby benefiting the poor farmers who purchase water from well owners (IWMI, 2012; Mukherji, 2008; Mukherji et al., 2012). The erstwhile Planning Commission of India had bought into this model of agricultural growth based on intensive well irrigation for triggering the second "Green Revolution" in eastern India (Planning Commission, 2013).

A few had argued that the farmers in West Bengal (WB) have been shying away from irrigated agriculture because of two important constraints induced by the wrong policies of the government in the past: the full cost that the state electricity

Water Policy Science and Politics. https://doi.org/10.1016/B978-0-12-814903-4.00012-9

board has been charging from farmers for providing power connections for agricultural pump sets and making it mandatory for farmers to obtain license from the state groundwater department to apply for power connection for agrowells. They lament that the state electricity department had been charging irrationally for power connections to farmers (IWMI, 2012; Mukherji, 2008; Mukherji et al., 2012).

Mukherji claimed that the government of WB embarked on "rationalizing" charges for agripower connections and removal of restrictions on drilling of wells in the blocks that are categorized as "safe" for groundwater exploitation, based on her research, and this kick-started a second Green Revolution in that state (source: http://www.solutions-site.org/node/811; https://www.gatescambridge.org/multimedia/blog/second-green-revolution). As per the claim, the farmers now have to pay for a new connection, on the basis of connected load, and not on the basis of the length of the extension and the number of electric poles required. Having seen its "grand success," recommendations are now made by others to extend this policy to Bihar. However, for Bihar, a brew of wonder pills like NREGA, feeder-line separation, groundwater pumping and recharge, and solar pumps irrigation are also suggested for faster poverty reduction (Shah, 2016).

These arguments and claims, however, failed to capture any of the important factors responsible for the agrarian impasse and the potential impacts of public policies relating to irrigation on the functioning of groundwater markets in eastern India, particularly in West Bengal and Bihar for which policy prescriptions are rolled out by these scholars. This chapter brings out the flaws in their analyses of the agrarian problems in the region, which were viewed through the lens of "irrigation-cereal production," the likely outcomes of the policies vis-à-vis access equity in groundwater, and comes out with concrete suggestions on what needs to be done to break the agricultural stagnation in the region.

12.2. GROUNDWATER IN EASTERN INDIA

Eastern India can be divided into two distinct parts. The first part is a very large part of eastern Gangetic Plains comprising the state of Bihar, West Bengal, Assam, and eastern part of Uttar Pradesh, which is mostly underlain by deep alluvial aquifers. The second part is its southern part comprising the states of Jharkhand, Chhattisgarh, and Odisha, which is mostly underlain by hard-rock formations barring the exception of coastal Odisha. The groundwater potential in the first part of the region is very high. Out of a total amount of 149.87 BCM of net groundwater availability in the region, the total annual utilization is only 77.77 BCM (source: based on CGWB, 2012). The region has subhumid to humid climates, with rainfall varying spatially from a lowest of nearly 1000 mm in eastern UP to a highest of more than 3000 mm in many parts of Assam and West Bengal. The high rainfall, flat topography, permeable soils, and favorable climate ensure very high recharge from precipitation and other sources.

In addition to the recharge from rainfall during both monsoon and non-monsoon periods, the total renewable groundwater of the region includes considerable amount from other sources such as river flows, ponds and other wetlands, canal seepage, and irrigation return flows. The states of UP, Bihar, and West Bengal have many surface-irrigation schemes irrigating large areas during the monsoon and nonmonsoon season. The stage of groundwater development in the region varies from a lowest of 22% in the state of Assam to a highest of 72% in the state of Uttar Pradesh (Table 12.1).

The groundwater availability in the second part as per the official estimates of CGWB is given in Table 12.2. Of the total 33.64 BCM of net available groundwater, the utilization is only 9.57, which is merely 28.4%. As per the estimates, there is a large positive groundwater balance in the region annually (23.67 BCM). But this is far from the reality with regard to groundwater situation in this region as wells in the region are either failing or are experiencing yield reductions due to overexploitation (Kumar and Singh, 2008). The estimates of nonmonsoon discharge, which is used to arrive at the net groundwater availability from renewable groundwater, are highly erroneous, as they are based on ad hoc norms. As pointed out by Kumar and Singh (2008), the nonmonsoon discharge from the recharged aquifers is very high in the hilly and undulating areas and the CGWB heavily underestimates this parameter in its assessment as evident from the measured values of lean season flows in the rivers of this region. There is not much groundwater available in this part of eastern India, which can be tapped for irrigation. Hence, we should limit the discussion to the issues in eastern Gangetic Plains.

TABLE 12.1 Status of Groundwater Development in the States of Eastern India (Gangetic Plains)

Name of State	Renewable Groundwater Resources From Recharge	Net Groundwater Availability	Annual Groundwater Draft (MCM)	Stage of Groundwater Development (%)
Assam	30.35	27.81	6.026	22
Bihar	28.63	26.21	11.36	43
Uttar Pradesh	75.25	68.27	49.48	72
West Bengal	30.50	27.58	10.91	40
Overall	164.73	149.87	77.78	51.9

Note: CGWB estimates the net groundwater availability by subtracting the estimates of nonmonsoon groundwater discharge from the estimates of total renewable groundwater.
Based on Central Ground Water Board (CGWB), 2012. Ground Water Year Book-India 2011–12. Central Ground Water Board, Ministry of Water Resources, RD&GR, GoI, Faridabad.

TABLE 12.2 Status of Groundwater Development in the States of Eastern India (Southern part of Gangetic Plains)

Name of State	Renewable Groundwater Resources From Recharge	Net Groundwater Availability	Annual Groundwater Draft (MCM)	Stage of Groundwater Development (%)
Chhattisgarh	12.22	11.54	3.60	31
Jharkhand	5.96	5.41	1.61	30
Odisha	17.78	16.69	4.36	26
Overall	35.96	33.64	9.57	28.4

Based on Central Ground Water Board (CGWB), 2012. Ground Water Year Book-India 2011–12. Central Ground Water Board, Ministry of Water Resources, RD&GR, GoI, Faridabad.

12.3. CAN INTENSIVE WELL IRRIGATION REDUCE RURAL POVERTY IN EASTERN INDIA?

It is well known that agriculture revolution swept West Bengal during the 1980s with the value of crop outputs jumping from 42 billion to 75 billion rupees from 1980–83 to 1990–93, mainly as a result of land reforms undertaken in the state after 1977–78 and provision of adequate public supply of credit and other inputs to small farmers (Ramchandran et al., 2003). During the 1980–93, the state recorded the highest annual compounded growth rate in crop yields (of 4.8%) in terms of rupees per hectare of GCA (gross cropped area) and crop value (6%). Against this, the growth rate in crop yield was only 1.9% per annum during 1990–93 and 2000–03, and that in crop value was 2.4% per annum during the same period. Some attribute the recent stagnation in agricultural growth to lack of expansion in irrigation facilities (source: based on Mukherji et al., 2012; Shah, 2016).

One of the contentions of these writers was that well irrigation in WB and Bihar hadn't grown at the same pace as in some of the agriculturally prosperous states such as (erstwhile) AP, Karnataka, Punjab, and Haryana (Mukherji et al., 2013; Shah, 2016). In the case of West Bengal, they argue that it was because of the corrupt practices of the officials of the state electricity board of that state and its groundwater department (Mukherji et al., 2012). For Bihar, the observation is that electricity supply to agriculture is generally poor, and this is limiting intensive groundwater use for irrigation (Shah, 2016).

This is highly misplaced and untenable. First of all, it is incorrect to compare high-rainfall, subhumid regions, like WB and Bihar, to the low- to-medium-rainfall semiarid states like AP, Punjab, and Haryana. The farmers in the former often take two crops without irrigation, whereas in low-rainfall semiarid to arid regions, even the monsoon crop might require one to two supplementary irrigations. Moreover, the aquifers in (erstwhile) AP and Karnataka are hard rock in nature, with extremely limited groundwater potential, and therefore would require much higher well density to support the same extent of irrigation than WB and Bihar (Kumar et al., 2014).

Second, a point was made, using the statistics of number of new agricultural power connections in the state of WB, about rent seeking and corruption in SWID (IWMI, 2012; Mukherji et al., 2012). This again is noncomprehensible. The fact is that the new agricultural power connections in WB peaked in 1989 and the number has been declining since then to become almost nil in 2003. But some researchers conveniently put the blame on the state electricity company and the groundwater department, by saying that the restrictive policies such as the Groundwater Act of 2005, which was designed to control new wells and prepare an inventory of wells, and the high connection charges led to the electricity department virtually stopping new connection since 2003. The point that is ignored is that in a region underlain by thick alluvial strata and with high natural replenishment of groundwater and low intensity of withdrawal, incidence of well failures will be very low and, therefore, there will be little need for redrilling unlike many other regions that experience groundwater overexploitation.

These arguments do not hold much water. First of all, a policy that was introduced in 2003 cannot affect irrigation growth in the 1990s. Secondly, no data are provided to see whether the number of applications from the farmers for new power connections and the connection charges has changed over the years. The only piece of information provided to substantiate the claim is a sweeping statement about corruption that till September 2010, only 8500 out of a total of 23,000 applications received by the SWID officials were approved. This no way makes the case of rent seeking as application can be rejected on several grounds, including the applicant's farm located in one of the semicritical blocks (Kumar et al., 2014) and documentation furnished being incomplete.

It was argued that the new policy of rationalizing power-connection charges would lead to remarkable expansion in well-irrigated area, reducing poverty in WB and other eastern states. As per the estimates by one of these authors, even if half a million farmers in WB take electricity connections for pump sets, it would lead to an additional irrigated area of 3.7 M ha, and if 50% of this target is achieved, the additional area would be 1.85 M ha. However, we would see that even these calculations, based on basic statistics, are wrong. Similar claims are now made for Bihar that agricultural intensification based on well irrigation would reduce poverty there and that NREGA, solar pumps for irrigation, groundwater recharge, etc. would do the trick (IWMI, 2012; Mukherji et al., 2012). The most recent argument is to promote solar pump based irrigation in the eastern Gangetic Plains, mainly West Bengal and Bihar (Shah, 2016; Shah et al., 2016).

12.4. FUNDAMENTAL FLAWS IN THE POLICY PRESCRIPTIONS FOR AGRICULTURAL GROWTH

The policy prescription for agricultural growth and rural poverty reduction in West Bengal and Bihar is that intensive groundwater irrigation powered by either electric or solar pumps would increase cropping intensity, raise agricultural surplus, and reduce rural poverty in these states. There are many fundamental issues that these policy prescriptions raise. They are as follows: (1) Is there real potential for groundwater irrigation in West Bengal and Bihar to expand? (2) Are policy decisions favoring intensive groundwater use leading to poor farmers benefiting from highly subsidized power and moving out of the poverty trap? (3) What is the likely impact of the policy decision on the economy and social welfare? In order to address the questions raised above, we need to revisit a few facts about agriculture in these states, which were conveniently ignored by the researchers, who rush to make policy prescriptions.

12.4.1 How Large is the Potential for Intensive Groundwater Use for Irrigation?

As regards West Bengal, it is one of the most intensively cropped and intensively irrigated states in the country with cropping intensity ranging from a lowest of 140% to a highest of 258%. Against the net sown area of 5.25 M ha, the gross cropped area is 9.2 M ha. Against the net irrigated area of 3.11 M ha, the gross irrigation is to the tune of 5.5 M ha, with an irrigation intensity of 176%. This means much of the irrigated land receives irrigation in two seasons. The state already has 5 lakh plus agricultural tube wells, of which more than one lakh are operated by electric pump sets and the remaining by diesel engines (Kumar et al., 2014).

Though not as high as West Bengal, Bihar also has very high cropping and irrigation intensities, much higher than many of the intensively groundwater-irrigated regions in India. Out of the net sown area of 5.712 M ha, around 4.56 M ha has assured irrigation. The cropping intensity is around 138%; what is important to note is that the entire north Bihar is flood-prone, a factor that works against capital intensive farming. A total of 41% of the state is flood-prone. The devastating flood of 2007 damaged a total area of 1.66 M ha in north Bihar (Government of Bihar, 2008). The availability of arable land in Bihar is one of the lowest in the world, that is, 07 ha per capita, which is one-eighth to one-sixth of what agriculturally prosperous states like Punjab and Haryana have (source: based on Kumar, 2003). That said, as analyzed by Kumar et al. (2014), the amount of additional area that can be irrigated through new power connections to farms and better electricity supply would be decided by three important factors.

The first factor is the amount of cultivable land that is in rainfed due to the lack of irrigation water. As the statistics presented earlier, WB has only 2.1 M ha of land, which is cultivated but remains unirrigated due to some reasons.

The cultivated area lying unirrigated in Bihar is only 1.15 M ha, and this includes a large area of land that is prone to floods.

The second factor is the agroclimate in the region/area in question, mainly the rainfall and agrometeorologic factors. Obviously, farmers would irrigate only if the rainfall cannot meet the crop water requirements. Interestingly, the rainfall in WB is one of the highest in the country, with the mean annual values varying from 1500 to 3000 mm. In Bihar, the rainfall is high to very high (varying from a lowest of 1165 mm in agroclimatic zone III to a highest of 1382 mm in agroclimatic zone II), and as earlier indicated, a large area of the northern part of this state is flood-prone (source, http://www.bameti.org/pdf/agriculture_profile_of_the_state.pdf). So, large areas in both the states won't require irrigation at all (or would need just one to two supplementary irrigation in winter) and farmers would be able to take the second crop merely using the soil moisture from monsoon rains. So, it is quite unlikely that the presently unirrigated land in these states would ever come under irrigation. But according to some of these authors, a total of 3.7 M ha is the additional irrigation potential from energized pump sets in WB alone, and there would be crop intensification in Bihar.

A recent study for the World Bank and government of West Bengal on performance of minor irrigation systems in the state showed that agroclimate and land-holding size (which together determine the irrigation water demand) are the two most important determinants of the performance of MI systems managed by farmer groups. The higher the difference between ET0 and effective rainfall in an area (ET0-Pe) and design command area, the better was the overall performance of the schemes in physical, financial, economic, and institutional parameters (Deloitte, 2015).

The third point that calls for more elaborate discussion and greater understanding of irrigation economics is the viability of irrigation using own pump sets for the poor small and marginal farmer. It is again a fact that WB and Bihar have the largest percentage of marginal holding. Nearly 90% of the farmers in WB are marginal farmers holding less than 1.0 ha of land. The corresponding figure for Bihar is 83%. One additional problem, which Bihar has to cope with, is land fragmentation and a society riddled with caste structure.

Going by the pattern of ownership of wells and pump sets in India, which is heavily skewed toward large and medium farmers (Kumar, 2007), it is quite logical to believe that all those well owners in the state, who already have electric and diesel pump sets, are large and medium farmers, and most of those who are left out are small and marginal farmers. The statistics shows that the average area irrigated by an electric pump set in WB is around 15 ha. The well-owning farmers, apart from irrigating their own land, provide irrigation services to several small and marginal farmers, though at prohibitive prices. The marginal farmers prefer such an arrangement over owing an energized well because the latter is economically unviable. Even if power connections are free and the land requires irrigation desperately, these farmers would not find it easy to drill a tube well and install electric pump set and meters.

Under such circumstances, even if the small and marginal farmers decide to install electric pump sets, the only gain would be reduced cost of irrigation water. But as pointed out by Kumar et al. (2014), the larger issue that the policy change raises is even more serious. What welfare gains does a state, which is in the brink of financial bankruptcy, make from investing a sum of Rs. 50,000–100,000 (US$ 800–1600) for power connection for a single farmer in the form of subsidy, when the intended beneficiary hardly gets Rs. 1000 (US$ 16) as incremental annual income from irrigated paddy production? Knowing that this is not going to happen so soon (with the state's poor record of rural electrification), another curious suggestion was made to provide the far more expensive "solar pumps" to the poor small and marginal farmers in states like Bihar (Shah, 2016; Shah et al., 2016). Such prescriptions are however devoid of any economic analysis that considers the economic benefits against the full economic cost of providing pumps run on solar PV systems.

In a state, where the farmers are scared of keeping their diesel engines in their fields unattended due to the fear of theft, one can imagine how many would come forward to install solar pumps that cost a princely sum of Rs. 4.5–5 lakh (US$ 7100–8000), even if the government of the day gives 80%–90% capital subsidy. A much better policy prescription for the state would have been to subsidize micro diesel engines, which the small and marginal farmers would find not only economically viable but also affordable. Here, the subsidy burden on the government would be much less (Kumar et al., 2014). Leave alone the issue of economic viability, even the technical feasibility of decentralized solar PV systems in rural areas is not established yet in this region that is poorly endowed in terms of solar irradiance (Bassi, 2015, 2017).

However, arguments for such pervasive subsidies are continued to be made. A recent paper argued that solar pump irrigation, supported by government subsidy to the tune of US$ 1000 per kilowatt, could increase the irrigation potential of Ganges basin by around 28–32 M ha (Shah et al., 2016). However, no details are given to substantiate this tall claim about the additional irrigation potential to be created by solar pumps in this intensively farmed region.

12.4.2 Misleading Statistics and Flawed Arguments

One sees contradicting statistics while moving from one paper to another. Mukherji et al. (2012) claimed that even if 50% of the existing pumps in WB are electrified, the number of electric pump sets would increase from 1.2 to 6.0 lakh. But the state has a total of only 5 lakh pump sets. Therefore, the maximum increase in electric pump sets possible through replacement of 50% of the existing pumps would only be 2.0 lakh, instead of 4.8 lakh claimed by the authors. It was further claimed that an electric well irrigates nearly 7.9 ha of land and hence estimates that the replacement of 4.8 lakh diesel engines by electric pumps would lead to an added area of 3.7 M ha. This would be possible only if we assume that the area irrigated by diesel engines was virtually zero (Mukherji et al., 2012).

All these researchers predict substantial increase in yield of paddy and wheat in Bihar and paddy in WB, through well irrigation. But they also mention that the current yields aren't much different from what it has been 150 years ago. The marginal return from irrigation for kharif crops like paddy in a high-rainfall, water-abundant region would be insignificant. One doesn't need to be an agricultural economist to know this fact. Even the winter wheat requires only one to two irrigations in Bihar, because of the presence of residual moisture in the soil. The problem in large parts of Bihar is not moisture deficit, but floods. In fact, several scientific studies from Indian Council of Agricultural Research and International Rice Research Institute highlighted the point that the climatic (yield) potential of paddy and wheat is very low in Bihar due to poor solar radiation and ecological problems such as flooding further reduce the yield realization (Aggarwal et al., 2000).

The WB government decision to pursue complete metering of agricultural power connections and charge on pro rata basis was praised by two researchers in 2012 (Mukherji and Das, 2012). But in a another publication, the lead researcher of the earlier article argued that the best agricultural power pricing policy for West Bengal is to go for a judicious "mix" of pro rata pricing and flat rate, claiming that such a mode of pricing would promote competitive water markets with lowering of water prices and water-use efficiency. The IWMI report says "Metering of agricultural pump sets is already promoting water-use efficiency, but a collateral damage has been a contraction in water markets, and hardest hit have been small holder farmers as they often rely on these markets for access to irrigation" (IWMI, 2012, p 2).

But no empirical evidence is provided to support this argument. In fact, the price at which water is traded in the market is not decided by the cost of production of water, but the market conditions, largely determined by the number of potential sellers against the potential buyers. No matter what you do to reduce the cost incurred by well owners, the prices would not come down unless the monopoly power of the seller is reduced (Kumar et al., 2013). Further, research showed that the mode of pricing of electricity does not influence the monopoly prices being charged by well owners. At the same time, the flat-rate pricing puts large well owners in a very advantageous position as it brings down their implicit unit cost of pumping groundwater. It will deter small and marginal farmers from investing in wells and entering the market, as their implicit cost of production of water will be high and will be forced to sell water at a price dictated by large well owners. Pro rata pricing of electricity would promote equity in access to groundwater, if many farmers from the same area have access to electricity connections, reducing the monopoly power of well owners (Kumar et al., 2013).

All these authors are unanimous in their opinion about further exploitation of the "abundant" groundwater in WB and Bihar for poverty reduction (Mukherji et al., 2012; Shah, 2016; Shah et al., 2016). But they advocate recharge of groundwater using NREGA schemes (IWMI, 2012). One wonders about the reason for the obsession with recharge during monsoon in groundwater-abundant regions like WB and Bihar, when the aquifer is fully replenished. Mukherji (IWMI, 2012) even suggests excessive groundwater pumping during summer to create more space in the aquifer for increased recharge during monsoon, quoting past research studies. Not only that this is an expensive strategy (Khan et al., 2014), but it was actually suggested by two scholars in the mid-1970s as a strategy for flood control in the Ganges basin and not for augmenting groundwater recharge (see Revelle and Lakshminarayana, 1975) and later on further explored by Khan and others with the same objective (Khan et al., 2014).

In states that are endowed with numerous small water bodies, it doesn't make hydrologic sense to invest in groundwater recharge as a regional water-management strategy. Instead, if these wetlands are well managed, the poor farmers can tap water from these small water bodies for irrigation at much lower cost. But what is most interesting is that if the recommendation of these researchers for intensifying groundwater use turns into a reality, the wetlands would get dried up during the summer season. A more important concern is "are we so desperate that we have to resort to moribund schemes like the NREGA to plan interventions for "groundwater recharge," which requires rigorous scientific studies?"

12.5. CONCLUSIONS

Policy making should be based on hard empirical evidence and logical reasoning and cannot be merely based on perceptions. Those who advocate policy shift in agricultural, power, and groundwater sectors in WB and Bihar seem to be doing the latter. Macrolevel, physical, and socioeconomic constraints to agricultural growth such as limited arable-land availability, an excessively large rural population depending on agriculture and very small operational holdings of millions of farmers, and climatic and ecological constraints cannot be removed by mere tinkering with certain norms and procedures. Bad policies framed on the basis of misinformation would surely and certainly lead to bad outcomes.

There is no doubt that economies of states such as Bihar and WB have to grow faster, to lift the people out of "poverty trap." Unfortunately, the route to that growth will not be "agricultural." For "agricultural revolution," along with fertile soils and water, farmers should have access to a good amount of arable land to cultivate and favorable climatic conditions such as good solar radiation for better crop growth, along with good supply of inputs such as fertilizers and quality seeds. Only then, they would be able to generate surplus. Or else, these ideas will remain as pipe dream.

As regards climate, it doesn't favor eastern India so much for achieving high productivity of the prevailing farming system. The region has a subtropical-to-tropical climate, and the region also receives high to very high rainfall. The future of farming in the region has to be based on the agroecology. The paddy-wheat system of cultivation has been very dominant in the eastern IGP for several decades (Pathak et al., 2003; Sikka and Bhatnagar, 2006), whereas paddy is the main crop in Orissa and Jharkhand (Sikka, 2009). Intensive irrigated paddy production, particularly cultivation of irrigated *boro* rice, had also resulted in the Green Revolution in West Bengal in the 1980s. But there are serious constraints in enhancing the income from this cropping system through yield improvement, induced by climate (Aggarwal et al., 2000; Pathak et al., 2003) and agroecology (Ladha et al., 2000). For instance, Pathak et al. (2003) found that the solar radiation during the growing season was much higher for the western side of the IGP (Ludhiana, Punjab) as compared with the eastern side (24 Parganas, WB) and minimum temperature much lower. They used crop simulation models to show that this difference in weather conditions could result in a difference in climatic potential yield between the two regions, to the extent of 3.0 t/ha (7.7 t/ha in WB against 10.7 t/ha in Punjab) in the case of rice and 2.7 t/ha (5.2 t/ha in WB against 7.9 t/ha in Punjab) in the case of wheat.

There is also a problem with the cereal- and water-centric approach to agricultural growth. Studies clearly show that over the period from the 1970s to 2000s, the contribution of crop sector to India's agricultural growth has declined from 79% to 60%. The contribution of livestock sector grew almost three times and accounted for more than one-third of the agricultural growth during that period (Joshi and Kumar, 2014). Within the crop sector, the contribution of cereals had declined from 30.7% in the 1980s to 6.9% during 2000–10, whereas the contribution of horticulture sector (within the crop sector) has been growing remarkably during the same period from 39.5% in the 1980s to 53.5% during 2000–10 (Birthal et al., 2014). The weakening contribution of cereals to India's agricultural growth also meant that the contribution of eastern Indian states, which are dominated by paddy (paddy accounting for 55% of the total cropped area in the region) with very little area under noncereal crops and horticultural crops, to overall growth in India's agricultural GDP also declined[1] (Birthal et al., 2014; Joshi and Kumar, 2014). Birthal et al. (2014) notes that diversification toward high-value commodities (fruits, vegetables, and milk) is a sustainable source of agricultural growth and also provides an opportunity for smallholders to enhance their income and escape poverty, as the demand for high-value food crops is expected to accelerate.

Therefore, intensification of well irrigation is not an essential ingredient for breaking the agricultural impasse in eastern India. But the narrative being created by some irrigation researchers for the past two decades is that farmers in eastern India remain poor because of limited access to or high cost of irrigation water for growing paddy and wheat, and therefore, the concept of agricultural growth and poverty reduction through intensification of well irrigation was promoted. The model they want to endorse was groundwater markets through either subsidized electricity for well owners or free agricultural

1. During the period from 1980–89 and 2000–09, It dropped from 14.4% to 6.3% in the case of West Bengal, from 13.8% to 6.5% in the case of UP, from 8.8% to 4.7% in the case of Bihar, from 4.2% to 3.1% in the case of Odisha, and from 2.3% to 1.5% in the case of Assam (Joshi and Kumar, 2014).

power connections or subsidized solar irrigation pumps. They could convince the erstwhile Planning Commission of India to accept this as the model of agricultural growth for eastern India (Planning Commission, 2013) with no tangible results so far.

12.6. THE WAY FORWARD

For agriculture in eastern India to prosper, it needs farming systems that suit its agroecosystems. The region is characterized by large areas under wetlands, including (1) areas that are under paddy grown in submerged conditions and (2) areas that are inundated due to floods and tides and having numerous wetlands that have water year round. As regards the inundated areas, there are large low-lying areas in the coastal region of West Bengal that are likely to get inundated during ocean tides. They include coastal mangrove regions. There are also floodplains that get marooned due to river flooding, mostly in north Bihar. While the areas that get water from tidal exchange would be suitable for saltwater shrimp, the floodplains would be suitable for freshwater shrimp and numerous varieties of native fish, along with paddy. Extensive shrimp farming can be practiced in these areas with very little farm inputs. Also, many of the fruit trees that are traditionally grown in north Bihar such as mango, guava, banana, and lychee can be grown more extensively. In addition, there are numerous vegetables that grow well in this region.

In the inland paddy-growing areas, which are in the very high-rainfall region with shallow groundwater and alluvial plains, water from ponds and lakes could be used to irrigate the rice fields, and shrimp can be farmed in them. The private ponds owned by farmers in their backyard and in the farms can be used for growing inland fish varieties such as rohu, catla, and mrigal. These fish varieties grow to a marketable size of nearly 700 g to 1 kg in a year. In groundwater-irrigated areas, the tanks can be replenished during the summer season using pumped groundwater. In the canal-irrigated areas of coastal Orissa, coastal plains of WB, and Bihar (south and north), the ponds and tanks can be replenished by the release from canals.

The hard-rock areas of the eastern India, comprising western parts of WB, western Orissa, and Jharkhand, have tropical, semiarid climates with high rainfall. As can be inferred from the data presented in this chapter (Tables 12.1 and 12.2), these regions are not as water-rich as the alluvial plains of West Bengal and north Bengal. A different type of farming system needs to be introduced. In these regions, though paddy is the dominant crop during kharif season, farmers grow many other crops, especially during winter season. In plots that are not traditionally used for raising paddy, farmers can take up vegetable production, and treadle pumps and micro diesel pumps can be used to irrigate these crops. Since the areas receive high rainfall (above 1000 mm), small water-harvesting systems such as ponds can also be built and used for supplementary irrigation of vegetables and fruits, along with fish rearing. The regions that are most ideal for this are the tribal areas of western Orissa; south Bihar and the western districts of West Bengal are also ideal for this.

Recent field-based research by Indian Council of Agricultural Research (ICAR) in Bihar and Orissa shows that well-designed multiple-use systems can enhance the productivity of use of both land and water in eastern India remarkably. This involved integrating fisheries, prawn farming, and duckeries with paddy irrigation using local secondary reservoirs for the water (Sikka, 2009). Therefore, what is required is better knowledge about favorable farming systems and extension services to impart this knowledge to the farmers. Better infrastructure for marketing of the high-value produce (fish, vegetables, and fruits) is the other important requirement.

REFERENCES

Aggarwal, P.K., Talukdar, K.K., Mall, R.K., 2000. Potential yields of rice-wheat system in the Indo-Gangetic plains of India. Rice-Wheat Consort. Pap. Ser. 10, 16.

Bassi, N., 2015. Irrigation and energy nexus: solar pumps are not viable. Econ. Polit. Wkly. 50 (10), 2015.

Bassi, N., 2017. Solarizing groundwater irrigation in India: a growing debate. Int. J. Water Resour. Dev. https://doi.org/10.1080/07900627.2017.1329137.

Birthal, P.S., Joshi, P.K., Negi, D.S., Agarwal, S., 2014. Changing Sources of Growth in Indian Agriculture: Implications for Regional Priorities for Accelerating Agricultural Growth. IFPRI Discussion Paper 716, International Food Policy Research Institute, Washington D.C.

Central Ground Water Board (CGWB), 2012. Ground Water Year Book-India 2011–12. Central Ground Water Board, Ministry of Water Resources, RD&GR, GoI, Faridabad.

Deloitte, 2015. Study of Existing Minor Irrigation Schemes in West Bengal. Final Report Submitted to West Bengal Water Resource Planning Department, Deloitte, Kolkata.

Government of Bihar, 2008. Bihar Economic Survey-2008-09. Ministry of Finance, Government of Bihar, Patna.

International Water Management Institute (IWMI), 2012. Agricultural Water Management Learning and Discussion Brief, AGWAT Solutions, Improved Livelihood for Small Holder Farmers. . April.

Joshi, P.K., Kumar, A., 2014. Agricultural Growth in India: Performance and Prospects. YOJANA, pp. 50–56.

Khan, M.R., Voss, C.I., Yu, W., Michael, H.A., 2014. Water resources management in the Ganges basin: a comparison of three strategies for conjunctive use of groundwater and surface water. Water Resour. Manag. 28 (5), 1235–1250.

Kumar, M., Bassi, N., Sivamohan, M.V.K., Venkatachalam, L., 2014. Breaking the agrarian crisis in Eastern India. In: Kumar, M.D., Bassi, N., Narayanamoorthy, A., Sivamohan, M.V.K. (Eds.), Water, Energy and Food Security Nexus: Lessons from India for Development. Routledge, London, pp. 143–159.

Kumar, M.D., 2003. Food Security and Sustainable Agriculture in India: The Water Management Challenge. Working Paper 60, IWMI, Colombo, Sri Lanka.

Kumar, M.D., 2007. Groundwater Management in India: Physical, Institutional and Policy Alternatives. Sage Publications, New Delhi.

Kumar, M.D., Singh, O.P., 2008. In: How serious are groundwater over-exploitation problems in India? A Fresh Investigation into an Old Issue. Managing water in the face of growing scarcity, inequity and declining returns: exploring fresh approaches. 7th Annual Partners' Meet of IWMI-Tata Water Policy Research Program, ICRISAT, Patancheru, AP, pp. 298–317.

Kumar, M.D., Scott, C.A., Singh, O.P., 2013. Can India raise agricultural productivity while reducing groundwater and energy use? Int. J. Water Resour. Dev. 29 (4), 557–573.
Ladha, J.K., Fischer, K.S., Hossain, M., Hobbs, P.R., Hardy, B.(Eds.), 2000. Improving the Productivity and Sustainability of Rice-Wheat Systems of the Indo-Gangetic Plains: A Synthesis of NARS-IRRI Partnership Research (No. 40). International Rice Research Institute.
Mukherji, A., 2008. The paradox of groundwater scarcity amidst plenty and its implications for food security and poverty alleviation in West Bengal, India: what can be done to ameliorate the crisis? In: 9th Annual Global Development Network Conference, Brisbane, Australia, pp. 29–31.
Mukherji, A., Das, A., 2012. How Did West Bengal Bell the Proverbial Cat of Agricultural Metering?: The Economics and Politics of Groundwater. Water Policy Research Highlight 2, IWMI-Tata Water Policy Program, IWMI Field Office, Anand, Gujarat, India.
Mukherji, A., Shah, T., Banerjee, P.S., 2012. Kick-starting a second green revolution in Bengal. Econ. Polit. Wkly 47 (18), 27–30.
Mukherji, A., Rawat, S., Shah, T., 2013. Major insights from India's minor irrigation censuses: 1986–87 to 2006–07. Econ. Polit. Wkly. 48 (26–27), 115–124.
National Rainfed Area Authority (NRRA), 2011. Challenges of Food Security and Its Management. National Rain-fed Area Authority, New Delhi.
Pant, N., 2004. Trends in groundwater irrigation in eastern and western UP. Econ. Polit. Wkly. 39 (31), 3463–3468.
Pathak, H., Ladha, J.K., Aggarwal, P.K., Peng, S., Das, S., Singh, Y., Singh, B., Kamra, S.K., Mishra, B., Sastri, A.S.R.A.S., Aggarwal, H.P., 2003. Trends of climatic potential and on-farm yields of rice and wheat in the Indo-Gangetic Plains. Field Crop Res. 80 (3), 223–234.
Planning Commission, 2013. Twelfth Five Year Plan (2012–2017) Economic Sectors Volume II. Planning Commission, Government of India, New Delhi.
Planning Commission, 2014. Report of the Expert Group to Review the Methodology for Measurement of Poverty. Government of India, New Delhi.
Ramchandran, V.K., Swaminathan, M., Rawal, V., 2003. In: Agricultural growth in West Bengal. Paper Presented at the All India Conference on Agriculture and Rural society in Contemporary India, Barddhman, 17–20 December.
Revelle, R., Lakshminarayana, V., 1975. The Ganges water machine. Science 188 (4188), 611–616.
Shah, M., 2016. Eliminating poverty in Bihar: paradoxes, bottlenecks and solutions. Econ. Polit. Wkly 51 (6), 56–65.
Shah, T., Pradhan, P., Rasul, G., 2016. Water challenges of the Ganga Basin: an agenda for accelerated reform. In: Bharati, L., Sharma, B.R., Smakhtin, V. (Eds.), The Ganges Basin: Status and Challenges in Water, Environment and Livelihoods. Earthscan/Routledge, London, UK.
Sikka, A.K., 2009. Water productivity of different agricultural systems. In: Kumar, M.D., Amarasinghe, U. (Eds.), Water Sector Perspective Plan for India: Potential Contributions From Water Productivity Improvements. International Water Management Institute, Colombo, Sri Lanka, pp. 73–84.
Sikka, A.K., Bhatnagar, P.R., 2006. Realizing the potential: using pumps to enhance productivity in the Eastern Indo-Gangetic Plains. In: Sharma, B.R., Villholth, K.G., Sharma, K.D. (Eds.), Groundwater Research and Management: Integrating Science into Management Decisions, Groundwater Governance in Asia Series-1. International Water Management Institute, Colombo, Sri Lanka, pp. 200–212.

Chapter 13

Deciding the Types of Interventions for Agricultural Development in Different Agroecologies

13.1. INTRODUCTION

Agricultural areas in India are officially grouped into two categories, namely, rainfed areas and irrigated areas. Rainfed areas are those where irrigation is less than 30% of the net sown area, and irrigated areas are those with extent of irrigation exceeding 30% (NRRA, 2012). These "rainfed areas," accounting for 60% of the total 143 M ha of cultivated area, are characterized by low agricultural productivity in terms of crop yields as compared with irrigated areas (Joshi et al., 2005; NRRA, 2011, 2012; Shah, 2013). For the kind of agroclimatic conditions that exist in India, an area can become rainfed due to two distinct reasons: (1) The area receives a large amount of precipitation, or the precipitation is more or less even during the growing season, and the quantum exceeds the evapotranspiration (ET) rates of crops that do not require irrigation to mature, while a large amount of water remains in the natural system in the form of runoff, groundwater, both, or in situ moisture; and (2) the area has very little water from the natural system in the form of soil moisture, runoff, and groundwater recharge, and the daily ET rates are high that only low water-consuming and short-duration crops are grown in the rainy season, and a small percentage of the area in the next season is irrigated (Kumar et al., 2017).

The first category of areas includes almost the entire Kerala; some very high-rainfall areas of coastal West Bengal, coastal Maharashtra, and Western Ghats region (Karnataka, Goa, and Maharashtra); and most areas of the northeast. The second category of areas includes Rayalaseema region of Andhra Pradesh, western Rajasthan, most parts of Saurashtra, several areas of Telangana, northern Karnataka, and many parts of Tamil Nadu. Though the two categories of areas have mutually contrasting conditions, both come under "rainfed" because of the unscientific criterion and definition (Kumar et al., 2017).

The first category of regions produces high biomass productivity (ton per hectare) and therefore yield high farm return (Rs) per hectare, whereas the

Water Policy Science and Politics. https://doi.org/10.1016/B978-0-12-814903-4.00013-0

second category of regions suffers from very low productivity, and therefore, the returns are very low (Droogers et al., 2001). As both types of areas fall under the category of "rainfed areas," many researchers try to make a comparison between the two in terms of biomass yield (kilogram per hectare) and productivity (rupees per hectare) to make the scientifically untenable argument that the latter has enormous potential for enhancing crop yields that can be realized (Pathak et al., 2009; Sharma, 2012; Wani et al., 2009) through soil moisture conservation, erosion control measures, and small water harvesting structures for supplementary irrigation (NRRA, 2012; Sharma, 2012).

There is also a third category of areas where crops require irrigation, and runoff water is available in the natural system but not available for utilization in the absence of irrigation infrastructure. The examples are Chhattisgarh, Jharkhand, Odisha, and some parts of Madhya Pradesh, and they form a case of deferred investment in water infrastructure. A large amount of water from the river basins located in these states (such as Narmada, Mahanadi, Brahmani-Baitarani, and Godavari) still discharge into the ocean annually, and there is scope for further exploitation of surface runoff from these basins (source, GoI, 1999).

Since it is the "amount and pattern of precipitation and evapotranspiration rates, which actually decide whether a crop in that region can be grown under rainfed conditions or has to be irrigated, ideally, the 'rainfed area' definition should consider the effective rainfall in an area and evapotranspirative demand of the crops to ascertain whether crops in a particular area can be grown under rainfed conditions or not" (source, based on Droogers et al., 2001). But the official definition links "rainfed areas" to their access to irrigation facilities. This also means that a "rainfed area" can become "irrigated," with increase in irrigation water imports or building of new irrigation infrastructure. Similarly, regions depending entirely on groundwater for irrigation could go dry, in the context of prolonged droughts, thereby making an irrigated area "rainfed" (Kumar et al., 2017).

13.2. WHAT MAKES AN AREA "RAINFED"?

For a long time, some decision makers in the Planning Commission of India and Ministry of Rural Development were concerned with this question: what do we do for rainfed areas, when so much investment is made for irrigated areas (NRRA, 2011, 2012; Parthasarathy Committee Report, 2006; Shah, 2013; XII Plan Working Group, 2011). Many civil society organizations and research institutions have also taken up the cause of rainfed areas, raising the issue of low productivity and high incidence of poverty there (NRRA, 2012; Sharma and Scott, 2005). A civil society network "Revitalizing Rainfed Agriculture Network," which has the mandate of exploring solutions for rainfed areas in India, has been advocating decentralized watershed management for improving productivity of rainfed areas (Shankar, 2011).

For improving rainfed agriculture in India, watershed development and management interventions were implemented by government-NGO consortia, with a massive public investment of US $700 million annually (Sharma and Scott, 2005).[1] But, these interventions had very limited success in terms of watershed area treated and productivity impacts on agriculture. By 2006, only 31.1 out of the 86M ha of rainfed area in the country had been treated under various programs of the government (Parthasarathy Committee Report, 2006). As regards the low productivity impacts, the problems identified include limited and temporary productivity gains, poor revegetation of common land, fast dissipation of groundwater recharge benefits (Joshi et al., 2004; Kerr et al., 2004; Palanisami et al., 2002), and shortage of high-quality professional human resources to deal with the technical and social aspects of planning and implementation (Shah, 2013). In the wake of these outcomes, the policy makers should not have been concerned about physical target achievements and instead should have targeted on real rainfed areas in India. On the contrary, the thrust of the erstwhile Planning Commission was on implementing the program on a larger scale (see Shah, 2013, p. 44).

Given the monsoon weather pattern and the interannual variability in rainfall that is high in the low-rainfall areas (Sharma, 2012), it is difficult to obtain optimum crop yields from purely rainfed areas. Most crops require some irrigation, and the extent of irrigation required depends on the difference between the actual rainfall received during the season and potential evapotranspiration (PET), which is determined by climatic factors. Even in the highest rainfall areas (like in Kerala or West Bengal) where average annual rainfall magnitude exceeds 2500mm in most places, some seasonal crops (winter crops) or annual crops are irrigated in certain seasons, in lieu of the effective precipitation in that particular year or particular season, falling below the evapotranspirative requirement of crops.

In the low-rainfall regions (most of which automatically fall in the "rainfed area" category by official definition), in many years, even the rainy-season crops would require one or two supplementary irrigations to mature due to monsoon failure. These crops get irrigated by local sources such as wells, though this doesn't get captured fully in the statistics on irrigated area that is outdated. As regards the second (Rabi season) crops, farmers cannot grow them unless water is applied artificially. Hence, most of the areas, which officially get labeled as "rainfed," are actually those where crops can give optimum yields only with irrigation. But, the official definition of "rainfed area" does not touch upon this aspect. Not being irrigated extensively doesn't qualify such areas to be "rainfed." It only makes them highly vulnerable to production risks.

In the high-rainfall (rainfed) areas, it is possible to see some seasonal crops that are purely rainfed. An example is the *Aus* and *Aman* paddy in West Bengal. In the low-rainfall ("rainfed") areas, it is not even possible to see such rainfed crops, and the situation there is highly dynamic and keeps changing from year to year depending on the amount and pattern of rainfall. Quite likely, in dry

1. The interventions comprised soil moisture conservation, erosion control measures, and water harvesting structures.

years, even the kharif or monsoon crops in those areas will have to be irrigated. It is therefore clear that the second category of "rainfed" areas, where the rainfall is poor, badly requires irrigation for agricultural productivity enhancement.

13.3. WATERSHED DEVELOPMENT FOR RAISING PRODUCTIVITY OF "RAINFED AREAS": WRONG SOLUTIONS

The policy makers overlook the vast heterogeneity in agroecology of areas that are currently classified as rainfed and irrigated and use the outdated definitions and statistics (NRRA, 2011, 2012). They ignore the fact that most of the officially classified "rainfed areas" in India remain so because of poor rains and high aridity, which result in low annual runoff and groundwater recharge and reduced irrigation potential, and that only a small percentage of the "rainfed areas" actually get sufficient rains to be able to grow crops under rainfed conditions (Droogers et al., 2001). In the rest of the areas, which fall under low to medium rainfall and also subjected to high year-to-year variability in rainfall, in situ soil moisture conservation measures would not help ensure optimum yield of crops during drought years when the actual amount of rainfall during the season will be very low and evaporation high (Kumar et al., 2017), resulting in very negligible excess runoff (Kumar et al., 2008). Yet, there is too much focus on groundwater recharge works in watershed development schemes (Shah, 2013, p. 44).

The agriculturally prosperous regions such as Punjab and Haryana, which experience low to medium rainfall and high aridity in most parts, had achieved that status due to surface-water imports, which enabled irrigation expansion and shift from "rainfed area" to "irrigated area" category.[2] By the year 1970–71, Punjab and Haryana had developed an irrigation potential of 4.2 and 2.2 M ha against a full potential of 5.0 and 3.3 M ha, respectively, to be realized by the year 2025 (MoA & I, 1976). Most of these areas are part of the alluvial plains and do not have well-defined watersheds. Large-scale import of surface water through canals also enhanced recharge to groundwater, given the flat topography in most parts and permeable, alluvial soils. This fact is completely ignored by policy makers who plan for watershed development for areas with such climatic conditions. Though in the low-rainfall areas (with poor irrigation infrastructure) farmers were making their own arrangements for irrigation by digging wells, these investments were not sustainable. The farmers compete with each other drilling deep to pump underground water, and in the process, many lose out because of well interference.

The planners and the policy makers come up with watershed development as a "silver bullet" for these rainfed areas. In fact, since the early 1990s, it

---

2. As noted by Kumar (2007), had it not been for the surface-water imports for irrigation, the groundwater balance in Punjab would have been even more precarious than it is today.

had become the panacea for all problems in the so-called rainfed areas for agricultural improvements, poverty reduction, livestock development, employment generation, and stopping outmigration (Fan and Hazell, 2000). More importantly, the emphasis of the program was in the areas with low rainfall and high aridity, where such needs were quite visible, though watershed treatment works were also implemented in the areas of very high rainfall as well with less intensity. There were recent attempts by the National Rainfed Area Authority to use aridity index to prioritize districts within the areas officially classified as "rainfed" for taking up watershed development projects (see NRRA, 2012).

Concerns were raised from time to time about poor target achievement in watershed development in terms of area coverage (XII Plan Working Group, 2011; Planning Commission, 2005, p. 101) and low impacts of the interventions (Joshi et al., 2005; NRRA, 2012; Shah, 2013). In the early 1990s, the lack of funds in the exchequer could be used as an excuse for preferring this cheap solution for such areas that had multiple natural and socioeconomic problems, but it can no longer be an excuse. Today, around INR 48,000 crore (US$ 7.5 billion) is spent annually on Mahatma Gandhi National Rural Employment Guarantee Act (MGNREGA) works in rural areas, and a lot of this is proposed to be used for watershed development works (see Shah, 2013).

13.4. IMPACTS OF WATERSHED DEVELOPMENT PROGRAMS IN INDIA

Watershed development programs are useful in areas that generate significant runoffs and that have heavy topsoil erosion (James et al., 2015). In an area, if the available radiative energy and potential evaporation rates are fairly low, then for a given amount of precipitation, runoff is likely to exceed evapotranspiration. Similarly, runoff would be a smaller fraction of precipitation if available radiative energy is very high resulting in high evapotranspiration (Arora, 2002). The low-rainfall areas with high aridity have low runoff generation potential (Arora, 2002; Prinz, 2002). In areas with steep or undulating topography, erosion of topsoil is likely to be higher than areas with relatively flat topography, owing to greater speed of flow of runoff.

The foregoing analysis suggests that in low-rainfall areas with high aridity, the hydrologic impact of small water harvesting structures built under the programs will be limited in terms of storage enhancement. The high potential evaporation in such areas also increases the water losses from small water bodies and soil profile. Due to the high interannual variability in rainfall, which is a characteristic of the low- to medium-rainfall regions (Sharma, 2012; Kumar et al., 2006), in general, the runoff from catchments in such areas has poor dependability. The effectiveness of water harvesting interventions in such areas will be very low during drought years (Kumar et al., 2006, 2008; Glendenning and Vervoort, 2011) owing to disproportionately high reduction in runoff

(Prinz, 2002). In spite of poor hydrologic feasibility, large numbers of small water harvesting structures were built under watershed development program in such areas (Kumar et al., 2008).

Watershed development interventions will produce significant benefits in catchments with steep or undulating topography (Kumar et al., 2017). In a denuded watershed of the Siwalik Hills in Haryana, a package of practices comprising trenching, brushwood/stone check dams, debris detention basins, and planting of *Acacia catechu* and *Dalbergia sissoo* reduced runoff from 30% to 10.8% and soil loss from 150 to 2.8 t/ha in a span of 40 years (Sharda, 2004). The size of watersheds, land use, and type of treatment also influenced the hydrologic impacts of watershed treatment works. Experiments in small watersheds of eastern Himalayas with different types of watershed treatments showed runoff ranging from 6.33% to 47.75% while the soil loss ranging from 22.7 to as high as 281.6 t/ha (Sharda, 2005).

The approach of integrated watershed management is recommended for large mountain watersheds to ensure sustainable water supplies (Vaidya, 2016), and this approach has been implemented in large watersheds/river basin with the help of scientific planning (Leclerc and Grégoire, 2016; Stewart and Bennett, 2016) and appropriate institutions (Leclerc and Grégoire, 2016). Though watershed management programs are quite old in India, the lack of scientific planning, especially in determining the type and intensity of the structures, is a major issue in their implementation. Interestingly, the focus of watershed development has modified over the last 25 years from soil conservation to water conservation to now include a more participatory planning approach (Shah, 2013). Nevertheless, findings of evaluation studies estimating the benefits of distributional equity or magnitude of social impacts from watershed development are often unclear or disputed (Hope, 2007; World Bank, 2004).

The available scientific studies to evaluate the impact of catchment management interventions on their environmental conditions pertain to the following: the hydrologic and biophysical impacts of the interventions at the level of microcatchments in terms of changes in runoff collection efficiency, the increase in vegetation cover, the soil loss from the catchment (Panwar et al., 2012; Garg et al., 2012), the groundwater recharge (Garg et al., 2012), and the way climate variability influences the impacts of land use and water harvesting interventions on the catchment outflows (James et al., 2015). They all show positive impacts at the local level (Panwar et al., 2012; Pathak et al., 2013). But, the sample microcatchments are not representative of the hydrologic conditions (rainfall, slope, geohydrology, soil type, land use, and land cover) of the extended catchments of which these sample microcatchments are a part. In a nutshell, the impact assessments do not factor in the "scale effects" in hydrology. In other words, the hydrologic processes that matter for microwatersheds (say, e.g., the rainfall-runoff relationship and catchment outflows) will not be same as that for the large catchment in which these microwatersheds fall.

Generally, the runoff rate for large catchments is much lower than that of the microcatchments that constitute it (Kumar et al., 2017).[3] The importance of this scale effect was well demonstrated by the water audit undertaken for Andhra Pradesh Rural Livelihoods Project (APRLP). It showed how it is critical to revise (scale down) runoff rates when one moves from a microwatershed to a large watershed or basin in terms of scale of operation. The analysis of runoff data for experimental watersheds carried out by the Central Soil and Water Conservation Research and Training Institute (CSWCRTI) in Bellary and by Central Research Institute for Dryland Areas (CRIDA) in Anantapur showed that average annual runoff is typically 2%–15% of average annual rainfall and highly dependent on such factors as soil type, slope, land use, and/or vegetative cover and the presence of infield soil and water conservation measures. The analysis of gauging data from the Central Water Commission (CWC) showed that annual surface runoff at the macrowatershed or basin scale is in the range of 0.8%–7.5% of rainfall. The relatively low runoff coefficients challenge the widely used assumption that runoff in this region is always in the range 30%–40% of annual rainfall (Rama Mohan Rao et al., 2003).

This difference makes it difficult to draw useful inferences of microcatchment-level studies for planning catchment-wide interventions in a large catchment, as it is the average runoff for the large catchment that matters, when watershed interventions are carried out at a large scale covering a macrowatershed. NGOs and government agencies generally implement watershed interventions (especially soil and water conservation works) in the whole of a macrowatershed or subbasin based on runoff rates that occur in small hilly upper catchments that produce excessively high runoff. This leads to overdesigned structures.

Empirical evidence that illustrates the "scale effects" in small water harvesting, which concern downstream impacts of water harvesting in the upper catchment areas, is available from several studies (Batchelor et al., 2003; Glendenning and Vervoort, 2011; Gosain et al., 2006; Gupta, 2011; Kumar et al., 2008; Ray and Bijarnia, 2006; Talati et al., 2005) and watershed development (Batchelor et al., 2003; Syme et al., 2012).

As noted by Kerr (2002), the socioeconomic benefits commonly associated with watershed development, which include improved crop yields and farmer returns, increased access to water for domestic supplies, and employment generation, can vary for different resource user groups located across watersheds. Emerging global evidence suggests that there are limits and tradeoffs to modify watersheds due to complex hydrologic and social systems'

3. For instance, in the catchment of Kabani River in Kerala, which is the uppermost catchment of Cauvery river basin, the rainfall-runoff relationship suggests that more than 80% of the rainfall is converted into runoff (mimeo), with a runoff of around 2.40 m, whereas the runoff estimated for the basin as a whole on the basis of dependable yield and the drainage basin area is only 0.216 m (Kumar et al., 2008).

interactions (Calder, 2005). This is partly because watershed development interventions modify land-use impacts on water resources, which in turn may alter downstream water access, while augmenting upstream water supplies (Gosain et al., 2006).

The water audits in watersheds of Andhra Pradesh and Karnataka under APRLP and Karnataka Watershed Development (KAWAD) project, respectively, showed clearly that watershed development activities and consequent increase in groundwater extraction for irrigation had major impacts on the pattern of water use and access. The intensive treatment of drainage lines contributed to a major reduction in the utility of traditional tank systems located downstream. Though these changes in pattern of water use were positive from an irrigation perspective, if the nonirrigation uses of tanks are considered, it would appear that the "irrigation" benefits have come at a social and economic cost. During the last 10–20 years, the utility of many tanks has declined for activities such as washing, bathing, watering livestock, and pisciculture. In extreme cases, reduced tank inflows are having a negative impact on domestic water supplies, especially where tanks are an important source of recharge of aquifers used for urban supply (Batchelor et al., 2003; Rama Mohan Rao et al., 2003).

Another illustrative example of the scale effect is available from Madhya Pradesh. The Narmada basin in Madhya Pradesh witnessed its large-scale implementation on a mission mode for almost a decade. The planning of these schemes however did not involve any hydrologic or geo-hydrologic considerations (Talati et al., 2005). A study carried out by Talati et al. (2005) involving primary data collected from two subbasins of Narmada, namely, Hathni and Kundi, and two microwatersheds in one of the basins showed that after watershed treatment activities, the streamflows in the two subbasins were reduced significantly, while the recharge fraction, estimated as the ratio of the average water level fluctuation (during monsoon) and the annual rainfall, was increased. Further, it was observed that the recharge fraction for the treated watershed was higher than that of untreated watershed in the same basin for higher magnitudes of rainfall (Talati et al., 2005).

In order to measure the real impact of watershed treatment on streamflow generation and the effect of water harvesting on downstream flows, a regression was run between rainfall and streamflow for two time periods: pre water harvesting period (15 years) and time period including post-WDP period (20 years). The results of the regression analysis showed a lower value of coefficient for the period encompassing the post-WDP period. This meant that the runoff corresponding to a unit rainfall reduced after the watershed treatments. Hence, it confirmed the preliminary findings that runoff rate is reduced in post-WDP period (Talati et al., 2005).

Economic valuation of watershed management programs fails to link the measurable indicators of watershed impacts to planned interventions (Hope, 2007) and is not able to capture the significant negative externalities (World Bank, 2004) as shown by several empirical studies across the country (Batchelor et al., 2003; Calder, 2005; Hope, 2007; Kumar et al., 2008; Syme et al., 2012).

Over and above, there is failure to capture the lasting effects of project interventions on the social and natural systems (Barron and Noel, 2011). Failure to quantify the hydrologic gains from the structures, which are essential to compute the economic value created from the use of the additional water stored in the catchment, is another pressing problem (James et al., 2015).

Findings of the studies on socioeconomic impact of watershed development programs from different authors, particularly those relating to poverty impacts, often contradict (source, based on Hope, 2007; Buhl and Sen, 2006; Kerr, 2003), while the skewed distribution of benefits from projects, which achieved resource conservation productivity gains, toward large landholders is visible (Kerr, 2003).

For instance, a review carried out on the socioeconomic and poverty impact of the Indo-German Watershed Development Program by Buhl and Sen (2006) showed some reduction in poverty and hunger as a "medium-term" positive impact of the program. Notably, on none of the attributes (seven MGD goals), the program showed a "very positive" impact. On environmental sustainability also, the program showed a positive impact both on the "short term" and "medium term." Its impacts on reducing child mortality and improving maternal health have been nil (Buhl and Sen, 2006).

A study in Madhya Pradesh found that majority of the farmers planting kharif crops are no better off after the project in income terms with no significant variation among social, income, or land-stratified groups. The smaller group of farmers who grow winter crops fares even worse, on average, but significant variation was found across social groups and land ownership. However, own-project evaluations based on qualitative perceptions showed an 84% increase in kharif yield and a 60% increase in winter yield. But, these positive impacts were associated with wage labor from the project and hence were short term in nature, and longer term improvements in water access were not identified (Hope, 2007).

A positive social impact was seen in terms of a significant reduction in domestic water collection time for households with the highest collection times. But, these households were still facing considerable collection costs (e.g., physical, opportunity, and health) and remained excluded from a basic level of domestic water access (Hope, 2007). But, the author found it reasonable to argue that the estimated lower level of domestic water access might be related to new upstream water conservation structures capturing more water, as planned, without fully understanding downstream water implications (see, e.g., Calder, 2005; World Bank, 2004).

Analysis of productivity, conservation, and poverty alleviation impacts of watershed programs in Maharashtra by Kerr (2003) looked into poverty alleviation trade-offs of achieving the objectives of productivity gain and conservation. The study suggested that the projects most successful in achieving conservation and productivity benefits also resulted in skewed distribution of benefits toward larger landholders. Those who had relatively larger holdings were satisfied with the project, while many landless people strongly resented their loss of access to common lands (Kerr, 2002).

To sum up, while there is increasing recognition among scholars of the need to take a look at the hydrology and the land use in the larger catchment in order to decide on the degree/scale of agricultural intensification and rainwater harvesting in order to avoid undesirable effects of watershed management (Calder et al., 2008), there are no comprehensive studies that were undertaken to look at the hydrologic and socioeconomic impacts of water harvesting structures, afforestation, and soil moisture conservation activities carried out under the watershed development programs at appropriate scales for nearly two decades (James et al., 2015). The existing studies use the microwatershed as the unit of analysis of impact, without taking cognizance of the fact that interventions cause negative externalities for the downstream areas of the microwatershed (Reddy and Syme, 2014).

13.5. WRONG CRITERIA FOR SELECTING AREAS FOR WATERSHED DEVELOPMENT

The XII Plan document envisaged a major breakthrough in the productivity of rainfed areas for achieving national food security through a massive program of watershed restoration and groundwater recharge (Shah, 2013, p. 44). Watershed development programs were implemented in all types of areas that were "rainfed" as per standard official definition—from the excessively high-rainfall areas of per humid areas in Kerala (nearly 3000 mm a year in most parts) to the hottest and driest areas of hyperarid, western Rajasthan that receive very scanty rains (less than 200 mm in a year), and from the high-rainfall hills of Himachal Pradesh to the low-rainfall, semiarid, and arid alluvial plains of north Gujarat—using funds from drought-prone area program or desert development program (Parthasarathy Committee Report, 2006) and, lately, MGNREG scheme (Shah, 2013). Watershed development activities were implemented under the banners of National Watershed Development Program for Rainfed Areas (NWSDPRA) by the Ministry of Agriculture and desert development program (DDP), drought-prone area program (DPAP), and integrated watershed development program (IWDP) by the Department of Land Resources of the Ministry of Rural Development (Parthasarathy Committee Report, 2006).

The scheme conceptualization completely ignored the fact that the drought-prone areas are those that actually experience high year-to-year variability in the monsoon rains, erratic rainfall, high aridity, etc., and such areas are not areas that experience high erosion and soil loss due to speeding runoff. In fact, the hilly and mountainous areas that are also characterized by high rainfall, speedy runoff, and excessive soil erosion provide the ideal environment for watershed development and management, which can produce benefits of soil erosion check and moisture conservation (Kumar et al., 2017). However, these attributes were never considered for evolving criteria for selection of districts for watershed development, and instead, areas were selected on the basis of poverty rates, drought proneness, aridity, and irrigation coverage (see, e.g., NRRA, 2012). Moreover, in the name of participatory approach, plans were prepared at the village level without realizing the fact that water doesn't follow administrative boundaries.

There was failure of watershed development programs almost everywhere due to the faulty selection criteria. A metaanalysis showed that the program was effective in the areas with rainfall ranging from 900 to 1100mm, with poor impacts in areas with rainfall less than 500mm and between 500 and 700mm (Joshi et al., 2005). The study recommended increased fund allocation for research and development to design innovative strategies for improving the effectiveness of programs (see, e.g., Joshi et al., 2005) and the frequent changes in the program implementation guidelines (Shah, 2013), which failed to enhance the impacts (Reddy, 2006). Obviously, the reason for these differential impacts of watershed programs across hydrologic regimes was not examined (Kumar et al., 2017).

The bumper production obtained by farmers in the rainfed areas in some abnormally wet years was attributed to the success of WSD program implemented there, though it was actually due to high groundwater recharge and high soil moisture storage that enabled cropped area expansion and optimal watering for crops. The routine evaluation and impact assessment studies of WSD programs suffered from problems of attribution, due to the flawed methodologies used.[4] But such methodological flaws insulated the proponents of "watersheds and rainfed areas" from major embarrassment for quite some time (James et al., 2015).

When these myths about the remarkable positive impact of WSD programs slowly began to unravel, arguments such as "gender-sensitive" agricultural development, "social justice" and democratic decentralization, and local capacity building were used to defend the investments (see, e.g., Parthasarathy Committee Report, 2006). While these are noble objectives, one can certainly raise questions about the objectivity in measuring these achievements (Kumar et al., 2017). Even when deficiencies are detected in the program, revision of watershed guidelines is resorted to as a ritual. This is evident from the comment of Mihir Shah (2013) on the program: "The Eleventh Plan has proposed a radically new approach based on the suggestions of the technical committee on watershed programme (Parthasarathy Committee). However, in a comprehensive review of the programme carried out by the Union Minister for Rural Development in 2012, it emerged that a number of practical impediments were coming in the way of putting the new paradigm of watershed management into practice on the ground and the pace of the programme was found to be less than satisfactory. A Planning Commission Committee has now revised the

4. A commonly used methodology for analyzing the impact of watershed development programs was to look at the change in impact variables such as local water availability and crop yields in a sample watershed between post and pretreatment periods. One of the typical problems of attribution induced by this methodology came from the lack of ability to factor in the impacts of rainfall on local water availability, and the rainfall, farm inputs and crop technologies on crop yields while analyzing the effect of watershed development on agricultural productivity on a time scale (James et al., 2015).

guidelines to provide necessary flexibility and momentum to the programme, even while strengthening its innovative features. One of the key deficiencies of the programme was found to be the shortage of funds to deploy high quality professional human resources for both social and technical aspects for which a special allocation has now been proposed" (Shah, 2013, p. 44). While phrases such as "radically new approach," "new paradigm," and "innovative approach" are profusely used, practically, nothing has changed on the ground in terms of enhancing the impact of the interventions.

13.6. CLASSIFICATION OF AREAS FOR AGRICULTURAL DEVELOPMENT PLANNING

The ideal classification of agricultural areas for development planners should be (1) cropped areas requiring irrigation and (2) cropped areas not requiring irrigation. Within the first category, there can be subcategories depending on the number of seasons of the year during which crops require irrigation to mature. The real rainfed areas are the latter, that is, areas where crops can mature without irrigation. This categorization will have to be done on the basis of hydrometeorological parameters and agroclimatic conditions. In the rainfed areas with moderate to steep slopes, watershed treatment activities for erosion control and soil moisture conservation could help enhance crop yields. Proper definition of "rainfed areas" would improve the resource allocations per unit area of land by allowing diversion of the resources to favorable areas (Kumar et al., 2017).

In the regions where crops require irrigation to mature, there are some areas with medium to high rainfall, where sufficient water in the natural system exists and can be tapped for irrigation through major, medium, and minor irrigation projects. But, the rest of the places with low-to-medium rainfall and high aridity and major expansion in irrigated area will be possible only through import of water from distant water-rich catchments (Kumar et al., 2017).[5] At present, in the absence of water-transfer schemes, groundwater is intensively used in such water-scarce regions to protect kharif crops and often raise the winter crops, and the resource is overexploited (Reddy and Chiranjeevi, 2016). Surface-water import for irrigation will also help address the issue of sustainable and equitable management of groundwater in such regions. This can be complemented by interventions to improve water-use efficiency in agriculture.

In the second type of areas, which do not require irrigation by virtue of the high to excessively high rainfall and subhumid to humid climate, depending on the agroecology, watershed development projects can be taken up. For instance,

5. The recent attempt by the government of Andhra Pradesh to transfer water from Godavari to Krishna and then to Rayalaseema region through the Pattiseema project is an illustrative example of such a model (Kumar et al., 2017). Water transfer from Sardar Sarovar reservoir on Narmada River to water-scarce areas of Central and North Gujarat, Kachchh, and Saurashtra regions are another good example (Jagadeesan and Kumar, 2015).

some regions of West Bengal and Jharkhand and the hilly areas of Uttarakhand experience seasonal water scarcity despite having above 1000mm rainfall as a large part of the rainfall during the monsoon season gets converted into runoff in the high slopes and goes uncaptured, without contributing to groundwater recharge. Speedy runoff also causes soil erosion problems (Kumar et al., 2017).

What is clear from the foregoing analysis is that we need to do away with the practice of using WSD programs as a standard solution for all impoverished areas. The selection of areas for watershed development should consider hydrologic, topographical, and climatic factors. Further, the interventions have to integrate hydrologic and biophysical considerations, with socioeconomic aspects for them to be effective and sustainable (Reddy and Chiranjeevi, 2016). It is important to assess the catchment for runoff and sediment transport before deciding on the degree of interventions (James et al., 2015).

13.7. DECIDING ON THE TYPES OF AWM INTERVENTIONS IN THE WATERSHEDS/CATCHMENTS

In the earlier section, we have seen what kind of areas would benefit from watershed development and water resources development and what kind of areas would require exogenous water for meeting irrigation demands of crops. Within the second category of areas, there could be many possibilities depending on the balance between renewable water availability and consumptive use of water. We anticipate five different scenarios in terms of hydrologic budget for the catchment, which takes into account water availability (inflows), stocks (static water resource), water supplies, current consumptive use (outflows), and potential water demand. These components will be largely decided by the hydrologic regime such as rainfall, aquifer conditions, streamflows, and occurrence of hydrologic extremes; socioeconomic conditions, namely, cropping pattern, irrigation pattern, and consumptive water use in agriculture and other sectors; climatic conditions; and overall water situation in the basin/catchment in terms of inflows and outflows. The selection of the types of agricultural water management interventions will be based on these considerations.

For instance, if the current inflows in the catchment (total precipitation, P) exceeds the total water supplies tapped naturally or artificially from various sources, namely, groundwater, surface runoff, and soil moisture to meet various competitive uses, and inflows also exceed the total consumptive water use ($ET_a + E$) yet there are unmet water demands in the catchment, then water resources development would help to meet these water demands.[6] To elaborate, if there is streamflow going out of the catchment, then runoff harvesting would

6. Such additional water demands can be for expanding the irrigated area, for intensifying cropping, or for bringing more area, which were earlier barren, under tree plantation or forest cover. In this case, the additional water is for meeting the evapotranspirative demand of the increased crop/vegetation cover.

help. But, if there is groundwater buildup in the catchment, which eventually would flow out as base flow and join the streams in the lower parts, then there will be scope for increased groundwater pumping to augment the supplies, which would take care of the unmet demands. This would reduce the downstream flows in the catchment (scenario 1).

If the current inflow in the catchment (P) is more or less equal to the total consumptive water demand at present or future ($P=ET+E$) and if all the inflow is tapped through various supply sources to meet the demands (with no water flowing out of the catchment), then the management intervention should aim at freeing some water for environmental flows downstream (scenario 2). In rural watersheds, agriculture will be the only sector that can free water for the environment, through water-use efficiency improvements, provided the area under irrigation does not expand. If we have to achieve irrigated area expansion, import of water from external sources would be required.

But, if the inflow from the catchment falls short of the total consumptive water demand ($ET+E>P$) and if all the inflows from various supply sources are tapped ($S=P$), with no water flowing out of the catchment ($ET_a+E=P$), then building water storage and impounding structures wouldn't help in improving the water balance in the catchment. It would only increase the water-spread area, thereby increasing the chances of evaporation of the stored water. The agricultural water management priority here should be to import water from external sources along with improving WUE in agriculture, so as to expand the area or meet the unmet demands from other sectors. The WUE improvement will be through reducing the catchment "outflows" (ET_a+E) in the form of beneficial consumptive uses—consumptive water uses in irrigated crops (T) and consumptive use of water for rainfed crops (T)—and nonbeneficial consumptive uses, the evaporation from water bodies and moist soils in the cultivated land.

There are also situations in which the total consumptive water use in the catchment (i.e., outflow) is higher than the inflows ($ET_a+E>P$) wherein the deficit is met through mining of groundwater with the result that the supplies equal the demand ($S=ET+E$) (scenario 4), a situation that is encountered in many watersheds falling in semiarid and arid regions of India. Here again, the abovementioned strategy to improve WUE can be adopted but might only help reduce the extent of groundwater mining. Large input of external water would be required here to stop groundwater mining and to expand the irrigated area.

Scenario 5 is one in which the total catchment yield exceeds the existing demands ($P>ET+E$) and would still have excess water, even after meeting any increase in demand, which is likely to occur in the future, resulting from expansion in cultivated land and increased water consumption in domestic, livestock, and other uses. Also, the available supplies from various sources are sufficient to meet the existing demands ($S=ET+E$). In such situations, the catchment is likely to remain "open." The "typical water budgets" for the four situations are presented in a matrix below (Table 13.1).

TABLE 13.1 Hydrologic Budget for Typical Watersheds and the Possible Agricultural Water Management Interventions

Inflow-Outflow	Catchment Water Supply-Demand Balance	Utilizable Supplies From Catchment Versus Inflows	Potential Agricultural Water Management Interventions
Outflow far less than catchment inflow ($ET_a+E<P$)	Catchment demand exceeding the utilizable supplies ($PET+E>S$)	Catchment inflow exceeding the supplies	WRD would augment the supplies to meet the deficits but would reduce the d/s flows WUE improvements without increase in area would free some water from agriculture thereby maintaining the d/s flows
Outflows equal to inflows ($ET_a+E=P$)	Catchment's utilizable supplies equal the demand ($S=PET+E$)	Entire inflow is tapped and no additional renewable water in the catchment	Reduction in nonbeneficial evaporation, reduction in beneficial ET, reduction in nonrecoverable DP in irrigation to free water from agriculture for d/s ecological uses
Outflows equal to inflows ($ET_a+E=P$)	Catchment's utilizable supplies less than the demand ($S<PET+E$)	Entire catchment inflow is harnessed; no additional renewable water to harness	External water input for irrigated area expansion Reduction in nonbeneficial evaporation, reduction in beneficial ET, reduction in nonrecoverable DP in irrigation to balance demand and supplies

(Continued)

TABLE 13.1 Hydrologic Budget for Typical Watersheds and the Possible Agricultural Water Management Interventions—cont'd

Inflow-Outflow	Catchment Water Supply-Demand Balance	Utilizable Supplies From Catchment Versus Inflows	Potential Agricultural Water Management Interventions
Outflows more than inflows ($ET_a + E > P$)	Catchment demand met from utilizable supplies ($PET + E = S$)	Entire catchment inflow is tapped; supply deficit is met through groundwater mining	External water input for irrigated area expansion Reduction in nonbeneficial E, reduction in beneficial ET, reduction in nonrecoverable DP to check mining of aquifers
Outflows far less than inflows ($ET_a + E < P$)	Entire catchment demand met from utilizable supplies ($PET + PE = S$)	Catchment inflow far exceeding utilizable supplies or water resource harnessed	Increase beneficial ET to meet future consumptive demands, increase soil infiltration and groundwater recharge through WHS, streamflow to downstream areas would reduce marginally

ET_a = actual evapotranspiration, E = evaporation (from water bodies and barren soil), PE = potential evapotranspiration, P = catchment inflow (precipitation), and S = utilizable water supplies.
Adapted from James, A.J., Kumar, M.D., Batchelor, J., Batchelor, C., Bassi, N., Choudhary, J., Gandhi, D., Syme, G., Milne, G., Kumar, P., 2015. Catchment Assessment and Planning for Watershed Management. Vol I. Main Report. PROFOR, World Bank Group, June 2015.

Now, a watershed can move from scenario 1 to scenario 2 if the consumptive water demand increases to touch the inflows and there is proportional increase in supplies by developing all water resources in the watershed to meet this demand. This means there is increase in either ET, E, or both. In this case, the outflows become equal to the inflows, with no water left in the catchment either in the form of renewable groundwater or runoff. If further increase in demand for water in the catchment is not met due to the absence of any stocks, then we would be in scenario 3 (like the situation in some watersheds of hard rock peninsular India), and if met through mining of groundwater, we would be in scenario 4.

The different types of interventions that can be considered for generating catchment management scenarios and the situations in which they can be considered are discussed below:

- *Water harvesting systems*: This can be introduced as an intervention in Case 1 and Case 5. The impact of water harvesting systems will be on the supply side of the water balance and therefore will have to be affected in terms of "change in storage." Since this essentially will have to come from the natural catchments, it would mean reduction in the flows downstream. The impact of building water harvesting systems in the catchment will also have to be affected through the introduction of new source of reservoir evaporation ("outflow") as well.
- *Micro irrigation and mulching for irrigated horticultural crops and row crops*: This can be introduced as an intervention in Cases 2, 3, and 4. The impact of micro irrigation and plastic mulching on water demand management will have to be affected in "catchment outflows" by reducing the consumptive use of irrigated crops. But, at the same time, in the case of MI, the deep percolation to shallow aquifer would be much less than that under traditional method of irrigation or nil.
The row crops for which drip system can be used are castor, cotton, fennel, groundnut, tomato, onion, maize, and fruit trees, plantation crops (coconut, etc.), and several of the vegetables for which the interplant spacing is more than one foot. Here, the total area under irrigated crops, which are amenable to MI system and mulching, needs to be assessed.
In order to assess the impact of water-saving technologies on overall water balance, it is important to know the physical impacts of these technologies in terms of real water saving, which comes from reduction in consumptive water use per unit of crop land. Such reduction can come from the following: (1) reduction in nonrecoverable deep percolation and (2) reduction in nonbeneficial evaporation from the soil covered by canopy and barren soil (see Allen et al., 1998; Kumar and van Dam, 2013).
- *Mulching for rainy-season crops:* This can be introduced as an intervention in Cases 2, 3, and 4. The impact of mulching on water demand of rainfed crops will be in terms of reduction in soil evaporation (E) values of the respective crops and increase in total water availability for meeting the transpirative demands (from the soil profile) (Xie et al., 2005). Under mulching, the water demand of the crop would be equal to transpiration instead of evapotranspiration. In terms of demand, it means that the supplementary irrigation can be reduced.
The row crops, which are amenable to mulching and which are grown during the rainy season in the catchment, need to be identified.
For crops, which are purely rainfed, we assume that a total of 50 mm of water could be saved through mulching. This value is quite reasonable. For instance, a study in northwest China on the effect of plastic mulching on

crop water use, yield, and soil evaporation for spring wheat (unirrigated) showed that soil evaporation is reduced by 0.49 mm per day, while the crop consumptive use (transpiration) is increased (by 49 mm) resulting in 82% higher yield and 51% higher water-use efficiency (Xie et al., 2005). In water balance models, this can appear as an increase in water supplies (green water) during normal years and reduction in pumping for the irrigated rainy-season crops.

- *Afforestation and tree and grass plantation*: This can be a catchment management intervention under Case 5, provided the runoff from the catchment carries excessive sediments. While grasses will have an effect on erosion control, it may not increase the ET "outflows" from the catchment and soil moisture storage as significantly as the deep-rooted trees, though this differential effect depends on the leaf area index and tree density (Oliveira et al., 2005). More importantly, unlike trees, grasses do not survive during the dry seasons in hot tropics, thereby bringing down the ET losses during the season to zero. During dry season, the ET losses through trees under natural conditions will be lesser than that of wet season, as in the former case, because soil drying leads to closing of stomata, thus reducing transpiration (source, www.forestry.gov.uk, Forestry Commission).
- The water for meeting ET demand of trees can come partly from precipitation "interception," partly from the moisture in the active root zone, partly from the unsaturated soil zone, and partly also from shallow groundwater. While its impact on overall yield of the catchment would be negative, depending on how the increased demand is being met from the hydrologic system, the impact will be seen on either runoff, groundwater, or both. If the vadose zone and topsoil contribute to evapotranspiration of trees, then the impact of afforestation will be on both groundwater system and runoff, whereas if shallow groundwater contributes to ET, then the most significant impact will be on base flows and groundwater. The higher the leaf area index, the higher will be the transpiration (Hamilton and King, 1983; Oliveira et al., 2005). On the other hand, litter cover on the forest floor increases infiltration rate significantly (Hamilton and King, 1983). Nevertheless, the large canopy cover will have some effect on the microclimate in terms of increasing the humidity and reducing temperature and incident solar radiation. While all these factors would reduce ET rates for the vegetation per unit area, reduced solar radiation affected by tree shade will adversely affect biomass outputs of standing crops (James et al., 2015).

13.8. CONCLUSIONS

As per the current official definition, areas with very low irrigation coverage and characterized by low to medium rainfall and high aridity currently fall in the "rainfed areas" category, along with areas with very high to excessively high rainfall. Agricultural water management interventions based on watershed treatment, which depend on local rainfall, are bound to fail in the former during droughts. Even the rainy-season crops actually require external water input for survival, and these small systems fail to provide that water. The reason is that during (meteorological) drought years, the amount of runoff generated from rainfall will be disproportionately lower than the reduction in rainfall, owing to faster moisture depletion from the soils as a result of high aridity and longer dry spells. The outflows (in the form of evapotranspiration, soil evaporation, and evaporation from water bodies) from such catchments are either higher than or equal to the inflows.

These areas require external water input in the form of irrigation for crops to survive, depending on the catchment water balance situation in terms of the difference between actual ET and catchment inflows. The effort of the planners should be to make those areas sufficiently irrigated so that they can reduce the vulnerability of their production systems to droughts. Along with irrigation, we need to focus on water-use efficiency improvement in irrigated crops through drips and mulching and rainfed crops through mulching. Simultaneously, research will be required on the development of crop varieties in the "low- to medium-rainfall" and "medium- to high-rainfall" areas with high aridity. But, the focus there should be on drought-resistant varieties as a short- to medium-term strategy, while the long-term strategy should be to develop sustainable irrigation.

In the medium- to high-rainfall regions, where the catchment inflows exceed the outflows yet there are unmet demands for water, water resources development including watershed development can be taken up to increase the water supplies and improve soil moisture regime to match the total water demand.

In the high-rainfall to excessively high-rainfall areas with low aridity, if topography is favorable, watershed treatment activities should be taken up with grass plantation and afforestation that increase evapotranspiration and reduce soil erosion. If the topography is not favorable, afforestation can be taken up in command land. In private lands, the focus should also be on developing crop varieties that can stand floods (in inland areas) and floods and soil salinity in the coastal plains. It will be meaningless to invest in expensive irrigation systems there, as they are bound to fail due to low irrigation demand.

To sum up, there are very few areas in India that offer favorable conditions for watershed development programs vis-à-vis the hydrometeorology, agroclimate, and topography. The program should be targeted at such areas, instead of implementing it lock, stock, and barrel in drought-prone areas and deserts in the garb of doing justice to ecologically deprived regions, with no positive effects. The latter types of areas require irrigation for enhancing agricultural productivity, and therefore, the investments required are much bigger than what a watershed development program would ideally attract.

REFERENCES

Allen, R.G., Pereira, L.S., Raes, D., Smith, M., 1998. Crop Evapotranspiration, Guidelines for Computing Crop Water Requirements. FAO Irrigation and Drainage Paper 56, Food and Agriculture Organization of the United Stations, Rome, Italy. 1999.

Arora, V.K., 2002. The use of the aridity index to assess the climate change effect on annual runoff. J. Hydrol. 265 (2002), 164–177.

Barron, J., Noel, S., 2011. Valuing soft components in agricultural water management interventions in meso scale watersheds: a review and synthesis. Water Altern. 4 (2), 145–154.

Batchelor, C., Rama Mohan Rao, M.S., Manohar Rao, S., 2003. Watershed Development: a solution to water shortages in semi-arid India or part of the problem? Land Use Water Resour. Res. 3, 1–10.

Buhl, S., Sen, R., 2006. A Preliminary Poverty Impact Assessment. Indo-German Development Cooperation, Natural Resource Management, GTZ, Delhi.

Calder, I., Gosain, A.K., Rama Mohan Rao, M.S., Batchelor, C., James, G., Bishop, E., 2008. Watershed development in India, 2. New approaches for managing externalities and meeting sustainability requirements. Environ. Dev. Sustain. 227–240. https://doi.org/10.1007/s10668-006-9073-0.

Calder, I.R., 2005. The Blue Revolution, Integrated Land and Water Resources Management. Earthscan Publications, London.

Droogers, P., Seckler, D., Makin, I., 2001. Estimating the Potential of Rain-Fed Agriculture. Working Paper 20, International Water Management Institute, Colombo, Sri Lanka.

Fan, S., Hazell, P., 2000. Should developing countries invest more in less-favoured areas? An empirical analysis of rural India. Econ. Polit. Wkly. 35 (17), 1455–1464.

Garg, K.K., Karlberg, L., Barron, J., Wani, S.P., Rockström, J., 2012. Assessing impacts of agricultural water interventions in the Kothapally watershed, Southern India. Hydrol. Process. 26 (3), 387–404.

Glendenning, C.J., Vervoort, R.W., 2011. Hydrological impacts of rainwater harvesting (RWH) in the case study catchment: The Arvari River, Rajasthan, India. Part 2: Catchment-scale impacts. Agric. Water Manag. 98, 715–730.

Gosain, A.K., Rao, S., Debajit, B., 2006. Climate change impact assessment on hydrology of Indian river basins. Curr. Sci. 90 (3), 346–353.

Government of India (GoI), 1999. Integrated Water Resources Development: A Plan for Action. Volume I. Report of the National Commission on Integrated Water Resources Development. Ministry of Water Resources, Govt. of India, New Delhi.

Gupta, S., 2011. Demystifying "tradition": the politics of rainwater harvesting in rural rajasthan. Water Altern. 4 (3), 347–364.

Hamilton, L.S., King, P.N., 1983. Tropical Forested Watersheds: Hydrologic and Soil Responses to Major Uses or Conversions. Westview Press, Boulder, Colorado.

Hope, R., 2007. Evaluating social impacts of watershed development in India. World Dev. 35 (8), 1436–1499.

Jagadeesan, S., Kumar, M.D., 2015. The Sardar Sarovar Project: Assessing Economic and Social Impacts. Sage Publications, New Delhi.

James, A.J., Kumar, M.D., Batchelor, J., Batchelor, C., Bassi, N., Choudhary, J., Gandhi, D., Syme, G., Milne, G., Kumar, P., 2015. Catchment Assessment and Planning for Watershed Management. Vol I. Main Report. PROFOR, World Bank Group. June 2015.

Joshi, P.K., Pangare, V., Shiferaw, B., Wani, S.P., Bouma, J., Scott, C.A., 2004. Socio-economic and Policy Research on Watershed Management in India: Synthesis of Past Experiences and Needs for Future Research. Global Theme on Agro-ecosystem, Report No. 7, International Crops Research Institute for Semi-Arid Tropics, Patancheru.

Joshi, P.K., Jha, A.K., Wani, S.P., Joshi, L., Shiyani, R.L., 2005. Meta-Analysis to Assess Impact of Watershed Program and People's Participation. Comprehensive Assessment Research Report 8, Comprehensive Assessment Secretariat, Colombo, Sri Lanka.

Kerr, J., 2002. Watershed development, environmental services, and poverty alleviation in India. World Dev. 30 (8), 1387–1400.

Kerr, J., 2003. Price policy, irreversible investment, and the scale of agricultural mechanization in Egypt. In: Lofgren, H. (Ed.), Research in Middle East Economics, Volume 5: Food, Agriculture and Economic Policy in the Middle East and North Africa. JAI Press, Oxford, UK, pp. 161–185.

Kerr, J., Pangare, G., Pangare, V., 2004. Watershed Development Projects in India: An Evaluation. Research Report 127, International Food Policy Research Institute, Washington, DC.

Kumar, M.D., 2007. Groundwater Management in India: Physical, Institutional and Policy Alternatives. Sage Publications, New Delhi.

Kumar, M.D., Ghosh, S., Patel, A.R., Singh, O.P., Ravindranath, R., 2006. Rainwater harvesting in India: some critical issues for basin planning and research. Land Use Water Resour. Res. 6 (1), 1–17.

Kumar, M.D., Patel, A.R., Ravindranath, R., Singh, O.P., 2008. Chasing a mirage: water harvesting and artificial recharge in naturally water-scarce regions. Econ. Polit. Wkly. 43 (35), 61–71.

Kumar, M.D., van Dam, J.C., 2013. Drivers of change in water productivity and its improvements at basin scale in developing economies. Water Int. 38 (3), 312–325.

Kumar, M.D., Reddy, V.R., Narayanamoorthy, A., Bassi, N., James, A.J., 2017. Rainfed areas: poor definition and flawed solutions. Int. J. Water Resour. Dev. https://doi.org/10.1080/07900627.2017.1278680.

Leclerc, M.-C., Grégoire, M., 2016. Implementing integrated watershed management in Quebec: examples from the Saint John river watershed organization. Int. J. Water Resour. Dev. https://doi.org/10.1080/07900627.2016.1251884.

Ministry of Agriculture and Irrigation (MoA & I), 1976. Report of the National Commission on Agriculture, Part V Resource Development. Ministry of Agriculture and Irrigation, Govt. of India, New Delhi.

National Rainfed Area Authority (NRAA), 2011. Challenges of Food Security and Its Management. National Rain-fed Area Authority, New Delhi.

National Rainfed Area Authority (NRRA), 2012. Prioritization of Rainfed Areas in India. Study Report 4, National Rainfed Area Authority, New Delhi. 100 pp.

Oliveira, R.S., Bezerra, L., Davidson, E.A., Pinto, F., Klink, C.A., Nepstad, D.C., Moreira, A., 2005. Deep root function in soil water dynamics in cerrado savannas of central Brazil. Func. Ecol. Br. Ecol. Soc. 19, 574–581.

Palanisami, K., Suresh Kumar, D., Chandrasekharan, B. (Eds.), 2002. Watershed Management: Issues and Policies for 21st Century. Associated Publishing Company, New Delhi. 341 pp.

Panwar, P., Pal, S., Bhatt, V.K., Tiwari, A.K., 2012. Impact of conservation measures on hydrological behaviour of small watersheds in the lower Shivaliks. Indian J. Soil Conserv. 40 (3), 257–262.

Parthasarathy Committee Report, 2006. From Hariyali to Neeranchal: Report of the Technical Committee on Watershed Programmes in India. Department of Land Resources, Ministry of Rural Development, Government of India, New Delhi.

Pathak, P., Sahrawat, K.L., Wani, S.P., Sachan, R.C., Sudi, R., 2009. Opportunities for water harvesting and supplemental irrigation for improving rainfed agriculture in semi-arid areas. In: Wani, S.P., Rockström, J., Oweis, T. (Eds.), Rainfed Agriculture: Unlocking the Potential. CABI Publishers, UK, pp. 197–221 (Comprehensive Assessment of Water Management in Agriculture Series-7).

Pathak, P., Kumar, A.C., Wani, S.P., Sudi, R., 2013. Multiple impact of integrated watershed management in low rainfall semi-arid region: a case study from eastern Rajasthan, India. J. Water Resour. Prot. 5, 27–36.

Planning Commission, 2005. Maharashtra State Development Report. State Plan Division, Planning Commission, Yojana Bhawan, Government of India, New Delhi.

Prinz, D., 2002. In: The role of water harvesting in alleviating water scarcity in arid areas. Keynote Lecture, Proceedings, International Conference on Water Resources Management in Arid Regions, 23–27 March, 2002, Kuwait Institute for Scientific Research, Kuwait. vol. III, pp. 107–122.

Rama Mohan Rao, M.S., Batchelor, C.H., James, A.J., Nagaraja, R., Seeley, J., Butterworth, J.A., 2003. Andhra Pradesh Rural Livelihoods Programme Water Audit Report. APRLP, Rajendranagar, Hyderabad, India.

Ray, S., Bijarnia, M., 2006. Upstream vs downstream: groundwater management and rainwater harvesting. Econ. Polit. Wkly. 41 (23), 2375–2383.

Reddy, V.R., 2006. Getting the implementation right: can the proposed watershed guidelines help? Econ. Polit. Wkly. XLI (40), 7–13.

Reddy, V.R., Chiranjeevi, T., 2016. The changing face of rainfed agriculture: need for integration of science, policy and institutions. In: Ramasamy, C., Ashok, K.R. (Eds.), Vicissitudes of Agriculture in the Fast Growing Indian Economy: Challenges, Strategies and the Way Forward. Academic Foundation, New Delhi, pp. 281–312.

Reddy, V.R., Syme, G.J., 2014. Integrated Assessment of Scale Impacts of Watershed Interventions: Assessing Hydro-geological and Bio-physical Influences on Livelihoods (edited). Elsevier Inc.

Shah, M., 2013. Water: towards a paradigm shift in the twelfth plan. Econ. Polit. Wkly. 48 (3), 40–52.

Shankar, P.S.V., 2011. Towards a paradigm shift in India's rainfed agriculture. Innov. Dev. 1 (2), 321–322.

Sharda, V.N., 2004. In: Water resource development through conservation measures in micro-watersheds in hills. Presented at the National Symposium on Enhancing Productivity and Sustainability in Hill and Mountain Agro-ecosystem, August 10–11, 2004, CSWCRTI, Dehradun, India.

Sharda, V.N., 2005. Integrated watershed management: managing valleys and hills in the Himalayas. In: Sharma, B.R., Samra, J.S., Scott, C.A., Wani, S.P. (Eds.), Watershed Management Challenges: Improving Productivity, Resources and Livelihoods. IWMI, ICAR and ICRISAT, Colombo, Sri Lanka.

Sharma, B.R., 2012. Unlocking Value out of India's Rainfed Farming Areas. Water Policy Research Highlight-10, IWMI-Tata Water Policy Program, Anand, Gujarat, India.

Sharma, B.R., Scott, C.A., 2005. Watershed management challenges: introduction and overview. In: Sharma, B.R., Samra, J.S., Scott, C.A., Wani, S.P. (Eds.), Watershed Management Challenges: Improving Productivity, Resources and Livelihoods. IWMI, ICAR and ICRISAT, Colombo, Sri Lanka, pp. 1–21.

Stewart, J., Bennett, M., 2016. Integrated watershed management in the Bow River basin, Alberta: experiences, challenges, and lessons learned. Int. J. Water Resour. Dev. https://doi.org/10.1080/07900627.2016.1238345.

Syme, G., Reddy, V.R., Pavelik, P., Croke, B., Ranjan, R., 2012. Confronting scale in watershed development in India. Hydrogeol. J. 20 (5), 985–993.

Talati, J., Kumar, M.D., Ravindranath, R., 2005. Local and Sub-Basin Level Impacts of Local Watershed Development Projects: Hydrological and Socio-economic Analysis of Two Sub-Basins of Narmada. Water Policy Research Highlight 15, IWMI-Tata Water Policy Research Program, Anand, Gujarat, India.

Vaidya, R.A., 2016. Governance and management of local water storage in the Hindu Kush Himalayas. Int. J. Water Resour. Dev. https://doi.org/10.1080/07900627.2015.1020998.

Wani, S.P., Rockström, J., Oweis, T., 2009. Rainfed Farming: Unlocking the Potential. Comprehensive Assessment of Water Management in Agriculture Series. CABI Publishing, Wallingford, England.

World Bank, 2004. Managing Watershed Externalities in India. The World Bank, Washington, DC.

Xie, Z., Wang, Y., Li, F., 2005. Effect of plastic mulching on soil water use and spring wheat yield in arid region of northwest China. Agric. Water Manag. 75 (2005), 71–83.

XII Plan Working Group, 2011. Natural Resources Management and Rainfed Farming. Report of the XII Plan Working Group.

Chapter 14

Impacts of Microirrigation Systems: Perception Versus Reality

14.1. INTRODUCTION

Over the years, the perceptions about the benefits of microirrigation systems among the scientific community had changed. It is now being viewed more as a technology to improve crop yields, save labor, and improve the efficiency of use of the water applied in the field and less as a water-saving technology.[1] India has the largest drip-irrigated area in the world (nearly 2 M ha), though it is only less than 4% of the country's net irrigated area (Kumar, 2016). Surely, those who raise questions about the water-saving benefits of microirrigation systems would surely receive attention. The reason is that in spite of the technology being there for almost two and a half decades in India, one question remains unanswered: "whether drip/sprinkler irrigation really saves water and if so how much."

While the above question is very broad and can be called "ambiguous," to make it more clear and specific, one can even ask: in this particular area, if we use drip irrigation for crop "A," how much would be the water saving? The reason for the ambiguity is that there is general awareness that water-saving benefits of drip irrigation are also dependent on crop type (Kumar, 2016). Yet, there are issues that need to be sorted out such as water saving at what level (i.e., whether field level or farm level or system level). This is because there are different notions of water saving (Seckler, 1996; Perry, 2007). Unfortunately, even the best water scientist or engineer won't be able to answer this question in one statement.

In the past, agricultural scientists and irrigation engineers used estimates based on the notions of engineering efficiency in irrigation, which used the amount of water consumed by the crop against the amount of water released

1. A few years ago, one of the important recommendations of water management conferences in India eventually was "there is a need to promote adoption of water saving devices in irrigation such as drips and sprinklers through mass awareness." In those years, views such as this, "drips and sprinklers will not save much water, but may only help the farmers raise yields," would be confronted with a lot of reactions. But, times have changed.

Water Policy Science and Politics. https://doi.org/10.1016/B978-0-12-814903-4.00014-2

from the reservoir/irrigation source. Therefore, the solutions for improving irrigation efficiency were quite straightforward—lining canals used for water conveyance and improving field efficiencies in irrigation. Over the years, advances in hydrology and crop sciences have only made water resources management far more challenging. Today, *water-saving impact of drip irrigation* is one of the most controversial and widely debated topics in the water (research) world, because of the ambiguity involved (Perry and Steduto, 2017).

These issues notwithstanding, the policy makers in the irrigation, agriculture, and energy sectors in India are often overenthusiastic about the resource conservation impacts of water-saving technologies and therefore advocate heavy incentives for farmers adopting microirrigation systems for irrigating crops. This chapter discusses the complex concept of "water saving" through microirrigation systems based on the different notions of "water saving" in agriculture, reviews the relate concept of "rebound effect" relevant to microirrigation system adoption in agriculture to analyze its impacts at the system and basin level, examines the limited global evidence on the water-saving impacts of drip systems to identify the determinants of change in actual water use post adoption, and analyzes how the "rebound effect" gets played out in different situations in India.

14.2. THE COMPLEX CONCEPT OF WATER SAVING THROUGH MICROIRRIGATION

Two concepts have made the analysis of water saving from water-efficient irrigation technologies such as drip and sprinkler complex. The first one concerns "scale effect," which suggests that water saving through the use of efficient irrigation technologies is scale-dependent—at what scale it is measured (whether plot level or farm level or irrigation system level or catchment level). To put it, in practical terms, the argument is that the extent of water saving achieved through the use of efficient irrigation devices would reduce as one moves from the field (where the system is installed) to the farm to the watershed or the catchment (Allen et al., 1998; Seckler, 1996; Perry, 2007).

The underlying premise is that the water, which is lost under traditional method of irrigation from the cropped field (plot) through field runoff due to excessive irrigation, is available to the nearest plot in the farm and the water that is lost in deep percolation is available as recharge to the shallow aquifer. The water that is lost from the farm through runoff is picked up from the drainage line by farmers using pumps, and the water that is lost in percolation is again available as recharge in the shallow groundwater, which in turn is pumped out by well irrigators. The bedrock of the concept of "scale effect" is "irrigation return flow."

Unfortunately, though this concept is widely used internationally to challenge those who promote drip irrigation as a panacea for water problems, there are no easy ways to actually compute the extent to which the deep percolation contribute to the "return flows" (aquifer recharge), for practical problem solving (Kumar and van Dam, 2013; Perry and Steduto, 2017). The fact that technologies such as drips are widely used in dry regions, where water table is generally deep, it becomes all the more important to find out how much of the percolating water reaches the water table (so as to be available for reuse).

The effect of geohydrology on return flows from irrigated fields had not been explored. The moot point is not all the percolating water could end up in the aquifers as return flows in such dry regions (as the unsaturated strata cannot act like a simple conduit for carrying water from the top soil to the saturated strata). Some of the water would eventually get evaporated from the top soil strata (1–3 m), some water would be picked up by the deep-rooted trees, and some water would remain in the unsaturated zone as hygroscopic water at high surface tension and will not be available for extraction by pumps or plants. If drip systems are adopted for crops in such areas, there could be some possible real water saving, due to the reduction in deep percolation (Kumar et al., 2008; Kumar and van Dam, 2013).

Naturally, if the water table is very shallow (say in the range of 10–30 ft), almost the entire water from deep percolation could end up in the shallow aquifer depending on the climatic condition of the area. This water will be available for reuse. Hence, there would only be "notional saving" of water (at the societal level) through adoption of drip technology, though the individual farmer would gain. It will not save much water in real terms, as there is no "loss of water" through percolation in the former case. This will ideally be the case in almost the entire Gangetic Plains that have shallow groundwater table.

The water balance, however, would altogether change if the area we are dealing with has saline aquifers (like in certain parts of the Indira Gandhi Nahar project command in western Rajasthan), as in such a case, the water that percolates deep will have to be treated as "water lost" as it will not be available for reuse. In nutshell, things are not so black and white when it comes to water saving through reduction in deep percolation.

Another source of water saving through drip technology is from the reduction in nonbeneficial consumptive use (evaporation from the soil profile, transpiration of weeds, etc.). There is no dispute on this among scientists. For many distantly spaced crops (orchards and row crops such as cotton, castor, fennel, and watermelon), there could be evaporation from the soil that is not covered by canopy, under flood (or small border) method of irrigation (Kumar and van Dam, 2013). The evaporation from unvegetated soil surface would increase when the area moves from humid and subhumid to arid and hyperarid (Allen et al., 1998). Even for field crops, soil evaporation would be significant in the initial stages of crop growth, as canopy cover would only be for a small fraction of the cropped area. While plant transpiration would be small, a major portion of the consumptive use will be from soil evaporation, which is nonbeneficial. Drip irrigation can ensure that water is applied only to the soil closer to the roots of the plants. So, the entire cropped area will not be wetted if farmers practice proper irrigation scheduling. However, the same argument won't hold water if the technology is sprinkler, as the entire cropped area is wetted.

Therefore, how much real water saving can be obtained per unit of land also depends on the crop type, technology (drip or sprinkler), and climate. What we also need to reckon with is the fact that the areas we deal with (for introducing microirrigation technologies) mostly are semiarid to arid and even hyperarid and not cold and humid areas. More importantly, in many such areas, the groundwater table is very deep due to overdraft and falls continuously. For instance, in semiarid to arid North Gujarat, an area that is known for groundwater intensive use for irrigation and that has also witnessed large-scale adoption of microirrigation systems (drips and sprinklers), groundwater level is in more than 40 m belowground (source, Government of India (GoI), 2012, pp. 9–12). In such areas, return flows to the pumped aquifer will be practically absent.

The drip irrigation baiters, however, continue their opposition saying that there will be no change in evapotranspiration (ET) achieved through the use of drip irrigation and therefore no water saving in real terms. They treat every crop like wheat, paddy, and alfalfa, with 95% full canopy cover, and bundle "transpiration" (T), evapotranspiration (ET), and consumptive use (CU), under "ET" to win arguments, while knowing fully well that they are different. It is quite likely that for the same level of ET, drip adopters get more yield as a greater portion of the ET gets used up as "transpiration," with reduction in soil evaporation (E). Also, it is possible that with the same level of ET, there is more CU under traditional method, as a result of some water being lost in nonrecoverable deep percolation. Under both the scenarios, the water productivity in relation to the amount of water consumed (depleted) in crop production will be higher as compared with conventional method of irrigation.

Obviously, as is evident from the foregoing discussions, the details are important. Unfortunately, not many researchers (neither the ardent proponents of drip technology nor the "drip baiters") want to get into the details and understand the context and instead are keen to generalize and reach their own conclusions. The outcome is that the conclusions with respect to the water-saving impact of efficient irrigation technologies are diametrically opposite. This was illustrated by a series of articles that appeared in Water International (first by Frederiksen and Allen, 2011 in *Water International*, 36 (3)), a response by Gleick et al. (2011) and separate responses to the same by Frederiksen et al. (2012) on the water-saving impacts of drip irrigation, with diametrically opposite views.

There is a great merit in some raising caution about overestimating the benefits of drip irrigation that does not take cognizance of the local context (Kumar and van Dam, 2013). The "return flows" from flood irrigation (wherever happens) from gravity canals at times produce much greater benefits (sustaining well irrigation, raising water table, and improving the quality of marginal quality groundwater) than what the same water would have generated if supplied through the canals to larger area. All these would cease to exist when the water is supplied through pipes and drips (Chakravorty and Umetsu, 2003). But, these comparisons are relevant in the case of traditional irrigation from canals and not wells. It is inappropriate to treat well irrigation in the same manner as gravity irrigation from canals (which take water from outside) while quantifying such benefits. In many instances, the cost of pumping extra water that is used to over-irrigate fields outweighs the return flow benefits. So, the argument about "return flow" benefits becomes convoluted when applied to well irrigation, while they are quite valid for gravity irrigation.

But there are limits to water saving and water productivity improvements through water-saving technologies like drips and sprinklers, even when the context is favorable for water saving. Expósito and Berbel (2017) analyzed the impact of irrigation water use in a closed river basin of Guadalquivir (in the Iberian Peninsula of Spain) on water productivity and examined how they have affected the "closure" process of the river basin. Following a period of expansion in irrigation, an administrative moratorium was declared on new irrigation water rights, with the right per hectare of irrigation coming down from around 9800 to 3400 m^3 from 1987 to 2012. Their analysis showed that a significant increase in mean irrigation water productivity was achieved in the premoratorium period (1989–2005), through a dominant use of drip and sprinkler irrigation technology and creation of new irrigated areas using the saved water devoted to high-value crops. The second phase (2005–12) was characterized by slower growth in the mean productivity of irrigation water, in spite of area under drip irrigation jumping from 45% to 66% of the irrigated area during 2004 and 2012. This was primarily the result of a significant reduction in water use per area (deficit irrigation). Findings show that water productivity gains reaching a plateau in the basin, since technological innovations (such as new crops, deficit irrigation, and water-saving and conservation technologies), have reached the limits of their capacity to save water and create new value.

14.3. THE REBOUND EFFECT

The second concept, which has become dominant in the debate on the water-saving benefits of microirrigation, is "rebound effect." The extent of water saving at system and basin level because of the widespread adoption of efficient irrigation systems is debated (Sanchis-Ibor et al., 2015; Ward and Pulido-Velazquez, 2008). This debate revolves around the question of "what farmers do with the saved water." Many believe that the aggregate impact of drips on water use would be similar to the impact on water use per unit area of land. Several others believe that with reduction in water applied per unit area of land, the farmers would divert the saved water for expanding the area, subject to favorable conditions with respect to water and equipment availability, and power supplies for pumping water, and under such circumstances, the net effect of adoption on water use could be insignificant at the system level (Kumar et al., 2008).

Others argue that with adoption of WSTs, there is a greater threat of resource depletion (Molle et al., 2004; Perry, 2007; Ward and Pulido-Velazquez, 2008), as in the long run, either the return flows from irrigated fields would decline or area under irrigation would increase, if conditions are favorable for cropped area expansion (Molle et al., 2004; Perry, 2007) or crop consumptive use (ET) would increase with drip-irrigated area replacing area under traditional method of irrigation (Ward and Pulido-Velazquez, 2008). However, both views fail to make a distinction between consumptive water use in crop production (depletion) and crop consumptive use (ET) (see Allen et al., 1998 for definitions) and how they could be vastly different in different contexts.

Nevertheless, some do argue that with the adoption of efficient irrigation technologies, the aggregate water use may decrease or increase depending on a variety of conditions such as agricultural system characteristics, aquifer characteristics, previous conditions of irrigation infrastructure, and induced changes in crop rotations (Berbel et al., 2015; Sanchis-Ibor et al., 2015).

While this is the case in many semiarid and arid regions, there are quite a few situations where the entire land is irrigated, but the resource (groundwater) is fast depleting (central Punjab and alluvial North Gujarat in India and Indus basin irrigation system in Pakistan). If it is so, such theorizing doesn't lead us anywhere.

Internationally, now, there are several studies that deal with this vexed issue, that is, what happens after large-scale adoption of microirrigation in a region in terms of water withdrawal and use (while some of it is actual empirical study, some are based on modeling, with the latter involving too many simplistic assumptions)? Such arguments of area expansion have merit when water is scarce and land is available in plenty, and there are no other constraints facing the farmers. In fact, a modeling study in Mijares basin in Valencia (Spain) by International Commission on Irrigation and Drainage (ICID) showed that in the region in question, the contextual factors did not allow expansion in area under irrigation or crop intensification, and as a result, large-scale adoption of drips actually led to significant reduction in consumptive water use (Sanchis-Ibor et al., 2015).

In another study, Ward and Pulido-Velazquez (2008) modeled the basin level impact of adoption of drip irrigation on actual water use in agriculture (in a basin in North America), and their modeling results showed that increase in area under drips (which replaced traditional method of irrigation) resulted in greater depletion of water in the basin. This happened because the modelers assumed that ET for drip-irrigated crop is higher than that of flood-irrigated crop. Their only logic was that the crop yield for drip-irrigated crop is higher than that of the latter, and therefore, ET has to be higher for the former. The modeling ignored the fact that under traditional irrigation practice, there could be consumptive uses that do not result in biomass production. Such simplistic assumptions in scientific studies, which make far-fetched conclusions, do not help anyone.

14.4. LOOKING BEYOND TECHNOLOGY

An important concern with drip irrigation is that many times, backed by poor knowledge of irrigation scheduling under the new technology, farmers end up overirrigating their fields using drips, as they get both free water and electricity (Palanisami et al., 2011). So, no social benefits are realized, against the huge public investments in the form of subsidies. This is really worrisome for the government and the society at large. There is growing evidence on this from some parts of India. A study carried out in Aurangabad District of Maharashtra involving comparative analysis of water use efficiency of irrigated sugarcane under drip system and under straight furrows showed no difference in the average values of water use efficiency (kilogram per cubic meter of water) between the two methods (Kumar et al., 2012). As pointed out by the study, one reason for this is that the farmers are not saving much cost by optimizing water application to the crops, in lieu of the zero marginal cost of using water and electricity, and many of the drip adopters were found to be applying water in the excessive irrigation regime that resulted in reduced yield.

Another important point that is missed out in the analysis is that often adoption of drip irrigation systems is associated with crop shift. With drip irrigation, farmers also go for high-value fruits, vegetables, and flowers. The reason is that the capital cost of installing drip systems cannot otherwise be offset by the incremental returns from the low-value crops they cultivate under traditional method owing to yield improvements (Schoengold and Zilberman, 2005). These crops are distinctly different from the traditional crops such as wheat, paddy, pearl millet, mustard, and sorghum in terms of plant architecture and the way they are cultivated in the field. For most fruit crops, the distance between plants is very large—ranging from 1.0 to 10.0 m (for mangoes). For many vegetables, the spacing between plants is 30–40 cm. The ET requirement of these crops will be different from that of traditional crops and would be further less, if irrigated with drips, as soil evaporation could be suppressed.

14.5. HOW FAR ARE THESE STUDIES RELEVANT FOR INDIAN CONDITIONS?

The arguments built on studies in other parts of the world have, however, missed certain critical variables determining area under irrigation and aggregate water use after drip system adoption. They are groundwater availability vis-à-vis power supply availability, crops chosen, and amount of land and finances available for intensifying cultivation. The most important of these factors is the overall availability of groundwater in an area and the power supply vis-à-vis water availability in the wells.

If power supply restrictions limit groundwater pumping, then farmers are unlikely to expand the irrigated area after adopting water-saving technologies. In many states, power supply to agriculture sector is only for limited hours, limiting the farmers' ability to expand irrigation in those places where water availability in the aquifer and demand for water is more than what the limited hours of power supply can pump. Such areas include the alluvial belts of Punjab, western Rajasthan, North Gujarat, and deltaic region of coastal Andhra Pradesh.

Since the available power supply is fully utilized during winter and summer seasons, farmers will be able to irrigate only the existing command even after adopting MI system. This is because the well discharge would drop when the microirrigation systems start running. The only way to overcome this is to install a booster pump. As electricity charges are based on connected load in most states, farmers have least incentive to do this. In the alluvial areas of North Gujarat, Punjab, and coastal deltaic region of AP, farmers will not be able to expand the area under irrigated crops. Hence, there will be some water saving due to drip adoption at the basin level.

On the other hand, if the availability of water in wells is less than what the available power supply can extract, the farmers are tempted to adopt MI systems. This is the situation in most of the hard-rock areas of peninsular India, central India, and Saurashtra region of Gujarat as the saved water could be used to expand the irrigated area. In such situation, there will be low aggregate reduction in the amount of water pumped from the aquifer.

In Michael region of central India, farmers use low-cost drips to give presowing irrigations to cotton, before monsoon, when there is extreme scarcity of groundwater. This helps them grow cotton in larger area as water availability improves after the monsoon (Verma et al., 2004).

The other factor is the availability of extra arable land for cultivation. In areas where land-use and irrigation intensity is already high—examples are central Punjab and coastal deltaic region of Andhra Pradesh—farmers might still adopt water-saving technologies for cash crops to raise yields or for newly introduced high-valued crops to increase their profitability. In such situations, adoption would result in reduced aggregate water demand. However, the fact that needs to be reckoned with is that in such situations, the economic incentive to adopt microirrigation systems is quite poor (Kumar et al., 2008).

The third factor is the crops chosen. Often, MI technologies follow a set cropping pattern. All the pockets in the country where adoption of drip irrigation systems has undergone a "scale," orchard crops are the most preferred crops (Narayanamoorthy, 2004). Even in the humid tropics of Kerala, farmers use drip irrigation for irrigating crops such as coconut, arecanut, banana, and nutmeg (Chandran and Surendran, 2016).

Farmers bring about significant changes in the cropping systems with the adoption of drips. With the adoption of orchards for drips, farmers are found to permanently abandoning cereal crops (Singh and Kumar, 2012). They generally start with small areas under orchards and install drips and after recovering the initial costs tend to bring the entire land under orchards. This is because orchards require special care and attention, and putting the entire land under orchards makes farm management decisions easier.

The irrigation water requirement of the cropping system consisting of field crops such as paddy, wheat, and pearl millet/sorghum combinations is much higher than that of fruit crops such as pomegranate, gooseberry, sapota, and lemon. Also for water-intensive crops such as mango, the irrigation water requirements during the early stages of growth would be much less than that of these field crops owing to wide plant spacing. Therefore, even with expansion in cropped area, the aggregate water use would drop, at least in the initial years. A research study carried out on the physical impacts of MI technologies in North Gujarat found that with MI adoption, several of the traditional cereal crops were replaced by cash crops amenable to MI systems. While the irrigation water use rates for individual crops reduced, the aggregate cropped area also reduced. As a result of these changes, groundwater use for irrigation reduced significantly at the farm level by nearly 7527 m^3 per farm (Singh and Kumar, 2012).

14.6. RESEARCH ON WATER-SAVING IMPACTS OF MICROIRRIGATION IN INDIA

The research on water-saving impacts of microirrigation in India is nearly two-and-a-half decade old. So far, they had primarily focused on field/plot level impacts of the system for individual crops, based on the data on the volume of water applied to the crops. The difference in water applied between the two methods (i.e., microirrigation and traditional method of irrigation) has been treated as reduction in water use due to microirrigation (see Narayanamoorthy, 2004, 2010; Palanisami et al., 2011). The actual water consumed in crop production (consumptive use, which is the sum of evapotranspiration, nonbeneficial evaporation, and nonrecoverable deep percolation) under traditional method of irrigation against the same for microirrigation has not been attempted empirically by any study, though some researchers have distinguished the fundamental differences between the two while also giving proper reasons for treating both equally in their respective local-specific situations (see Singh and Kumar, 2012; Kumar et al., 2012). As a result, many of research studies looked at notional water saving, and as pointed out by Dhawan (2000) in the context of drips, this has resulted in overestimation of social benefits of using such systems or underestimation of the social costs. The arguments by Prof BD Dhawan made in 2000 still hold good for most of the studies happening in this topic.

At the next level, investigations into the overall change in water use at the farm level due to changes in cropping pattern and area expansion owing to the adoption of microirrigation systems have not been attempted by most of the researchers who worked on the topic, with the exception of Singh and Kumar (2012) that looked at the impact at the farm level. Because of these two shortcomings, intuitively, the real water-saving impacts of microirrigation systems have been overplayed by most researchers and therefore the social benefits. As a result, heavy capital subsidy for microirrigation systems is being provided by the state and central governments. From a public policy perspective, what is even worse is that little attention is being paid to the fact that in the absence of proper knowledge of irrigation scheduling, farmers might end up overirrigating the crops even after adopting MI systems with the result that no significant benefits would be accrued from MI adoption.

14.7. CONCLUDING REMARKS

The analysis and illustrative examples presented in this chapter reinforced the fact that water-saving impact of drip irrigation is highly location-dependent, an argument earlier made by Kumar et al. (2008) and Kumar and van Dam (2013). The ongoing polemics about the impact of drip irrigation can be settled if we invest in high-quality empirical research involving field measurements of water consumption and compare the same with (hydrobiophysical economic) modeling results in different climatic and geohydrologic environments and cropping conditions to see whether such modeling results are reliable and where there is a need for making these models robust. In any case, the governments need to exercise caution while launching large-scale projects to promote microirrigation, involving big subsidy support to farmers. Besides settling the issue of whether there is real water saving through the use of MI, the local-specific conditions and their implications for water saving and "rebound effect" need to be fully understood. Mechanisms need to be in place to reward farmers who actually reduce water consumption after microirrigation system adoption.

There are also limits to water-saving and water productivity improvements through sprinklers and drips, as illustrated by the example of Guadalquivir river basin in Spain. We also need to look beyond technologies for water saving. Irrigation scheduling is very important to achieve the intended benefits of microirrigation adoption, as farmers can continue to overirrigate their fields even post MI system adoption. Another point that is missed in the discussion on water saving is that a lot of the reduction in water use might be achieved through crop shift, which technology adoption is often associated with, resulting from change in consumptive use of water by the crop.

But policy makers are often overenthusiastic about the water-saving benefits of microirrigation, without having much support of field evidence to that effect. The knowledge gap that exists with regard to water-saving impact of microirrigation systems is clearly evident. Part of the reason is that the advanced concepts in agricultural water management (beneficial consumptive use; nonbeneficial consumptive use; recoverable nonconsumptive use (recharge); and nonrecoverable, nonconsumptive use) are yet to find even acceptance among Indian academia to an extent that matter. Many of the research outputs produced by agricultural universities on the spread, impacts, and economics of microirrigation systems in India even in the recent past are merely "run-of-the-mill" productions that fail to use any new concepts in agricultural water management and drip irrigation economics, in spite of having got the opportunity to cover nine different states representing many agroclimatic and other physical conditions (see, for instance, Palanisami et al., 2011).

In the recent past, large-scale drip irrigation projects were implemented in some areas in India in the farmers' fields, with government funding. The most recent one is from Karnataka, wherein the project is expected to cover an area of 450,000 acre of irrigated sugarcane in the state, with a government assistance of Rs. 10,000 per acre. The impacts of these projects on consumptive water use in irrigation and real water saving need to be analyzed through proper water accounting studies to assess the social benefits from such interventions. Needless to say, the heavy government subsidies for capital equipments can be justified only if there is significant social benefit accrued through real water saving.

REFERENCES

Allen, R.G., Willardson, L.S., Frederiksen, H., 1998. In: de Jager, J.M., Vermes, L.P., Rageb, R. (Eds.), Water use definitions and their use for assessing the impacts of water conservation. Proceedings ICID Workshop on Sustainable Irrigation in Areas of Water Scarcity and Drought, Oxford, England, September 11–12, pp. 72–82.

Berbel, J., Gutiérrez-Martín, C., Rodríguez-Díaz, J.A., Camacho, E., Montesinos, P., 2015. Literature review on rebound effect of water saving measures and analysis of a Spanish case study. Water Resour. Manag. 29, 663–678.

Chakravorty, U., Umetsu, C., 2003. Basinwide water management: a spatial model. J. Environ. Econ. Manag. 45 (2003), 1–23.

Chandran, K.M., Surendran, U., 2016. Study on factors influencing the adoption of drip irrigation by farmers in humid tropical Kerala, India. Int. J. Plant Prot. 10 (3), 347–364.

Dhawan, B.D., 2000. Drip irrigation: evaluating returns. Econ. Polit. Wkly., 3775–3780. October 14.

Expósito, A., Berbel, J., 2017. Agricultural irrigation water use in a closed basin and the impacts on water productivity: the case of the Guadalquivir River Basin (Southern Spain). Water 9, 136. https://doi.org/10.3390/w9020136. www.mdpi.

Frederiksen, H.D., Allen, R.G., 2011. A common basis for analysis, evaluation and comparison of off stream water uses. Water Int. 36, 266–282. https://doi.org/10.1080/02508060.2011.580449.

Frederiksen, H.D., Allen, R.G., Burt, C.M., Perry, C., 2012. Responses to Gleick et al. (2011), which was itself a response to Frederiksen and Allen (2011) (Correspondence). Water Int. 37, 183–197. https://doi.org/10.1080/02508060.2012.666410.

Gleick, P.H., Christian-Smith, J., Cooley, H., 2011. Water use efficiency and productivity rethinking the basin approach. Water Int. 36, 784–798. https://doi.org/10.1080/02508060.2011.631873.

Government of India (GoI), 2012. Groundwater Year Book-India 2011–12. Central Ground Water Board, Ministry of Water Resources, Govt. of India.

Kumar, M.D., 2016. Water saving and yield enhancing micro irrigation technologies: theory and practice. In: Viswanathan, P.K., Kumar, M.D., Narayanamoorthy, A. (Eds.), Micro Irrigation Systems in India: Emergence, Status and Impacts. Springer:Singapore, pp. 13–26.

Kumar, M.D., Niranjan, V., Puri, S., Bassi, N., 2012. Irrigation Efficiencies and Water Productivity in Sugarcane in Godavari River Basin, Maharashtra (Report Submitted to the World Wild Fund for Nature). Institute for Resource Analysis and Policy, Hyderabad.

Kumar, M.D., Turral, H., Sharma, B.R., Amarasinghe, U., Singh, O.P., 2008. In: Kumar, M.D. (Ed.), Water saving and yield enhancing micro irrigation technologies in India: when do they become best bet technologies? Managing Water in the Face of Growing Scarcity, Inequity and Declining Returns: Exploring Fresh Approaches, Vol. 1, Proceedings of the 7th Annual Partners' Meet of IWMI-Tata Water Policy Research Program, ICRISAT, Hyderabad, pp. 13–36.

Kumar, M.D., van Dam, J., 2013. Drivers of change in agricultural water productivity and its improvement at basin scale in developing economies. Water Int. 38 (3), 312–325.

Molle, F., Mamanpoush, A., Miranzadeh, M., 2004. Robbing Yadullah's Water to Irrigate Saeid's Garden: Hydrology and Water Rights in a Village of Central Iran. Research Report 80, International Water Management Institute, Colombo, Sri Lanka.

Narayanamoorthy, A., 2004. Drip irrigation in India: can it solve water scarcity. Water Policy 6 (2), 114–130.

Narayanamoorthy, A., 2010. Can drip method of irrigation be used to achieve the macro objectives of conservation agriculture? Indian J. Agric. Econ. 65 (3), 428–438.

Palanisami, K., Mohan, K., Kakumanu, K.R., Raman, S., 2011. Spread and economics of micro-irrigation in India: evidence from nine states, review of agriculture. Econ. Polit. Wkly. XLVI (26 & 27), 81–86.

Perry, C.J., 2007. Efficient irrigation; inefficient communication; flawed recommendations. Irrig. Drain. 56, 367–378.

Perry, C.J., Steduto, P., 2017. Does Improved Irrigation Technology Save Water? A Review of Field Evidence. Discussion Paper on Irrigation and Sustainable Water Resources Management in the Near East and North Africa. Food and Agriculture Organization of the United Nations, Cairo, Egypt.

Sanchis-Ibor, C., Macian-Sorribes, H., García-Mollá, M., Pulido-Velazquez, M., 2015. In: Effects of drip irrigation on water consumption at basin scale (Mijares River, spain). 26th Euro-Mediterranean Regional Conference and Workshops, Innovate to Improve Irrigation Performances, 12–15 October 2015, Montpellier, France.

Schoengold, K., Zilberman, D., 2005. The Economics of Water, Irrigation and Development. February 2005, Available from, http://www.ctec.ufal.br/professor/vap/Handbook.pdf.

Seckler, D., 1996. The New Era of Water Resources Management: From "Dry" to "Wet" Water Savings. Issues in Agriculture No. 8, Consultative Group on International Agricultural Research. April 1996.

Singh, O.P., Kumar, M.D., 2012. Hydrological and farming system impacts of agricultural water management interventions in north Gujarat, India. In: Kumar, M.D., Sivamohan, M.V.K., Bassi, N. (Eds.), Water Management, Food Security and Sustainable Agriculture in Developing Economies. Routledge/Earthscan, London, pp. 116–137.

Verma, S., Tsephal, S., Jose, T., 2004. *Pepsee* systems: grass root innovations under groundwater stress. Water Policy 6, 303–318.

Ward, F.A., Pulido-Velazquez, M., 2008. Water conservation in irrigation can increase water use. PNAS 105 (47), 18215–18220.

Chapter 15

Implications of Rising Demand for Dairy Products on Agricultural Water Use in India

15.1. INTRODUCTION

Globally, income growth is resulting in high calorie intake by population, especially in the developing countries. In most countries, it is reflected in greater consumption of animal products, such as meat, fish, fruits and vegetables, and nonconventional food items (soft drinks). China is a bright spot. But India has been an exception. The biggest impact of income growth, in this largely vegetarian country, is on the consumption of dairy products—milk, butter, curd, and cheese (Alexandratos and Bruinsma, 2012). The country also recorded a 700% increase in milk production in the past 60years—from a mere 17 m. t—while the population grew by 235%, raising the per capita milk supply significantly.

Delgado (2003) predicted that by 2020, the per capita demand for milk in India would become 288 g/day, growing at an annual rate of 3.5% (Delgado, 2003). However, if the available data are any indication, the per capita consumption of milk in India has already crossed this by the year 2011–12 itself, given the fact that the availability was 296 g per capita per day in that year, that the exports are negligible, and that the demand and supply match. In fact as pointed out in a recent paper by Rajeshwaram et al. (2014), the annual growth in per capita availability of milk has been very high since 2005–06. The fact that the wholesale price index (WPI) for milk has been increasing since April 2006 at a rate of 10.5% in spite of the increase in availability, one can say that this price trend is a clear response to the ever-increasing demand for milk in the market for both direct consumption and its processing and manufacturing of its various derivatives. This trend is visible in the farm gate price as well. As reported by Rajeshwaram et al. (2014), the average price paid by Gujarat Cooperative Milk Marketing Federation (GCMMF) has been on a continual upward trend, increasing by 30% in the 5-year period from 2001–02 to 2006–07, whereas it increased by 90% in the next 5-year period ending 2011–12.

Dairying farming is an important subsector of farming sector in India that has been contributing remarkably to India's agricultural GDP growth in the recent years (Rada, 2013). Dairy farming in India is managed by many millions

Water Policy Science and Politics. https://doi.org/10.1016/B978-0-12-814903-4.00015-4

of small farmers who hold one to two animals, who also do conventional crop cultivation. Hence, it has a huge role in increasing the profitability of farming and improving the viability of farming sector.

The impact of milk production on the water environment in India, however, has been least studied, except for the work of Singh (2004) and Singh et al. (2004), which mainly covered semiarid regions of Gujarat. The general belief is that consuming vegetarian diet will create low water footprint as compared with having a meat-based diet. This is largely true for regions with temperate climate, for which most of the available data on water intensity of milk production are available. Hence, there is too little discussion about the impact of growing dairy farming sector on India's water resources. In this chapter, we look at the drivers of milk demand in India, the water intensity of dairy production in different climatic conditions, the key determinants of dairy production, and their overall implications for the future of dairy production and water use in that sector. The chapter also discusses the ways to promote improved water-use efficiency in milk production.

15.2. THE INCOME VS PRICE ELASTICITY OF MILK CONSUMPTION

The average income elasticity of milk consumption is very high for households in developing countries, especially for the poor households (the lower 20% category), with an elasticity of 2.15% (Gerosa and Skoet, 2012). This means every 1% increase in per capita income would lead to 2.15% increase in the consumption of milk. But what is more significant is the negative price elasticity (−3.0) of milk consumption, which is one of the highest in the world (Dastagiri, 2004). Any increase in the price of milk would lead to a disproportionately higher reduction (three times more) in the consumption of milk. Therefore, it is important that prices are kept stable and not volatile, given the importance of milk in the overall food security—dairy calorie intake and nutritional value.

The late Dr. Verghese Kurien, whose life and work were dedicated to the dairy sector and dairy farmers in India, was often criticized for not acting in the interest of the farmers, by not allowing them to tap the international market and get more attractive price for their produce. But the fact remains that the farmers and the consumers were protected from the price volatility they would have otherwise been subjected to under such circumstances, by creating an excellent system for procurement of farmers' produce, thereby protecting them from exploitation by private players.

So, the ideal situation is that we have an (physical and institutional policy) environment wherein the Indian farmer is able to produce milk and get remunerative prices on the one hand and the consumers are able to meet their daily demands, without being hit by inflation. This is because allowing large-scale exports, which benefit the farmers at times, might affect the domestic consumers as they would be forced to pay high prices in lean season, whereas allowing import of cheap milk from countries like Germany, France, Belgium, and Holland would destroy Indian milk producers and dairy business, while it may help the domestic consumers get milk at affordable prices in the market, though only in the short run. Therefore, the government has to tread carefully with regard to the policies it pursues in the dairy sector with the aim of bringing out maximum social welfare and long-term sustainability of the dairy sector in the country.

The government has a target of increasing the production of milk to 191 million ton by the year 2020 (Rajeshwaram et al., 2014), which stood at 132.4 m. t in 2011–12. This is nothing in comparison with the growth since 1950–51, when the total production was a mere 17 m. t. The question is whether business as usual can continue under the scenario of growing resource scarcity. There is a general perception that the humid and water-rich areas are more conducive to milk production, as intuitively the dairy animals can feed on natural vegetation (wild grasses, tree leaves, shrubs, etc.) generally available in plenty in such areas, with cattle feed and some nutritional supplements of oil cakes, etc. It is true that in such areas, dairy farming will have much lesser impact on the environment in terms of resource depletion, as compared with water-scarce areas, if practiced with the same level of intensity.

As found in a study carried out by Kumar and Singh (2008), water productivity (inverse of water intensity) in relation to "blue water" is very high for milk production in such high rainfall and humid regions. The reason is that the animals there feed on wild grasses or fodder crops largely grown under rainfed conditions. The results are discussed in the next section.

15.3. WATER PRODUCTIVITY VARIATIONS IN MILK PRODUCTION ACROSS REGIONS

To understand the water footprint of dairy production in a region, it is important to know the physical productivity of water in milk production. The physical productivity of water in milk production was estimated for two types of livestock in North Gujarat and three types of livestock in western Punjab and central Kerala. The input data used for this were the average daily milk yield, the average daily quantities of dry and green fodder and cattle feed for the livestock (kg), and the daily drinking water use by the livestock (m^3), all estimated for the entire animal life cycle, and the physical productivity of water for different types of green and dry fodder (kg/m^3) was estimated using the standard formula (for details, see Kumar (2007) or Singh (2004)). Subsequently, the water productivity in milk production in economic terms was estimated using the average net return from milk production using the gross return and average cost of production of milk.

Comparing the result for western Punjab and North Gujarat (Table 15.1) shows that the physical productivity of water for both buffalo and crossbred cow is much higher in western Punjab, when compared with north Gujarat. Further, the difference in economic productivity is much higher than that in physical

TABLE 15.1 Milk Yield and Physical and Economic Productivity of Water in Milk Production in Two Semiarid Regions

	Punjab			North Gujarat		
Variables	**Buffalo**	**Crossbred Cow**	**Indigenous Cow**	**Buffalo**	**Crossbred Cow**	**Indigenous Cow**
Average milk yield (L/day)	3.25	4.46	2.98	3.12	5.33	N.A
Water productivity (WP) (L/m^3)	1.79	2.53	3.68	0.31	0.49	N.A
WP in Milk Production (Rs./m^3)	7.06	17.44	16.41	0.190	0.17	N.A

Based on Singh, O.P., 2004. Water productivity of milk production in North Gujarat, Western India. Proceedings of the 2nd Annual Conference of Asia Pacific Association of Hydrology and Water Resources (APHW 2004), Suntec City, Singapore, 2004 and Kumar, M.D., Malla, A.K., Tripathy, S., 2008. Economic value of water in agriculture: comparative analysis of a water-scarce and a water-rich region in India. Water Int. 33 (2), 214–230.

productivity. The high physical productivity of water in milk production in case of western Punjab could be attributed to the lower volume of embedded water in the inputs used for cattle owing to higher physical productivity of both green and dry fodder. In the case of western Punjab, it was found that only green fodder such as winter jowar (fodder) and kharif bajra (fodder) and dry fodder available from residues of paddy (hay) and wheat (straw) were used. Since paddy and wheat have very high yields in the region, the physical productivity of dry fodder is very high. The cumulative effect of both these factors is in reducing the amount of embedded water, whereas in the case of North Gujarat, alfalfa, a highly water-intensive irrigated green fodder, was found to be most common.

The difference in feeding practices can be seen from Table 15.2. Though the amount of green and dry fodder quantities is less in the case of North Gujarat, alfalfa (figures in brackets) accounts for nearly 70% of the green fodder for both buffalo and crossbred cow. Further, the quantum of cattle feed used for dairy animals in North Gujarat is much higher than that of western Punjab. The much higher water productivity in economic terms was due to (i) the lower cost of

TABLE 15.2 Comparison of Daily Average Feed and Fodder Consumption per Milch Animal in Western Punjab and North Gujarat

Feed/Fodder	Animal Type	Bathinda (Western Punjab)	Mehsana (North Gujarat)
Green fodder (kg/day)	Buffalo	19.46	12.98 (9.25)
	Indigenous cow	12.92	Nil
	Crossbred cow	14.41	12.96 (9.07)
Dry fodder (kg/day)	Buffalo	7.94	5.48
	Indigenous cow	5.07	Nil
	Crossbred cow	4.33	6.44
Concentrate (kg/day)	Buffalo	2.28	5.21
	Indigenous cow	1.2	Nil
	Crossbred cow	1.4	5.36
Drinking water (L/day)	Buffalo	55.8	59.10
	Indigenous cow	52.6	Nil
	Crossbred cow	60.2	49.10

From Kumar, M. D., Malla, A.K., Tripathy, S., 2008. Economic value of water in agriculture: comparative analysis of a water-scarce and a water-rich region in India. Water Int. 33 (2), 214–230 and Singh, O.P., 2004. Water Productivity of Milk Production in North Gujarat, Western India. Proceedings of the 2nd Annual Conference of Asia Pacific Association of Hydrology and Water Resources (APHW 2004), Suntec City, Singapore, 2004.

production of milk, owing to the lower cost of production of cattle inputs such as dry and green fodder, resulting in much higher net returns, and (ii) the lower volume of embedded water in cattle feed and fodder. The difference in the cost of inputs mainly comes from that of irrigation water. In North Gujarat, pumping depths are much higher than that of Punjab. This results in very high capital and variable cost of irrigation owing to expensive deep tube wells, high-capacity pump sets, and very high electricity charges.

Palakkad receives very high rainfall and is less arid than North Gujarat and western Punjab, with resultant low evaporation and evapotranspiration. The region has a lot of naturally grown grasses that provide nutritious fodder for livestock. They also get dry fodder from residues of crops, particularly paddy. The advantage of such regions is that not only the consumptive use of water by fodder crops would be very less but also most of this water needs would be directly met from precipitation. The study in Palakkad shows that green grass accounts for 84%–95% of the total green fodder fed to livestock (Table 15.3).

TABLE 15.3 Average Feed and Fodder Fed to Livestock in Palakkad, Kerala (kg/day/animal)

Name of Feed and Fodder	Average Daily Input (kg)		
	Buffalo	Crossbred Cow	Indigenous Cow
A. Green fodder	*16.00*	*15.59*	*12.17*
1. Local green grass	13.37	14.05	11.59
6. Maize	2.64	1.54	0.58
B. Dry fodder	*11.75*	*11.39*	*10.63*
1. Paddy straw	11.75	11.39	10.63
C. Concentrate	*3.37*	*3.34*	*2.59*
1. Balanced cattle feed	1.57	1.73	1.12
2. Cotton seed cake	0.38	0.44	0.25
7. Wheat bran	0.43	0.66	0.28
8. Rice bran	0.99	0.51	0.94
D. Drinking water (Lt.)	*0.034*	*0.029*	*0.023*

Source: Rajesh, R., Tirkey, R., 2005. Water Intensity of Milk Production in India: Analysis From Two States. MTS Report Submitted to Institute of Rural Management, Anand and IWMI-Tata Water Policy Programme, Anand.

TABLE 15.4 Total Water Use and Water Productivity in Milk Production, Palakkad, Kerala

	Kerala		
Particulars	Buffalo	Crossbred Cow	Indigenous Cow
1. Green fodder (m^3)	0.16	0.10	0.04
2. Dry fodder (m^3)	4.73	4.59	4.28
3. Concentrate (m^3)	4.67	4.06	3.87
4. Drinking water (m^3)	0.034	0.029	0.023
5. Total water used (m^3)	9.60	8.77	8.21
Milk production (L/day)	2.46	3.49	2.36
Irrigation water productivity (IWP) (L/m^3)	0.26	0.40	0.29
Effective IWP in milk production (L/m^3)	0.50	0.74	0.51
IWP in milk production (Rs./m^3)	0.51	0.90	0.74
Effective IWP in milk production (Rs./m^3)	1.00	1.88	1.55

Source: Rajesh, R., Tirkey, R., 2005. Water Intensity of Milk Production in India: Analysis From Two States. MTS Report Submitted to Institute of Rural Management, Anand and IWMI-Tata Water Policy Programme, Anand.

This has a remarkable impact on water use for green fodder fed to cattle. It was found to be in the range of 40–160 L/day per animal (Table 15.4). As a result, the effective water productivity in milk production (physical) was higher as compared with the semiarid North Gujarat, 0.50, 0.74, and 0.51 L/m^3, respectively, for buffalo, crossbred cow, and indigenous cow (Table 15.4). Though the actual irrigation water productivity in milk production is much lower than these figures, a significant chunk of the water used up is the embedded water in cattle feed. It was found to be 48.7%, 46.2%, and 47.1% of the total water used for milk production, for buffalo, crossbred cow, and indigenous cow, respectively (see Table 15.3). Since local water resources are not used for their production and are available from imports, they are not considered while estimating (effective) irrigation water productivity.

Further, the cost of production of fodder was found to be negligible, when compared with that of cattle feed. The water productivity in economic terms was also relatively high when compared with North Gujarat. The estimated effective irrigation water productivity was Rs. 1.0/m^3, Rs. 1.88/m^3, and Rs. 1.55/m^3 for buffalo, crossbred cow, and indigenous cow, respectively (see Table 15.4) (Rajesh and Tirkey, 2005). Groundwater depletion due to agricultural withdrawal is not a problem in this region. But the amount of land available for dairy farming is a major constraint for increasing dairy production. While per capita land availability is high in semiarid regions, it is extremely low in humid and subhumid regions.

15.4. PRODUCTIVITY VS PRODUCTION

But the distinction between productivity and production potential is hardly ever made in the discourse on water-environment debate. What happens in reality is just opposite to what is desirable from an environmental sustainability point of view. In the water-rich areas, dairy farming is not as intensive as water-scarce areas.

In water-scarce Punjab, the per capita availability of milk is 1032 g/day, whereas in water-rich and humid Tripura, it is just 109 g/day. It is 200 g/day in Kerala (source: Dept. of animal husbandry, dairying and fisheries, Ministry of Agriculture 2015–16). What stops Tripura and Kerala from increasing its milk production to meet its own demand and makes it dependent on milk imports? It is the shortage of land—not for grazing but for producing the cereals and the fodder, which the animal needs. In Punjab and Haryana, the vast amount of land under crop cultivation (paddy and wheat) supports dairy farming, which is integral to the region's farming enterprise. The other factors that determine dairy production potential are access to grazing land and wasteland.

The data on per capita gross sown area, per capita pasture land, and per capita wasteland in eight major Indian states (estimated on the basis of data on gross cropped area, area under grazing land, and area under wasteland and the population (Census 2011) figures of each state) are given in Fig. 15.1. It shows that the per capita land available from common lands (wasteland and pasture land) and

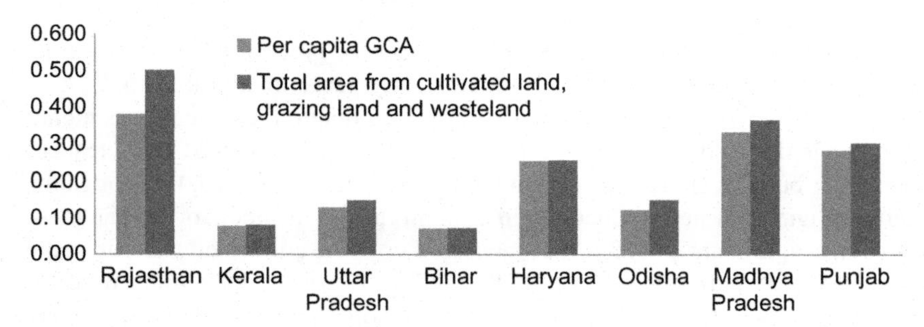

FIG. 15.1 Per capita land area under different classes available for livestock (ha).

cultivated area in arid Rajasthan is 0.50 ha. The per capita land availability is 0.256 ha for Haryana, 0.36 ha for Madhya Pradesh, and 0.30 for Punjab. Against these, the figure is only 0.081 ha for Kerala, whereas in Tripura, the arable land availability is as low as 0.068 ha per capita (not in Fig. 15.1) against 0.30 ha in Punjab, which is the largest milk producer in the country in per capita terms.

The large amount of groundwater resources depleted annually for irrigation in Punjab cannot be merely attributed to paddy-wheat production system alone. Part of it has to go to dairy, as milk production is heavily dependent on straw and hey from wheat and paddy.

Dairy production is very intensive in North Gujarat, another region, known worldwide for impressive dairy sector and large dairy processing plants. The per capita milk availability is 545 g/day. But dairy farming in that region (comprising Mehsana, Banaskantha, Patan, Sabarkantha, and Ahmedabad) thrives on its scarce and rapidly depleting groundwater resources (Kumar, 2007). The nutritious but highly water-intensive alfalfa is grown intensively in this hot and arid region for feeding the high-yielding (crossbred) cows and buffalo.

15.5. MAINTAINING THE DELICATE BALANCE BETWEEN FOOD SECURITY AND ENVIRONMENT

As regards maintaining the delicate balance between food security and environment, armchair critics of modern irrigated agriculture suggest that such water-scarce regions should not go for highly water-intensive crop and dairy production, though such arguments are being paraded by concerned citizens widely even in places like California (when it was hit by droughts). The reality is that the water-scarce regions today export milk, which has a huge amount of embedded water, to other parts of the country (Singh et al., 2004). We also need to remember that within agriculture, the only subsector that has been growing consistently in India over the past 4–5 decades is dairy sector, thanks to the technological and institutional innovations in that sector by AMUL under the great leadership of Dr. Verghese Kurien.

Anything we do in that sector will have long-term implications for agricultural growth, food security, and farming system resilience. In fact, more regions, where conventional crop production is becoming risky due to climatic uncertainties and water scarcity, are taking up dairy farming as an alternative, given the stable and remunerative prices and the greater ease in maneuvering inputs as compared with growing conventional crops. For instance, in case of fodder shortage, the farmers would be able to import dry fodder from distant places. Often, governments provide fodder banks during scarcity periods to help livestock farmers. Only the water for voluntary consumption will have to be managed.

Textbook prescriptions are often made with regard to saving water in agriculture using technological interventions. It is as though farmers have all resolved to save water in their farms and are just waiting for a device to help them in doing that. But no matter which technology is used for irrigating crops, farmers are concerned with the returns per unit of land/animal owned, and not water.

Under the current system of water and energy pricing in agriculture, water saving does not result in cost saving for the farmer, and therefore, there is no private incentive to save water (source: based on Kemper, 2007; Kumar and van Dam, 2013; Schoengold and Zilberman, 2005), unless there is sufficient evidence that reducing water application would result in yield improvement and increase in gross revenue that would offset the investment for water-efficient irrigation technology (Kumar and van Dam, 2013). Even if they optimize watering to crops, they would do so just to expand the area to maximize their farm returns. In the process, they would use the same amount of water as under conventional method, depending on the relative scarcity of water and land.

Over the years, the provincial governments in the country have made water and electricity so cheap that, leave alone the societal cost, the direct economic cost of using these resources is not borne by the farmer in most cases.

In the case of irrigation water, the water charge is levied on the basis of the area irrigated and the crop type and not on the basis of volume of water supplied to the farms. The amount of subsidy per unit area varies remarkably across states. If we leave states such as Mizoram, Jammu and Kashmir, Himachal Pradesh, and Meghalaya, which are not known for investments in public irrigation, and Jharkhand whose irrigation schemes are ongoing, the estimated subsidy was found to be varying from a highest of Rs. 103,962/ha in Andhra Pradesh to Rs. 2995/ha in the case of Chhattisgarh. The investment in per hectare terms is very high in AP, and one of the reasons for this is the very high working expenses, owing to the large number of (river) lift irrigation schemes built recently by the erstwhile government of AP, incurring substantial cost for energy for lifting water. The second highest irrigation cost in per hectare terms is in Maharashtra (Rs. 68,498/ha), which has the largest number of irrigation projects in the country, followed by Gujarat (Rs. 63,543/ha) (source: Kumar, 2017).

As regards electricity, in many Indian states, electricity to farm sector for pumping well water is supplied free; in some cases, the electricity charge is based on connected load of the pump (annual electricity charge is calculated on the basis of the installed capacity of the pump), and at the aggregate level, electricity supply is heavily subsidized. Only in two to three states, the electricity supply for groundwater pumping is metered, with subsidized unit rates. West Bengal has the longest experience in metering power supply to agriculture and charge rates as high as Rs. 4.15/kWh.

As Table 15.5 indicates, the extent of subsidy made available to the farmers varies from state to state. As per the data for the year 2013–14, only Tripura state has an agricultural power tariff that is higher than the average cost of power supply. In all other states, there is power subsidy and varies from Rs. 1.75 in Assam to Rs. 8.67 in Jharkhand. However, the average cost of power supply also varies across states. It is highest in Jharkhand (Rs. 9.42) and lowest in Sikkim (Rs. 3.19). In the states where power subsidy prevails, the extent of subsidy varies from a highest of 100% in Tamil Nadu, Punjab, Himachal Pradesh, and Puducherry to a lowest of 28% in Assam (Kumar, 2017).

TABLE 15.5 Power Subsidies in Agriculture in Indian States

Name of State	Power Supply Cost (Rs./kWh)	Agricultural Power Subsidy (Rs./kWh)	Extent of Subsidy in Farm Power (%)
Undivided AP	5.63	5.19	92.14
Assam	6.29	1.75	27.82
Bihar	7.85	3.74	47.67
Chhattisgarh	4.11	2.57	62.51
Gujarat	4.96	2.78	56.10
Haryana	6.46	6.00	92.80
Himachal Pradesh	5.26	5.25	99.78
J&K	6.74	5.06	75.01
Jharkhand	9.42	8.67	91.96
Karnataka	5.05	1.98	39.24
Kerala	5.97	4.24	71.03
MP	5.39	1.88	34.88
Maharashtra	5.84	3.26	55.78
Meghalaya	5.24	0.00	0.00
Punjab	5.78	5.78	99.99
Rajasthan	6.98	5.17	74.11
Tamil Nadu	6.46	6.46	100.00
Uttar Pradesh	7.06	4.82	68.25
Uttarakhand	5.10	2.82	55.18
West Bengal	6.13	1.98	32.26
Arunachal Pradesh	8.24		
Goa	3.72	2.66	71.44
Manipur	8.55	7.14	83.51
Mizoram	7.41	0.00	0.00
Nagaland	9.06		
Puducherry	4.06	4.03	99.22
Sikkim	3.19		
Tripura	5.10	−0.44	−8.70

Source: Kumar, M.D., 2017. Market Analysis: Desalinated Water for Irrigation and Domestic Use in India. Prepared for Securing Water for Food: A Grand Challenge for Development in the Center for Development Innovation U.S. Global Development Lab Submitted by DAI Professional Management Services.

So, at times, even field-level efficiency improvement is not achieved with such technologies. Therefore, on the technology front, the focus needs to be shifted to introducing crops that can give higher returns per unit of land but are less water-intensive, along with the drip technology.

That said, we also need to be concerned about food and nutritional security of the country's population. The fact is that these water-scarce regions produce surplus milk and export to deficit regions, which include those that are water-rich. So, there are no regions that can meet the shortfall in production resulting from such a strategy of change in farming system. Hence, the repercussions of such strategies on national food and nutritional security and livelihoods of dairy farmers will be immense.

India can't afford to think of meeting a mere 2% of its milk demand (2.7 m. t) through imports, as it is large enough to shake the global dairy market. But as per some back of the envelope calculations, dairy farming in India should be consuming anywhere near 250–270 BCM of water annually, including the rain-water directly available from the soil profile (*green water*). This is more than the storage available in all (large and small) surface reservoirs put together in India. This is 25%–30% of the total annual consumptive water use (*blue water and green water*) in agriculture in the country. While dairying is part of integrated farming system in most parts of India, in the recent years, it has also emerged as a major source of income for farmers a significant share of farmers with very small holding (less than 0.01 ha) (MoAFW, 2016).

Therefore, it is crucial that water management in agriculture seriously looks at the options for improving water-use efficiency in milk production, which covers all components of dairy production system. Since there is already a major shift toward high-yielding breeds of cows and buffalo, the only option available to make dairy production more efficient is to create incentive to use water and other inputs economically through fiscal and market instruments. Let there be experiments with pricing of water and electricity and rationalizing of water allocation/energy supply to farm sector. Once we do that, farmers are smart enough to find ways to achieve high efficiency in production, be it efficient way of growing alfalfa, rice, jowar, and bajra or selection of water-efficient crops as fodder or efficiency in harvesting and feeding practices or by choosing the right kind of animal breed to suit the climate and ecology.

If they find crop and milk production not giving enough returns to offset the costs, they would shift to water-efficient ones, after trying with changes in feeding practices—with greater use of dry fodder and low-water-consuming fodder crops. The power of market instruments is hardly ever appreciated by our water policy makers. Shortfall in production caused by such shifts can create market shocks, with the result that the demand-supply imbalance would address the profitability concerns of the farmers. This is similar to what has happened in the "pulse" sector. The shortfall in supply and the rising prices of pulses had once again made pulse production very attractive for the farmers.

In Australia, the Southern Murray-Darling basin communities could withstand a decade-long drought during 1999–2009, through water trading. During the droughts, tradable water rights were introduced by the basin authorities in a progressive manner through water right reforms. The farmers who were raising perennial crops and wine growers purchased "water rights" from those who were earlier growing cereals but found it unviable to grow those crops, with major cuts in their allocation. In the process, both the buyers and sellers were benefited. The output of a general equilibrium model developed to analyze the economic impact of water trading was that the basin produced larger amount of crops in value terms (AUD 220 million per annum) but used much less amount of water (NWC, 2010).

The measures suggested above are not "easy to implement." In fact, they require lot of innovations in crafting institutions and nurturing them and changing the legal framework governing water use in our country, and this would be a long and arduous process. However, they would be far more effective than the plethora of "solutions" that appear very simple and implementable and are implemented today. As Saleth and Dinar (1999) pointed out, the opportunity cost of not going for these difficult measures is what should drive the future investments in institutional development for resource management and allocation. Ultimately, the aim should be to make both water and milk flow in amble measure.

15.6. CONCLUSIONS

India's dairy sector is poised to grow leaps and bounds, driven by the rising demand for dairy products on the one hand and the greater access to production technologies and market infrastructure on the other. The implications of dairy production on India's water resources are huge, with an estimated total consumption of nearly 250–270 BCM. The importance of water as a critical input for dairy production is hardly known to dairy professionals and policy makers in the animal husbandry sector in India. It is generally considered that the water requirement for dairying is equal to the sum of water consumption and the water required for washing animals. The amount of embedded water in dairy inputs such as feed and fodder is largely ignored. The regions that give high water productivity, owing to abundant vegetation and favorable climate, do not offer high production potential in lieu of the limited availability of arable land.

On the other hand, the regions that have high production potential have low productivity in relation to water due to the hot and arid climate. The prospects of India engaging in large-scale import of dairy products are unlikely, given its implications for the livelihood security of the tens of millions of small-scale milk producers in the country. Future focus should therefore be on improving the productivity of water use, in regions where the production potential is high (in terms of access to cultivated land and grazing land), through the use of fiscal instruments, as water-use efficiency improvement will not happen with adoption of water-saving irrigation devices.

15.7. WHAT CAN AFRICA LEARN FROM INDIA'S EXPERIENCE?

Many African countries have been experiencing some growth in the consumption of milk and other dairy products in the recent decades owing to population growth and a modest increase in per capita consumption (4% during 1990–2004), while the annual growth in production of milk (3.1% during 1990–2004) is not keeping with it. In many countries of north and sub-Saharan Africa, the annual growth in milk production ranged from 0.5%–2.5% (in 12 countries) to 2.5%–5% (in four countries) to more than 5% (in two countries). Egypt in the north and Ethiopia in SSA recorded more than 5% annual growth in milk production during this period. Given the economic pressure to reduce the imports, many countries are likely to follow policies to promote dairy development in a big way (Ndambi et al., 2007). Moreover, livestock keeping is found to significantly change the economic dynamic of SSA small holder farmers. A recent study involving field data of 13,000 small holder farms from 19 SSA countries showed a steep increase in food availability with increase in per capita holding in terms of number of total livestock units (Frelat et al., 2016).

From the perspective of food security, it is important that countries of sub-Saharan Africa increase the production of milk so as to meet the current domestic demand of dairy products and to keep the prices of these dairy products lower enough to make it affordable to greater proportion of the population in future. Semiintensive farms owned by the state and cooperative and small family-owned intensive farms are becoming widespread across Africa. Semiintensive and intensive dairy farming, particularly the latter (most of the latter coming up near the cities), would eventually consume a lot of water directly for irrigating crops meant for green fodder, in the hyper arid and semiarid regions of North and sub-Saharan Africa. This can increase the pressure on the limited freshwater resources available in the case of North African countries (Egypt, Libya, and Algeria) and the limited utilizable freshwater in the case of sub-Saharan Africa.

REFERENCES

Alexandratos, N., Bruinsma, J., 2012. World Agriculture Towards 2030/2050: The 2012 Revision. ESA Working Paper No. 12-03, Agricultural Development Economics Division, FAO, Rome.

Dastagiri, M.B., 2004. Demand and Supply Projections for Livestock Products in India. Policy Paper 21, ICAR (NCAP), New Delhi.

Delgado, C.L., 2003. Rising consumption of meat and milk in developing countries has created a new food revolution. J. Nutr. 133 (11), 3907–3910.

Frelat, R., Lopez-Ridaura, S., Giller, K.E., Herrero, M., Douxchamps, S., Djurfeldt, A.A., Erenstein, O., Henderson, B., Kassie, M., Paul, B.K., Rigolot, C., Ritzema, R.S., Rodriguez, D., van Asten, P.J.A., van Wijka, M.T., 2016. Drivers of household food availability in sub-Saharan Africa based on big data from small farms. Proc. Natl. Acad. Sci. USA 113 (2), 458–463.

Gerosa, S., Skoet, J., 2012. Milk Availability—Trends in Production and Demand and Medium Term Outlook. ESA Working Paper No. 12-01, A. D. E. Division, Food and Agriculture Organization of the United Nations, Rome.

Kemper, K.E., 2007. Instruments and institutions for groundwater management. In: Giordano, M., Villholth, K. (Eds.), Agricultural Groundwater Revolution: Opportunities and Threats to Development. CAB International, Wallingford, Oxfordshire.

Kumar, M.D., 2007. Groundwater Management in India: Physical, Institutional and Policy Alternatives. Sage Publications, New Delhi.

Kumar, M.D., 2017. Market Analysis: Desalinated Water for Irrigation and Domestic Use in India. Prepared for Securing Water for Food: A Grand Challenge for Development in the Center for Development Innovation U.S. Global Development Lab Submitted by DAI Professional Management Services.

Kumar, M.D., Singh, O.P., 2008. Groundwater stress due to irrigation in semi-arid and arid regions: is dairying a boon or a bane? In: Kumar, M.D. (Ed.), Managing Water in the Face of Growing Scarcity, Inequity and Declining Returns: Exploring Fresh Approaches, Volume 1, Proceedings of the 7th Annual Partners' Meet of IWMI-Tata Water Policy Research program, ICRISAT, Patancheru, Andhra Pradesh, pp. 202–213.

Kumar, M.D., van Dam, J.C., 2013. Drivers of change in agricultural water productivity and its improvements at basin scale in developing economies. Water Int. 38 (3), 312–325.

Ministry of Agriculture and Farmers Welfare (MoA & FW), 2016. State of Indian Agriculture 2015–16. Department of Agriculture, Cooperation and Farmers Welfare, Ministry of Agriculture and Farmers Welfare, Govt. of India, New Delhi.

National Water Commission (NWC), 2010. The Impacts of Water Trading in the Southern Murray–Darling Basin: An Economic, Social and Environmental Assessment. NWC, Canberra.

Ndambi, O.A., Hemme, T., Latacz-Lohmann, U., 2007. Dairying in Africa—status and recent developments. Livest. Res. Rural. Dev. 19 (8), 2007.

Rada, N.E., 2013. In: Agricultural growth in India: examining the Post Green Revolution Transition. Selected Paper Prepared for Presentation at the Agricultural & Applied Economics Association's 2013 AAEA & CAES Joint Annual Meeting, Washington, DC, August 4–6, 2013.

Rajesh, R., Tirkey, R., 2005. Water Intensity of Milk Production in India: Analysis From Two States. MTS Report Submitted to Institute of Rural Management. Anand and IWMI-Tata Water Policy Programme, Anand.

Rajeshwaram, S., Naik, G., Dhas, R.A.C., 2014. Rise in Milk Price-A Cause for Concern on Food Security. Working Paper # 472, Indian Institute of Bangalore.

Saleth, R.M., Dinar, A., 1999. Water Challenge and Institutional Responses (A Cross Country Perspective). Policy Research Working Paper Series 2045, The World Bank, Washington, DC.

Schoengold, K., Zilberman, D., 2005. The Economics of Water, Irrigation and Development. February 2005, Available from http://www.ctec.ufal.br/professor/vap/Handbook.pdf.

Singh, O.P., 2004. In: Water productivity of milk production in North Gujarat, Western India. Proceedings of the 2nd Annual Conference of Asia Pacific Association of Hydrology and Water Resources (APHW 2004), Suntec City, Singapore, 2004.

Singh, O.P., Sharma, A., Singh, R., Shah, T., 2004. Virtual water trade in dairy economy. Review of agriculture. Econ. Polit. Wkly. 39 (31), 3498–3503.

FURTHER READING

Government of India (GoI), 2014. Annual Report (2013–14) on the Working of State Power Utilities & Electricity Departments. Power & Energy Division, Planning Commission, Government of India, New Delhi.

Kumar, M.D., Malla, A.K., Tripathy, S., 2008. Economic value of water in agriculture: comparative analysis of a water-scarce and a water-rich region in India. Water Int. 33 (2), 214–230.

Chapter 16

Water Policy Making: What Other Developing Economies Can Learn From India

16.1 INTRODUCTION

Policy making in the water sector should be driven by the objective of addressing problems related to water such as water scarcity, water pollution, and droughts and floods in a manner that the resource is able to perform all its three functions, namely, social good, economic good, and environmental good. In the developed countries, many of the problems related to water such as scarcity of water for human uses, shortage of water for economic production functions (crop production, dairy farming, and manufacturing uses), pollution of water bodies, environmental water stress in rivers, and flood hazards either do not occur frequently or are not severe, by virtue of having adequate water infrastructure and capable water institutions (Kumar et al., 2008a, 2016). Therefore, there is a need to focus on countries that score low on water security.

An index was developed by Kumar et al. (2008a), which is a derivative of the water poverty index, to express the water situation of a country, and is called sustainable water use index. Going by the values of the index, around 20 out of the 125 countries for which SWUI values were estimated had its values lower than 40 (less than 50% of the maximum value of 80), and all except two of them (Cambodia and Haiti) are in sub-Saharan Africa and are economically very poor. Out of 41 countries having the index values in the range of 40–50, 14 are from sub-Saharan Africa and many from Latin America and Central and South Asia. For India, the value of the index was 46.4, and for China, it was 44.1. Hence, if we want to understand the implications of Indian experience with water policy making, it makes logical sense to look at sub-Saharan Africa, which is yet to achieve the level of development India has.

The region comprising around 50 countries is largely agrarian. Yet, it has the unique distinction of having the lowest level of cereal yields and agricultural productivity growth rates (von Braun, 2007; FAO, 2006). The region is the most water stressed in the world (HRD, 2006). Yet, a small fraction of the

Water Policy Science and Politics. https://doi.org/10.1016/B978-0-12-814903-4.00016-6

utilizable water resources of the region have so far been tapped (Falkenmark and Rockström, 2004; Xie et al., 2014). Much of the region is generally considered to suffer from what is called "economic water scarcity," and a very few countries actually experience physical scarcity of water (Xie et al., 2014).[1] Poverty reduction is closely linked to water development for irrigated agriculture in these low-income countries in the absence of other economic opportunities in rural areas (UN Water/FAO, 2007). Yet, the region is yet to see significant investments in water resources development, including irrigation development (Xie et al., 2014). The region suffers from inadequate human resource capacities in water sector, apart from poor finances (Falkenmark and Rockström, 2004). It is thus also the most food insecure region in the world (IFPRI, 2011; Weismann, 2006) depending largely on donor aid and food imports (von Braun, 2007).

In regard to agriculture, food security, and poverty scenario, in the 1960s, most of the Asian continent barring the far-eastern economic giants, particularly India and China, looked the same way sub-Saharan Africa stands today. But, Asian countries made significant strides in terms of maintaining high growth in agricultural productivity and production, lifting hundreds of millions of people out of poverty. Irrigation development has played a crucial role in rural poverty reduction, and with irrigation, the region moved away from a "low resource use-low productivity" regime to largely an "intensive resource use-moderate to high productivity" regime. While in the semiarid and arid parts of Asia groundwater played a significant role in revolutionizing irrigation, the latter also brought along with it problems of "aquifer mining" supported by the lack of adequate planning, legal framework, and governance, threatening the long-term sustainability of irrigated agriculture (UN Water/FAO, 2007).

Sub-Saharan Africa, which is riddled by conflicts, political instability, poor governance, corruption, politics of exclusion, high rural poverty (Ong'ayo, 2008), and weak human resource capacities (Falkenmark and Rockström, 2004), is not comparable with India when it comes to government policy making. Yet, there are regions within India, which are as poor and food insecure as some countries of sub-Saharan Africa. Because of the heterogeneous agroclimate and socioeconomic conditions prevailing across the country, Indian experience provides important learnings for sub-Saharan Africa. The lessons would address questions like the following: what should be the long-term strategies for water, food, and energy security and poverty reduction? This chapter tries to draws important lessons on water management strategies for countries of sub-Saharan Africa, based on the findings of analyses presented in the previous chapters. Finally, the fledgling economies

---

1. Economic scarcity of water refers to a condition where the investments in water resources and relevant human capacity are not substantial enough to meet water demands in an area where the population does not have the financial means to make use of an adequate water source on its own.

of sub-Saharan Africa and parts of South and Southeast Asia need to manage the political process of policy making well to obtain the desired outcomes in terms of sustainable water management. In this regard, some policy directions for the region are also offered.

16.2 FOOD INSECURITY AND WATER SCARCITY IN SUB-SAHARAN AFRICA

16.2.1 Where Does Sub-Saharan Africa Stand in Terms of Food and Water Security?

Most of the sub-Saharan African countries are highly food insecure. According to FAO (2015), one-third of the people in sub-Saharan Africa are undernourished in 2014–16, and the reduction in extent of undernourishment was found to be only 30 percentage points over a period of 25 years (FAO, 2015). According to the global hunger index data published by IFPRI in 2001 (IFPRI, 2011), out of the 26 countries that have alarming to extremely alarming hunger index scores, 22 are in the African continent (IFPRI, 2011). A total of 217.8 million people were estimated to be undernourished in 2014–16 (FAO, 2015).

An analysis of progress in reducing hunger, expressed in terms of reduction in GHI scores among countries, shows that the progress has been relatively less in sub-Saharan Africa as compared with many countries in Southeast Asia and Latin America. The reduction in GHI scores ranges from 0.0%–24.9% for some in Southern and Central Africa to 25.0%–49.9% for some others. A few countries in sub-Saharan Africa showed increase in hunger. At the same time, the reduction in GHI score has been much higher (above 50%) for many countries of Latin America (Brazil, Uruguay, and Chile), Middle East, Central Asia (Turkey), and China. The achievement of India in reducing hunger was less as compared with more populous countries like China. The GHI score for the country went down from 30.4% to 23.7% during the period from 1990 to 2011. What is even more alarming is the fact that both the percentage and aggregate number of undernourished people increased in sub-Saharan Africa during the period from 1990–92 to 2014–16 (FAO, 2015).

As per the report of the joint monitoring program of WHO and UNICEF, nearly 322 million people from sub-Saharan Africa lack access to safe drinking water, and only 53% of the urban population and 28% of the rural population in these countries had access to improved sanitation facilities (WHO/UNICEF, 2006). What is more alarming is the great unevenness in access to water supply and sanitation across countries and between districts/regions within countries.

As per a report by the London School of Hygiene and Tropical Medicine, which used data from 138 national surveys of 41 countries of sub-Saharan Africa, access to improved drinking water and sanitation is extremely uneven within individual countries in sub-Saharan Africa. Access to clean water is highly variable, ranging from a low 3.2% in some districts of Somalia to as high as 99% in Namibia's urban centers. Adequate sanitation facilities are equally inconsistent. Improved sanitation ranged from 0.2% in parts of Chad to close to 100% in Gambia. The rural households in the districts—with the lowest levels of access within a country—were 1.5–8 times less likely to use improved drinking water, 2–18 times less likely to use improved sanitation, and 2–80 times more likely to defecate in the open, compared with rural households in districts with the best coverage.

16.2.2 The Root Cause of Food Insecurity in Sub-Saharan Africa

Poor agricultural growth is one of the main reasons for high rate of food insecurity in sub-Saharan Africa. Cereal yields are the lowest in sub-Saharan Africa among all the regions of the world. During the three decades from 1967 to 1997, the increase in cereal yields there has been negligible, while yields doubled in South Asia to touch 2000 kg/ha during the same period (Rosegrant et al., 2001, based on FAOSTAT). Though cereal yields in the region had been growing at an annual rate of 2.2% during 2000 and 2010 and harvesting area growing at an annual rate of nearly 2%, resulting in an output growth of 4.1% per annum, its impact on per capita supply of cereals has not been very high, as population in the region grew at an annual rate of 2% during the same period (FAO, Statistics Division, FAOSTAT).

The impediments to agricultural growth are many, and there is a complex web of problems. But some of the most critical ones are poor investment in irrigation; poor adoption of modern agricultural technologies, including high-yielding varieties and farm machinery; poor workforce in agriculture; poor extension services in agricultural sector; and poor market infrastructure (FAO, 2009). These problems are compounded by high-rainfall variability (Gommes and Petrassi, 1994; Thornton et al., 2014) and frequent occurrence of droughts, though variability in rainfall is high in the drier regions and low in the wetter regions (Gommes and Petrassi, 1994). But, on the governance front, weak institutions, a nonvibrant civil society, and a weak agricultural research system characterized by inadequate human resource base (Kumar et al., 2008a) and rising food price are the problems. Out of the 24 countries that have low human development indices (less than 0.50), 23 are in Africa (HDR, 2009). Many of these problems can be averted through improving water security of the people in this region.[2]

Worldwide, experiences show that improved water security (in terms of its access to water, levels of use of water, overall health of water environment, and enhancing the technological and institutional capacities to deal with sectoral challenges) leads to better human health and environmental sanitation, food security and nutrition, livelihoods, and greater access to education for the poor (see, for instance, UNDP, 2006). This aggregate impact can be segregated with irrigation having direct impact on rural poverty (Bhattarai and Narayanamoorthy, 2003; Hussain and Hanjra, 2003); irrigation having impact on food security, livelihoods, and nutrition (Hussain and Hanjra, 2003), with positive effects on productive workforce; and domestic water security having positive effects on health, environmental sanitation, with spin-off effects on livelihoods and nutrition (positive), school dropout rates (negative), and productive workforce.

Currently, accessibility to safe water is very scant in sub-Saharan Africa, with only 22%–34% of populations in eight sub-Saharan countries having access to safe water. The UNEP projects that in the year 2025, as many as 25 African nations—roughly half the continent's countries—are expected to suffer from a greater combination of increased water scarcity and water stress. Dirty water and poor sanitation account for vast majority of the 0.8 million child deaths each year from diarrhea—making it the second-largest cause of child mortality. Diseases and productivity losses linked to water and sanitation amount to 5% of GDP in the sub-Saharan Africa (HDR, 2006).

The strong inverse relationship between sustainable water use index, which captures the overall water situation in a country (Kumar et al., 2008a; Kumar, 2009) and the global hunger index (GHI), developed by IFPRI for 118 countries (Wiesmann, 2006) provide a broader empirical support for some of the phenomena discussed above.[2] The estimated R square value for the regression between SWUI and GHI was 0.60. The coefficient was also significant at 1% level. It shows that with improved water situation, the incidence of infant mortality (below 5 years of age) and impoverishment reduces. In that case, improved water situation should improve the value of human development index, which captures three key spheres of human development such as health, education, and income status (Kumar, 2009).

Therefore, the root cause of the problems of food insecurity in sub-Saharan Africa lies in water insecurity (Kumar et al., 2008a; Shah and Kumar, 2008). There are recent attempts to link food insecurity to national income using the correlation between per capita GDP and food insecurity using time-series data for different regions (IFPRI, 2011; Wiesmann, 2006). This is mainly because of autocorrelation, that is, countries having high water security (expressed by us in terms of SWUI) also have high per capita income. The fact, as illustrated by Kumar et al. (2008a) and Shah and Kumar (2008), is that water security remained crucial to achieving progress for many countries in the form of high human development indicators and economic growth. Water security in these, semiarid to arid tropical countries will come only through water development (Falkenmark and Rockström, 2004) either through building of storages or through judicious exploitation of groundwater.

While climate variability has significant impact on agricultural outputs in sub-Saharan Africa through impacts at the plot level (though effect on crop yield) and aggregate level (through effect on cropped area and irrigated area and crop-livestock systems) (Thornton et al., 2014), it also impacts on food systems

2. In addition to these 118 countries for which data on GHI are available, 18 developed countries were included in the analysis. For these developed countries, we have considered zero values, assuming that these countries do not face problems of hunger.

(Codjoe and Owusu, 2011), human health (McMichael et al., 2006), and nutrition (Lloyd et al., 2011; Nelson, 2009).

16.2.3 Challenges in Achieving Food Security

Sub-Saharan Africa has less than 25% of its cultivable land under crop production (FAO, 2009). The region has the lowest-irrigated to rainfed area ratio of less than 3% (FAO, 2006, Fig. 5.2, p. 177). The drought-prone areas of sub-Saharan Africa are characterized by one of the lowest levels of agricultural productivity in the world, primarily due to water stress during crop growth. One of the reasons for poor utilization of arable land for cultivation is the lack of assurance in obtaining yields, in the wake of uncertain rainfalls and the absence of irrigation facilities.

Even the economic growth of this predominantly agrarian region is closely correlated with rainfall (Barrios et al., 2004; Foster et al., 2006). Irrigation is the key to improving water security, expanding crop land, and also raising crop yields in the drought-prone areas. But, one of the biggest challenges in reducing the region's vulnerability to droughts and associated problems of food insecurity and hunger is in developing this irrigation (FAO, 2009; UN Water/FAO, 2007). Currently, only two of the sub-Saharan African countries have extensive irrigation. They are South Africa and Madagascar, with 1.43 and 1.15 M ha of irrigated area (You et al., 2007). The total area under irrigation is only 4% of the cultivated area in sub-Saharan Africa, whereas the cultivated area itself is only 8% of the total geographic area of the region (You et al., 2010).

Apart from irrigation, improving access to safe water and improved sanitation will also go a long way in reducing child mortality through control of fatal water-borne diseases. It will help improve education, by reducing school dropout rates. It will increase the productive workforce by improving family health and nutrition. Therefore, design of water supply systems in these countries for enhancing food production should take into consideration this particular linkage.

In India, the initial impetus in irrigation development came from the public sector, mainly covering large medium and minor surface irrigation systems (Shah, 2009). But, later on with the advancement in drilling technology, massive rural electrification, institutional finance for well development, and heavy subsidies for electricity for agricultural use, groundwater development for irrigation took place in a big way in the countryside, with well irrigation becoming intensive in the semiarid and arid parts of the country (Kumar, 2007). Today, well irrigation surpasses surface irrigation and accounts for nearly two-thirds of the net irrigated area in India (Shah, 2009).

But, there is little reason to believe that the irrigation development trajectory would be more or less the same in sub-Saharan Africa, given the drastically different sociopolitical scenario and human resource capacities of the countries of that region. Surface irrigation development in sub-Saharan Africa is likely to happen at a slow pace even in the coming years, and too little can be done about changing it from a water sector perspective, unless the macrolevel issues of political instability, corruption in governments, institutional capacity, and finance are addressed to. These are in addition to the host of social and environmental issues that the surface water projects raise. The corruption in government is likely to reduce the donor confidence in many of the countries there.

Most of the leverage therefore lies in groundwater development through private sector initiatives (Kumar, 2012). But, unlike India, where drilling technology has come very handy in rural areas due to low cost and easy accessibility, well drilling is very expensive in most African countries (Kadigi et al., 2012). Apart from issues of high drilling costs, there are also technical issues. They concern uncertainty about resource condition. The scientific information about groundwater resource conditions is patchy at the local level, while some information about aquifer characteristics and recharge and abstraction are available at the regional level (Foster et al., 2006). Unless farmers are convinced about their ability to hit water, investments are unlikely to come, even if funds are available. An associated challenge is in rural electrification (Kumar, 2012). A recent study by IFPRI that assessed irrigation potential expansion in the region using hydroeconomic modeling showed that motor pump irrigation could increase irrigation in sub-Saharan Africa by around 30 M ha and treadle pump by around 24 and 22 M ha, small reservoirs by 22 M ha, and river diversion by 20 M ha (Xie et al., 2014), though the study hasn't provided any analysis of the economics of such irrigation development interventions. The other challenge is to change the investment climate. In addition to irrigation, there is a need for greater investment in public goods that support agriculture, such as research and extension, rural roads, storage facilities, education, and health (FAO, 2009).

16.3. PRACTICAL AGENDA FOR WATER MANAGEMENT FOR AGRICULTURAL PRODUCTION, FOOD SECURITY AND DRINKING WATER SUPPLY

16.3.1 Basin Level Planning of Surface Water Systems

In the early stages of irrigation development in independent India, the focus was on large irrigation systems involving reservoirs and barrages for water storage and diversion and canal systems for water distribution and delivery. The 1970s, 1980s, and 1990s saw expansion in groundwater irrigation, which occurred mostly with private investments from farmers. The past two decades have, however, seen an increasing focus on small water harnessing systems, with an accent on decentralization and community participation. This has emerged in response to growing water scarcity in semiarid and arid regions, with the supplies from both public irrigation systems and groundwater-based sources being unable to meet the growing demand for water in agriculture. The underlying assumption perhaps was that small structures could be managed easily through local efforts, without much help from public agencies, whose performance in managing large irrigation schemes, anyway, has not been impressive (Kumar, 2012).

But, these efforts haven't produced any laudable results due to the lack of scientific planning and overdoing, without any attention being paid on the basin hydrology. In fact, most of the structures were built in "closed" basins. The scale effects of small-scale water harvesting and watershed development at the macrolevel were ignored (Glendenning and Vervoort, 2011; Kumar et al., 2006; Syme et al., 2011). This has led to a situation of dividing the waters within the basin rather than augmenting the utilizable water resources. This had seriously impaired the economic viability of the structures also, apart from causing ecological problems of reduced streamflows in downstream parts (Kumar et al., 2006, 2008b). This is leading to a phenomenon, which Falkenmark and Rockström refer to as "overcrowing" in water and technical water stress.

Many African countries are also caught up in this movement for small water harvesting. There has been some work in eastern African countries (Oweis et al., 1999; Rockström et al., 2002), Mexico (Scott and Ochoa, 2001), and India to show the impact of water harvesting on crop water productivity. Rockström et al. (2002) have shown remarkable effect of supplementary irrigation through water harvesting on physical productivity of water expressed in kilogram per evapotranspiration, for crops such as sorghum and maize. However, the research did not evaluate the incremental economic returns due to supplementary irrigation against the incremental costs of water harvesting. It also does not quantify the real hydrologic opportunities available for water harvesting at the farm level and its reliability. The work by Scott and Ochoa (2001) in the Lerma-Chapala Basin in Mexico showed higher gross value product from crop production in areas with better allocation of water from water harvesting irrigation systems. But, their figures of surplus value product that takes into account the cost of irrigation were not available from their analysis. In arid and semiarid regions, the hydrologic and economic opportunities of water harvesting are often overplayed (Kumar and van Dam, 2013). The works in India on the effectiveness of rainwater harvesting systems are very much applicable to sub-Saharan Africa; as like India, most of sub-Saharan Africa also experiences high variability in rainfall and droughts (Kumar, 2012).

With high capital cost of WH systems, the small and marginal farmers would have less incentive to use it for supplementary irrigation, as incremental returns due to yield benefits may not exceed the cost of the system. This is particularly so for crops having low economic value (Kumar and van Dam, 2013) that dominate cropping in sub-Saharan Africa. The dominant crops of the region are cassava, maize, millet, cow pea, barley (all food grains), bean (vegetable) and cotton, groundnut, cocoa, and coffee. Among these crops, only a few of the high-valued ones require irrigation. The other crops such as oil palm, banana, sugarcane, potatoes and yam, rice, and wheat are grown in small area (You et al., 2007).

Since sub-Saharan Africa is still at the early stage of water development (Falkenmark and Rockström, 2004), there is a scope for using integrated catchment approach in basin-wide planning of small and large water resource systems, in order to avoid the phenomenon, which Falkenmark and Rockström refer to as "technoeconomic water scarcity," wherein the number of people to compete for a unit of water is excessively large. To enable sufficient scientific inputs for planning the water systems, such tasks should be entrusted with a competent scientific/technical agency, which is outside the jurisdiction of public irrigation enterprises. The small and large systems have to coexist. Water resources development should begin with small structures. The large water systems can be planned downstream of the small structures to harness the untapped water from these local catchments (Kumar, 2012). While community participation might be possible and also desirable for building and maintaining small water harvesting schemes, they should be built only in an area only if the local hydrology and topography permit. Care should be exercised to make sure that political interference in the planning decisions is minimal.

16.3.2 Treading Carefully on Groundwater Development

Available data on groundwater development in African countries show very low degree of groundwater development at the regional scale in sub-Saharan Africa (Foster et al., 2006; Siebert et al., 2010). But, planning for groundwater development cannot be driven by the macrolevel information about groundwater recharge and abstraction. Microlevel information about utilizable resource and abstraction and the stage of development would be essential. Further, given the complex characteristics of the geologic formation (Kadigi et al., 2012) and high spatial and temporal variability in rainfall, the reliability of the existing data on groundwater recharge is highly questionable. Therefore, attempt should be made to accurately estimate recharge from precipitation. Far more important is the requirement for robust methodologies for estimating the groundwater balance, which considers inflows and outflows from the aquifers. More importantly, the assessment of overexploitation should consider negative consequences of groundwater overuse, which are physical, economic, social, and ethical rather than merely considering the abstraction against utilizable recharge.

This is one of the crucial problems hindering serious groundwater resources planning for agriculture in India (Kumar et al., 2012; Sreenivasan and Lele, 2017). The current estimates seriously underplay groundwater overdevelopment problems in hard-rock regions of India as they fail to capture the undesirable effects such as sharp decline in water levels, rampant well failures, and reduction in well yields that have serious economic consequences (Kumar et al., 2012).

Groundwater resource estimation is a politically sensitive issue in India. Representatives of political parties want to put their constituency under "safe" category, even when they are overexploited or critically exploited. The reason is that most of the institutional financing for well development for an administrative unit is dependent on the status of groundwater development for that unit (Moench, 1992). As a result, the agency estimates are often subject to tampering, as people's representative would like to see that fund flows for drilling wells and purchase of pump sets in their constituency continues uninterrupted. We need to reckon with the fact that access to rural livelihoods is linked to access to groundwater. Therefore, groundwater resource estimation could be a big source of rent seeking by official agencies. In order to make it free from political interference, it is important that the task of resource evaluation and planning is assigned to independent agencies.

High energy costs would continue to pose a major obstacle in the development of the groundwater sources throughout Africa as most smallholders do not benefit from generous energy subsidies like farmers in India and other South Asian countries. Drilling costs are also high in the region compared with other countries, estimated to be on average USD 100/m, which is more than 10 times the cost in India (Wurzel, 2001; Kadigi et al., 2012).

Since optimum development of groundwater can help minimize the cost of abstraction of unit volume of water, mechanisms for regulating groundwater use will have to be thought through much before exploitation becomes intensive. If groundwater is also abundant in a region with shallow water table, low-cost irrigation systems such as treadle pumps could be promoted, taking advantage of the shallow water table conditions and rural labor force. Another option would be micro diesel pumps combined with very shallow tube wells. Nevertheless, the rich farmers might be able to install pump sets and use larger volume of groundwater. In areas with extremely limited groundwater resources, the traditional wells with rope and bucket or treadle pumps would be appropriate if water table is within 20–25 ft below the ground.

One of the important areas for future research is the potential of groundwater for irrigation development and water supplies in developing countries. In this regards, sub-Saharan African countries pose greater challenges in planning groundwater development, as the information available on aquifer recharge and abstraction are too little and patchy (Foster et al., 2006; Kadigi et al., 2012). The priority for countries in sub-Saharan Africa is to generate the knowledge and information about renewable groundwater and the stocks, before they embark on policies for large-scale development of this resource. The scientific challenge is great in view of the greater climatic variability across space and across years and variation in geologic formation (Foster et al., 2006; Kumar, 2012).

16.3.3 Conjunctive Management of Surface Water and Groundwater

Sub-Saharan Africa has Savanna Climate, with erratic rainfall and moderate population pressures. Water use to water resource endowment ratio remains low due to the lack of irrigation. Though theoretically there is a lot of unused potential, it is difficult to mobilize this water due to poor institutional preparedness, the lack of human resource capacity, and financial constraints (Falkenmark and Rockström, 2004).

Climatic variability is very significant in sub-Saharan Africa, characterized by year-to-year variation in annual rainfall (Thornton et al., 2014). Within the tropical semiarid region of sub-Saharan Africa, there are regional differences. Rainfall gradients are steep—as much as 100mm per 100km in West Africa—passing from 100mm in the northern region of the Sahelo-Saharan zone to over 1600mm in the Guinean zone. The duration of the rainy season also varies greatly, ranging from 1 month in the desert margin to more than 8 months in the Guinean coastal zone (Kumar, 2012).

The Indian experience provides lessons that the focus should be on balanced development of surface water and groundwater resources for irrigation in sub-Saharan Africa, failing that there could be several undesirable consequences.[3] To facilitate such a balanced development, actions would be required at two levels, first at the policy level and second at the level of planning, designing, and execution of irrigation schemes. The naturally water-scarce arid and semiarid regions should embark on surface irrigation using water imported from water-rich catchments from within the region or from water-abundant regions. Such imports might be possible in West Africa and to an extent in east Africa, due to the high-rainfall gradients. At the next level, the schemes should be operated in such a way that while providing irrigation water to crops, water delivery should simultaneously enable recharge of the local shallow aquifers through return flows. This would ensure sustainable groundwater use. By embarking on conjunctive management, the need for a high degree of water exploitation could be minimized.

In high-rainfall regions of Kenya and Ethiopia and southern parts of western Africa, small water harvesting systems such as minireservoirs with lift irrigation device should be promoted as a matter of policy. Lift devices could be used to draw water from these reservoirs to provide supplementary irrigation to crops during dry spell. High rainfall over longer time duration also means lower cost of harvesting unit volume of water. The region will not require full irrigation due to the availability of green water (Kumar, 2012).

16.3.4 Planning Multiple Use Water Systems

The cost of exploitation of water resources is going to be very high in Africa owing to a variety of reasons (Foster et al., 2006; Kadigi et al., 2012). One of them is the huge cost of drilling, and the other is the very high cost of fuel for pumping water (Kadigi et al., 2012). A typical irrigation system in sub-Saharan Africa would cost anywhere between US$ 600 and 100/ha for a traditional community-based system (water harvesting and flood recession) to US$

---

3. Some of them witnessed in India are water logging and salinity resulting from underutilization of groundwater resources in the canal command areas of Punjab, Bihar, and Haryana owing to the availability of cheap canal water against expensive groundwater pumping and overexploitation of groundwater in the semiarid and arid regions (large parts of alluvial Gujarat, Rajasthan, and Punjab and hard-rock areas of Tamil Nadu, Andhra Pradesh, Telangana, Maharashtra, and Karnataka), resulting from overdependence on these resources owing to the lack of surface water.

1000–1500/ha for an individual (small well and pump-based) system to US$ 3000–8000/ha for an intercommunity system (deep tube well, small dam, or diversion system) (You et al., 2010). Therefore, it is imperative that the values realized from their uses are higher than what one would obtain if the same is diverted for crop production alone.

At the same time, the willingness of people to pay for water services, be it for productive uses or domestic uses, is likely to remain low, due to the poor economic conditions and the lack of understanding of the benefits of using more water (in the case of productive uses like irrigation) or better-quality water (in the case of domestic uses). In poor neighborhoods, water systems that are not capable of meeting the multiple needs of the communities are unlikely to find importance in their day-to-day affairs, which would affect their willingness to pay for the services being rendered (GSDA, IRAP, and UNICEF, 2013).

As literature shows in amble measure, marginal increase in volume of water supplied from a single-use system or marginal improvement in the quality of its water could result in significant gains in the economic benefits realized from its uses, which can far exceed the costs.[4] If the amount of water supplied to rural households from a drinking water supply scheme is increased marginally to exceed the domestic requirements of the households, people might show greater willingness to invest in livestock rearing and kitchen gardens thereby increasing their economic outputs and improving livelihoods. This would also increase their willingness to pay for drinking water supply services and maintain the systems, as the water supply scheme, apart from meeting the basic survival needs, also helps the communities to sustain their livelihoods.

There are two distinct possibilities that exist in sub-Saharan Africa. First, the physical infrastructure of small and large irrigation schemes could be extended to cover rural water supply, given the fact that the access to improved water supply is quite low in this region. Research has already shown that the feasibility exists for building numerous small reservoirs in sub-Saharan Africa, and water from such reservoirs can also be diverted for rural drinking water supply after treatment for bacteriological quality (Xie et al., 2014). If the source of irrigation is surface water from rivers and lakes, the water might require some preliminary treatment using sand filters, etc., for the removal of physical contaminants and organic matter before being supplied for domestic uses. In the case of groundwater-based irrigation systems, they could be used for multiple purposes without much additional infrastructure. However, groundwater-based drinking water schemes should be built only in areas where aquifers are well endowed with large stock (static groundwater) and water quality is good, in order to ensure sustainability of the sources during droughts.

4. Using data from Sri Lanka, Renwick (2008) showed that increase in water availability from a rural water supply system to meet livestock rearing, kitchen garden, and small enterprise needs resulted in an incremental income of US$ 25–70 per capita per year (US$ 1 equals to INR65).

On the other hand, if rural water supply schemes are built, then provision could be made to increase the total per capita supplies to accommodate productive uses such as livestock rearing, kitchen garden, and tree planting in the villages. In this case, the cost of production of unit volume of water would be less than what it would be if water is only used for drinking and domestic purposes. In such cases, small reservoirs will have to be built at different points within the village for livestock drinking, if livestock demand is major. If kitchen gardens and homesteads are a priority for the community, additional connections will have to be provided to the households for directly taking water or watering there backward.

16.3.5 Farming System Approach in Promoting Microirrigation

In view of the fact that the access to water for agriculture is very poor in sub-Saharan Africa owing to very high cost of exploitation, many researchers and development professionals have advocated microirrigation systems as a technology to enable farmers to produce and to improve food security situation and reduce poverty with limited amount of water (Kadigi et al., 2012; Polak, 2004). Low-cost microirrigation systems such as bucket kit and drum kits were suggested for farmers with very small holdings and extremely limited access to water resources (Polak, 2004). Drip and sprinkler irrigation system are used in irrigating the citrus and sugarcane farms of South Africa and Swaziland and export flowers and vegetable crops in Kenya, Ethiopia, and Tanzania (Kadigi et al., 2012).

But such suggestions lacked a farming system perspective both at the level of individual farms and at the regional farming system level, which has to look at the risks associated with such farming models. The reasons are as follows. First of all, microirrigation systems are most amenable to fruit and vegetable crops and some of the cash crops like cotton, groundnut, and potato. These technologies are not yet proven to be technoeconomically viable for cereals (Kumar, 2016). Hence, growing these crops will not help farmers to meet their staple food needs. They might be able to sell these produce in the market and purchase food grains from the earnings. But that will depend on how successfully they are able to market these crops and how backward and forward linkages are established. Large-scale production of these crops can lead to price crash, if new markets are not developed along with, bringing tremendous hardship to the growers. Nevertheless, the issue is not limited to the marketing of produce alone. Most of these crops are highly susceptible to diseases. Crop protection measures are extremely important for the survival of these crops and for good harvest.

Also, the input costs are high for fruits and vegetables toward seeds and pesticides. The farmers need to have sufficient capital. So far, as fruits are concerned, there is an additional burden of long gestation period for most of them. The minimum time duration for horticultural crops is 1 year (for papaya and banana) and can sometimes be up to 3–4 years for crops like sapota, guava, lemon, and citrus. For farmers with marginal landholding, this might come as an added constraint. All these increase the farming risk.

An alternate scenario is that large numbers of farmers from a region succeed in adopting new farming systems based on market-oriented crops with the use of microirrigation technologies, and then, this can even motivate them to replace the traditional cereal crops in their farms with the high-valued cash crops for earning greater income. This was the trend found in North Gujarat region in India, where large-scale adoption of microirrigation systems along with fruits and vegetables occurred, with some shrinkage in the area under cereals such as wheat and bajra (Kumar et al., 2010). As sub-Saharan Africa is already heavily dependent on food imports, similar trends can cause regional food shortages and food inflation.

For both the scenarios, institutional mechanisms are more important to reduce the risk. In the first case, it will be in the form of proper extension services for agronomic inputs and credit services for purchase of expensive seeds and pesticides and equipment for the control of production environment (polyhouse, net house, etc.). In the second case, it will have to be in the form of market support, like better support prices and procurement systems for cereals to create incentive for farmers to continue growing them.

A recent work (Wanvoeke, 2015) involving extensive literature review and intensive field research carried out in Burkina Faso argues that the intensive work to promote adoption of low-cost drip irrigation did not meet with any success in sub-Saharan Africa. It argued that the use of the technology has not been sustainable and it only worked when there was significant external support as part of developmental interventions.[5] According to the author, smallholder drip irrigation has failed to spread over the last decade because the technology doesn't fit African farmers' needs. More attention should be put in how farmers use and perceive new irrigation technologies differently than other actors, instead of limiting the analysis to attempts to realize the potential of the technologies.

16.4 WATER POLICY MAKING IN AFRICAN COUNTRIES

Many African nations are young and do not have great democratic traditions. Unlike many developed countries and some of the fast-growing developing

5. In the context of Burkina Faso, Wanvoeke (2015) noted that most farmers using drip irrigation do so as "pilot farmers." He noted that though development agencies often use pilot farmers as proof of farmers' "use" or "adoption" of drip irrigation, pilot farmers seemed to engage with these projects not because of the performance of the technology in terms of water and labor saving, but because of other benefits that come with the promotional package such as agricultural inputs, water-lifting devices, microcredit, and infrastructure. Drip system may also serve as a tool to acquire prestige or to forge new alliances with funders and services providers. These "side benefits" are accrued in the initial phase itself. That explains why these "pilot farmers" initially engage with the project but quickly stop using drip kits as soon as the project ends.

countries like India and China, the scientific and academic research institutions are not mature enough to feed into government policy making. The policies of the national governments for the development sector in general and water sector in particular are driven by donor agenda and sometimes agenda of the international NGOs, which manage knowledge dissemination in the sector owing to their financial muscle and intellectual control over scientific and academic journals. Many of the ideas that resonate in the water sector as solutions to water problems are those promoted by the international development agencies and multilateral donors, irrespective of whether they are relevant to the socioeconomic, legal, and institutional context of African countries or not. They include decentralized small-scale water harvesting, small drinking water schemes managed by local communities, irrigation management transfer, integrated water resources management, and water rights system.

Nongovernmental organizations, once considered as altruistic groups whose aim was to impartially influence public policy with no vested interests, are now increasingly being perceived as groups that prioritize their own ideologies or that respond to the interests of their donors, patrons, and members rather than those of the groups they represent. As noted by Tortajada (2016), there are cases where actions of grassroots NGOs on natural resources management, particularly water resources, have been largely driven by donor's interests. The objectives have been to promote specific agendas that often match with their own ideologies, rather than one which caters to the needs of socially backward communities (Tortajada, 2016). The main problem has been that such ideologies are often nurtured by poor knowledge on the sciences involved and have not always considered how effective they are at the larger societal level (Kumar, 2010).

As solution to problems of water insecurity, several of the international NGOs have pushed for rooftop rainwater harvesting for domestic water supplies and small-scale runoff harvesting for drought proofing of rainfed production systems for dealing with problems as serious as drinking water scarcity and food insecurity, respectively, for sub-Saharan Africa, through sheer campaigning. The experience of India with rainwater harvesting for domestic water supplies and recharge and watershed development for enhancing agricultural productivity in rainfed areas would be a great learning experience for these countries in framing policies so that they do not have to repeat the mistakes that India made. What is most important is that unlike India, where formal water supply systems built and run by the provincial governments exist in rural areas, in many of these SSA countries, such public systems are largely absent. Therefore, embracing systems like rooftop water harvesting tanks as an alternative to conventional water supply systems can be highly risky, making the communities living in the rural areas highly vulnerable, especially when many of these countries are drought-prone and rainfall variability is high. Similarly, overemphasis on rainfed production systems, supported by rainwater harvesting and supplementary irrigation to improve domestic food security of farm households, can also be risky.

During the past couple of decades, the union and provincial governments in India had focused on decentralized management of rural water supply schemes, following constitutional amendments to give powers to local self-governments for performing many functions of the government including water supply. As a result, technologies chosen for rural water supply were such that they could be operated and maintained by the local village panchayats. Increasing preference for bore-well-based schemes was the result of this changed focus. The low-capital investment was another reason for this preference. This approach has resulted in a huge compromise on the sustainability of the schemes with wells drying up during summer and droughts as evident from the analysis of rural water supply schemes in Maharashtra (Bassi et al., 2014).

In sub-Saharan Africa, there is an increasing tendency among governments to develop local groundwater for provision of rural water supply through drilling of deep wells in view of the lower initial costs. During droughts, these sources can dry up quickly, especially in the complex hydrogeologic environments (Foster et al., 2006). With no surface reservoirs that can provide clean water, the village communities might be forced to use poor-quality water from traditional sources such as ponds for drinking and cooking and other domestic uses including livestock use or reduce their water use for personal hygiene, increasing their vulnerability to problems of public health.

Policies with regard to technology selection for rural water supply have to be driven by considerations of sustainability of water supplies and cost-effectiveness, rather than initial costs. Therefore, the utilities in sub-Saharan African countries need to generate data on the technological feasibility, cost-effectiveness, and management efficiency of different types of water supply systems in different physical and socioeconomic environments in order to determine policies relating to technology choices, levels of investment, and pricing of water.

Worldwide, large dams are under constant public scrutiny for their environmental performance. Hence, dam builders in developing countries are under great pressure to comply with the environmental safeguards and to reduce the social and environmental costs associated with such projects. This has happened because the costs associated with human displacement especially that of the indigenous people, forest submergence, and loss of biodiversity and aquatic life received worldwide attention, especially after the report of the World Commission on Dams in 2001.

At the aggregate level, sub-Saharan Africa has utilized only a small fraction of the water resource potential of the region for expanding irrigation, and there is large untapped potential, though the extent of utilization is high in a few countries (Svendsen et al., 2008). But it is quite obvious that the above considerations would induce new constraints for dam building in several of these countries for irrigation, hydropower generation, and flood control.

But what these countries can learn from Indian experience is that large-dam-based water systems also create positive outcomes for the sustainable development of water resources and energy, by raising the water table in the overexploited regions thereby preventing groundwater mining, producing clean energy (from hydropower), reducing energy requirement for pumping groundwater (Jagadeesan and Kumar, 2015; Vyas, 2001), and improving the quality of groundwater in areas that suffer from heavy mineralization (Jagadeesan and Kumar, 2015). These benefits are extremely significant for developing countries that suffer from environmental problems. Hence, the criteria for evaluating the performance of large water systems need to be more comprehensive taking into account many of the indirect (social, economic, and environmental) benefits, rather than stringent. The onus is on governments and academic institutions in these countries to adopt pragmatic methodologies to analyze the real benefits and cost of large water resource systems and convince the international donors that fund such projects.

In the context of sub-Saharan Africa, interannual variability in the climate and extreme climatic events are much more serious than climate change, affecting the biological systems and food systems, health, and nutrition. An analysis of climate variability (rainfall departure from mean values expressed in terms of standard precipitation index) and agricultural GDP growth of three African countries (Ethiopia, Mozambique, and Niger) showed strong positive correlations (Thornton et al., 2014). Water resources development policies for the region should take cognizance of this reality, which has significant implications for the effectiveness of small-scale local water systems that depend on local catchments. The reason is these systems are bound to fail when the area actually requires water in the event of a meteorologic drought, as water yield from the local catchments would be disproportionately low.

In many South Asian countries, particularly in India, "quick fix" solutions such as small water harvesting interventions are being pursued for a long time to deal with drought conditions with no positive effect (Glendenning and Vervoort, 2011; Kale, 2017; Kumar et al., 2006, 2008b) in the name of decentralized management of water. This had resulted in deferred investment for much needed large water infrastructure projects involving reservoirs and water distribution systems. African countries, which have low levels of water security and which experience high climatic variability, should make policy choices that encourage long-term solutions for achieving water security rather than as the goal rather than decentralized water management.

Rural electrification is very poor in many African countries and some countries of South and Southeast Asia, because of poor financial conditions of the power utilities in these countries and the high rates of rural poverty that suppress the electricity demand making power distribution systems in rural areas unviable. Less than 10% of sub-Saharan Africa's rural population has access to electricity. Electricity is also a costly resource for Africa. The average tariff for individual customers is 13 US$ cents per kilowatt hour, that is, close to that of OECD countries (14 c€/kWh in France) for a standard of living that is 15 times lower. The electricity charge is also substantially higher than the cost observed in other developing regions (Huet and Boiteau, 2017).

With climate change mitigation actions dominating the international development agenda, there is sudden focus on nonconventional renewable energy systems as a way to meet the energy needs of these developing countries. Here, again, the emphasis is on private development of power using decentralized systems (Huet and Boiteau, 2017). This is in spite of the enormous potential Africa offers for hydropower development, particularly small hydropower units (Klunne, 2013).

India is putting in a large amount of money to provide capital subsidies for purchase of solar photovoltaic systems that would replace conventional diesel engines that are often highly inefficient. Proposals are also being made to provide solar PV systems to farmers who are already connected to the electricity grid so as to reduce their dependence on fossil-fuel-based power and also to feed solar electricity into the grid.

Recent analysis carried out for India of comparative economics of solar PV systems and conventional diesel engines show that the former are not economically viable even when their clean energy benefits are taken into account (Bassi, 2017). In the context of Africa, such analyses need to be undertaken before major initiatives are launched to solarize the power sector in the rural areas especially for agriculture.

The challenge for the governments in sub-Saharan Africa will be to pursue a path of sustainable development of energy resources that are ecologically sound but at the same time cheap and affordable. Policies need to be framed to enable faster development of hydropower with sufficient safeguard for protection of the environment.

16.5 CONCLUSIONS

There is no magical wand or silver bullet for addressing water problems in the developing economies. Challenges are far greater in the poor-income countries of sub-Saharan Africa (Kumar, 2012). That said, wise management of water is going to be crucial for food and energy security and achieving improved access to safe water supply. The question is what water management for irrigation, water supply, and energy production really means in the context of these developing economies, which are mostly falling in semiarid and arid and sometimes humid tropics. In the African context, there is a need to understand climatic variability and its implications for the paradigm of water resources development and water management.

The greatest challenge for many developing countries will be to make their own water resources development and management choices that match with their hydrologic, climatic, socioeconomic, and institutional reality and their developmental goals. As regards the latter, these countries need to improve their "food security" and energy security and their water supply situation especially in the rural areas dramatically over the next decade or so if they have to achieve a reasonable degree of progress in human development and economic growth. The water development and management paradigms they choose should match with these developmental needs, and the policy choices they make for the water sector should reflect those needs.

The governments in these countries need to make informed choices for the water sector by facilitating knowledge generation on water resources development management through the creation of knowledge institutions that are independent of the government. The idea is that in such situations, they do not have to depend on local civil society groups and international NGOs that pursue agendas that merely fit into their developmental philosophy and ideological positions. There is also a need for strengthening the institutional regime that regulates the quality of academic research in these countries so that peddling of falsehood by the influential writers can be prevented to a great extent.

REFERENCES

Barrios, S., Bertinelli, L., Strobl, E., 2004. Rainfall and Africa's Growth Tragedy. Paper.

Bassi, N., 2017. Solarizing groundwater irrigation in India: the growing debate. Int. J. Water Resour. Dev. https://doi.org/10.1080/07900627.2017.1329137.

Bassi, N., Kumar, M.D., Niranjan, V., Kishan, K.S., 2014. The decade of sector reforms of rural water supply in Maharashtra. In: Kumar, M.D., Bassi, N., Narayanamoorthy, A., Sivamohan, M.V.K. (Eds.), The Water, Energy and Food Security Nexus: Lessons from India for Development. Routledge/Earthscan, London, pp. 172–196.

Bhattarai, M., Narayanamoorthy, A., 2003. Impact of irrigation on rural poverty in India: an aggregate panel data analysis. Water Policy 5 (2003), 443–458.

Codjoe, S., Owusu, G., 2011. Climate change/variability and food systems: evidence from the Afram Plains, Ghana. Reg. Environ. Chang. 11, 753–765.

Falkenmark, M., Rockström, J., 2004. Balancing Water for Humans and Nature: The New Approach in Eco-Hydrology. Earthscan, New York.

Food and Agriculture Organization (FAO), 2015. Regional Overview of Food Insecurity Africa: African Food Security Prospects Brighter Than Ever. Food and Agriculture Organization of the United Nations, Accra.

Food and Agriculture Organization of the United Nations (FAO), 2006. The AQUASTAT Database. FAO, Rome. www.fao.org/aq/agl/aglw/aquastat/dbase/index.stm.

Food and Agriculture Organization of the United Nations (FAO), 2009. The special challenge for Sub-Saharan Africa, high level expert forum. In: How to Feed the World 2050. Food and Agriculture Organization, Rome. 12–13 October.

Foster, S., Tuinhof, A., Garduño, H., 2006. Groundwater Development in Sub-Saharan Africa: A Strategic Overview of Key Issues and Major Needs. Case Profile Collection # 15, The World Bank, Washington, DC.

Glendenning, C.J., Vervoort, R.W., 2011. Hydrological impacts of rainwater harvesting (RWH) in the case study catchment: the Arvari River, Rajasthan, India. Part 2: Catchment-scale impacts. Agric. Water Manag. 98, 715–730.

Gommes, R., Petrassi, F., 1994. Rainfall Variability and Drought in Sub-Saharan Africa since 1960. Agro Meteorology Working Paper # 9, Food and Agriculture Organization of the United Nations, Rome.

Groundwater Survey and Development Agency/Institute for Resource Analysis and Policy and UNICEF, 2013. Multiple Use Water Services to Reduce Poverty and Vulnerability to Climate Variability and Change. Final Report of a Collaborative Action Research Project in Maharashtra, India, UNICEF, Mumbai, India.

Huet, J.-M., Boiteau, A., 2017. Rural electrification in Africa: an economic development opportunity? Private sector and development. In: PROPARGO Magazine. March 2017.
Human Development Report (HDR), 2006. Human Development Report-2006. United Nations, New York.
Human Development Report (HDR), 2009. Overcoming Barriers: Human Mobility and Development. United Nations, New York.
Hussain, I., Hanjra, M., 2003. Does irrigation water matter for rural poverty alleviation? Evidence from South and South East Asia. Water Policy 5 (5), 429–442.
International Food Policy Research Institute (IFPRI), 2011. 2011 Global Hunger Index: The Challenge of Hunger-Taming Price Spikes and Excessive Food Price Volatility. International Food Policy Research Institute, Concern Worldwide and Welthungerhilfe, Washington, DC.
Jagadeesan, S., Kumar, M.D., 2015. The Sardar Sarovar Project: Assessing Economic and Social Impacts. Sage Publications, New Delhi.
Kadigi, R.M.J., Tesfay, G., Bizoza, A., Sinabau, G., 2012. Irrigation and Water Use Efficiency in Sub Saharan Africa. Supporting Policy Research to Inform Agricultural Policy in Sub Saharan Africa and South Asia. Global Development Network.
Kale, E., 2017. Problematic uses and practices of farm ponds in maharashtra. Econ. Polit. Wkly. LII (3), 20–22.
Klunne, W.J., 2013. Small hydropower in Southern Africa—an overview of five countries in the region. J. Energy South. Afr. 24 (3), 14–25.
Kumar, M.D., 2007. Ground Water Management in India: Physical, Institutional and Policy Alternatives. Sage Publications, New Delhi.
Kumar, M.D., 2009. Water Management in India: What Works and What Doesn't. Gyan Books, New Delhi.
Kumar, M.D., 2010. Managing Water in River Basins: Hydrology, Economics, and Institutions. Oxford University Press, New Delhi.
Kumar, M.D., 2012. Water management for food security and sustainable agriculture: strategic lessons for developing economies. In: Kumar, M.D., Sivamohan, M.V.K., Bassi, N. (Eds.), Water Management, Food Security and Sustainable Agriculture in Developing Economies. Routledge/ Earthscan, London, pp. 211–220.
Kumar, M.D., 2016. Water saving and yield enhancing micro irrigation technologies: theory and practice. In: Viswanathan, P.K., Kumar, M.D., Narayanamoorthy, A. (Eds.), Micro Irrigation Systems in India: Emergence, Status and Impacts. Springer, Singapore, pp. 13–26.
Kumar, M.D., van Dam, J., 2013. Drivers of change in agricultural water productivity and its improvement at basin scale in developing economies. Water Int. 38 (3), 312–325.
Kumar, M.D., Ghosh, S., Patel, A., Singh, O.P., Ravindranath, R., 2006. Rainwater harvesting in India: some critical issues for basin planning and research. Land Use Water Resour. Res. 6 (2006), 1–17.
Kumar, M.D., Shah, Z., Mudgerikar, A., 2008a. Water, human development and economic growth: some international perspectives. In: Kumar, M.D. (Ed.), Managing Water in the Face of Growing Scarcity, Inequity and Declining Returns: Exploring Fresh Approaches. Volume 2, Proceedings of the 7th Annual Partners' Meet of IWMI-Tata Water Policy Research Program. ICRISAT, Hyderabad, pp. 842–858.
Kumar, M.D., Patel, A.R., Ravindranath, R., Singh, O.P., 2008b. Chasing a mirage: water harvesting and artificial recharge in naturally water-scarce regions. Econ. Polit. Wkly. 43 (35), 61–71.
Kumar, M.D., Bassi, N., Singh, O.P., Sharma, M.K., Srinivasu, V.K., 2010. Hydrological and Farming System Impacts of Agricultural Water Management Interventions for Sustainable Groundwater Use in North Gujarat. Final report submitted to Sir Ratan Tata Trust, Mumbai. Institute for Resource Analysis and Policy, Hyderabad, India.

Kumar, M.D., Sivamohan, M.V.K., Narayanamoorthy, A., 2012. The food security challenge of the food-land-water nexus in India. Food Sec. 4 (4), 539–556.

Kumar, M.D., Saleth, R.M., Foster, J.D., Niranjan, V., Sivamohan, M.V.K., 2016. Water, human development, inclusive growth, and poverty alleviation: international perspectives. In: Kumar, M.D., James, A.J., Kabir, Y. (Eds.), Rural Water Systems for Multiple Uses and Livelihood Security. Elsevier Science, Amsterdam, The Netherlands, pp. 17–47.

Lloyd, S.J., Kovats, R.S., Chalabi, Z., 2011. Climate change, crop yields and undernutrition: development of a model to quantify the impacts of climate scenarios on child undernutrition. Environ. Health Perspect. 119, 1817–1823.

McMichael, A.J., Woodruff, R.E., Hales, S., 2006. Climate change and human health: present and future risks. Lancet 367, 859–869.

Moench, M., 1992. Drawing down the buffer. Econ. Polit. Wkly. XXVII (13), A7–A14.

Nelson, G.C., 2009. Climate Change: Impact on Agriculture and Costs of Adaptation. International Food Policy Research Institute, Washington, DC.

Ong'ayo, O.A., 2008. In: Political Instability in Africa: where the problem lies and alternative perspectives. Presented at the Symposium 2008: "Afrika: een continent op drift," organized by Stichting Nationaal Erfgoed Hotel De Wereld, Wageningen, 19th September, 2008.

Oweis, T., Hachum, A., Kijne, J., 1999. Water Harvesting and Supplementary Irrigation for Improved Water Use Efficiency in Dry Areas. SWIM Paper, Colombo, Sri Lanka.

Polak, P., 2004. Water and the other three revolutions needed to end rural poverty. Water Sci. Technol. 51 (8), 134–143.

Renwick, M., 2008. In: Multiple use water services. GRUBS Planning Workshop, Nairobi, Kenya.

Rockström, J., Barron, J., Fox, P., 2002. Rainwater management for improving productivity among small holder farmers in drought prone environments. Phys. Chem. Earth 27, 949–959.

Rosegrant, M.W., Paisner, M., Meijer, S., Witcover, J., 2001. 2020 Global Food Outlook: Trends, Alternatives and Choices, A 2020 Vision for Food, Agriculture, and Environment Initiative. International Food Policy Research Institute, Washington, DC.

Scott, C., Ochoa, S., 2001. Collective action for water harvesting irrigation in Lerma-Chapala basin, Mexico. Water Policy 3, 555–572.

Shah, T., 2009. Taming the Anarchy: Groundwater Governance in South Asia. Resources for Future and International Water Management Institute, Washington, DC.

Shah, Z., Kumar, M.D., 2008. In the midst of the large dam controversy, objectives, criteria for assessing large water storages in developing world. Water Resour. Manag. 22, 1799–1824.

Siebert, S., Burke, J., Faures, J.M., Frenken, K., Hoogeveen, J., Döll, P., Portmann, F.T., 2010. Groundwater use for irrigation—a global inventory. Hydrol. Earth Syst. Sci. 14, 1863–1880.

Sreenivasan, V., Lele, S., 2017. From groundwater regulation to integrated water management the biophysical case. Econ. Polit. Wkly. 52 (31), 107–114.

Svendsen, M., Ewing, M., Msangi, S., 2008. Watermarks: Indicators of Irrigation Sector Performance in Sub-Saharan Africa. African Infrastructure Country Diagnostic, Summary of Background Paper # 4, International Food Policy Research Institute, Washington, DC.

Syme, G., Reddy, V.R., Pavelik, P., Croke, B., Ranjan, R., 2011. Confronting scale in watershed development in India. Hydrogeol. J. https://doi.org/10.1007/s10040-011-0824-0.

Thornton, P.K., Ericksen, P.J., Herrero, H., Challinor, A.J., 2014. Climate variability and vulnerability to climate change: a review. Glob. Chang. Biol. 20 (11), 3313–3328.

Tortajada, C., 2016. Nongovernmental organizations and influence on global public policy. Asia Pac. Policy Stud. 3 (2), 266–274.

UN Water/Food and Agriculture Organization of the United Nations (FAO), 2007. Coping With Water Scarcity: Challenges of the Twenty-First Century. 2007 World Water Day, 21st March, 2007.

United Nations Development Program (UNDP), 2006. United Nations Development Program Report, New York.

von Braun, J., 2007. The World Food Situation: New Driving Forces and Required Actions. International Food Policy Research Institute, Washington, DC.

Vyas, J., 2001. Water and energy for development in Gujarat with special focus on the Sardar Sarovar project. Int. J. Water Resour. Dev. 17 (1), 37–54.

Wanvoeke, M.J.V., 2015. Low Cost Drip Irrigation in Burkina Faso: Unraveling Actors, Networks and Practices. . Thesis submitted in fulfilment of the requirements for the degree of doctor at Wageningen University by the authority of the Rector Magnificus.

World Health Organization and United Nations Children's Fund (WHO/UNICEF), 2006. Meeting the MDG Drinking Water and Sanitation Target: The Urban and Rural Challenge of the Decade. Report of the Joint Monitoring Program, World Health Organization and United Nations Children's Fund, Geneva.

Wiesmann, D., 2006. Global Hunger Index: A Basis for Cross Country Comparison. International Food Policy Research Institute, Washington, DC.

Wurzel, P., 2001. Drilling Boreholes for Handpumps. Working Papers for Water Supply and Environmental Sanitation 2 Swiss Center for Development and Cooperation, St. Gallen, Switzerland.

Xie, H., You, L., Wielgosz, B., Ringler, C., 2014. Estimating the potential for expanding smallholder irrigation in Sub-Saharan Africa. Agric. Water Manag. 131, 183–193.

You, L., Wood, S., Sichra, U.W., 2007. Generating Plausible Crop Distribution and Performance Maps for Sub-Saharan Africa Using a Spatially Disaggregated Data Fusion and Optimization Approach. Discussion Paper 00725 Energy and Production Technology Division, International Food Policy Research Institute.

You, L., Ringler, C., Nelson, G., Wood-Sichra, U., Robertson, R., Wood, S., Guo, Z., Zhu, T., Sun, Y., 2010. What Is the Irrigation Potential for Africa? A Combined Biophysical and Socioeconomic Approach. IFPRI Discussion Paper 00993 Environment and Production Technology Division, International Food Policy Research Institute (IFPRI), Washington, DC.

FURTHER READING

Batchelor, C., Singh, A., Rao, R.M., Butterworth, J., 2002. In: Mitigating the potential unintended impacts of water harvesting. International Water Resources Association International Regional Symposium. Hotel Taj Palace, New Delhi, "Water for Human Survival". pp. 26–29.

Chapter 17

Conclusions

Through various chapters of this book, we have seen that policy choices made in water, agriculture, energy, and climate sectors in India are an outcome of political processes. The ever-increasing influence of civil society groups with strong ideological positions; the increasing presence of academics and researchers, who peddle falsehood; and the "academic-bureaucrat-politician nexus" are essentially the defining features of this process. Some civil society groups and academics tend to push certain paradigms and "successful models" of water management, which fit into a new outlook on water management that they constantly call for.

The following views characterize this outlook: (i) any water project, which involve submergence of forests and human displacement, should be completely avoided; (ii) rather than augmenting water supplies, water-use efficiency in irrigated agriculture should be enhanced significantly to manage the demand for water in that sector and to allocate more water for the other sectors; (iii) sufficient flows need to be maintained for the environment in all rivers; (iv) the performance of new schemes should be assessed rather in relation to their ability to improve equity in access to water than augmenting water supplies; and (v) new irrigation schemes, if at all required, should meet the growing needs of the farming enterprise, rather than contributing to the country's grain basket (Iyer, 2011, 2012; Shah, 2016).

They propagate several myths to argue that the new paradigm can offer viable alternatives to large water projects (see Iyer, 2011, 2012; Shah, 2016), one of them being the potential of rainfed agriculture to feed the population of the country (Iyer, 2011). In fact, the aforementioned "outlook" on water management is part of a larger narrative to gain legitimacy for the "alternatives."

There are major political-economic considerations involved in water management decision-making, and there is too little of scientific and technological considerations. It is very rare that the political interests and long-term interest of the society are aligned, and as a consequence, good water management decisions in terms of policies, programs, projects, and schemes happen only rarely. As we have seen through several of the chapters in this book, in most situations, the political interests and societal interests do not match. This is the harsh reality of developing countries like India. We would illustrate this by picking up each of the cases we have discussed.

Water Policy Science and Politics. https://doi.org/10.1016/B978-0-12-814903-4.00017-8

Results from the earlier analyses of global data sets presented in Chapter 2 have shown that improved water security improves economic conditions and reduces poverty and income inequality of countries through the human development route, and they were corroborated by microlevel evidence from intensive primary research carried out in India on the various positive impacts of large water storages on social and economic variables that correspond to human development, economic growth, and poverty reduction.

The analysis also showed that nations need not wait till their economic conditions improve to improve their water situation. Instead, they should start investing in infrastructure and institutions for water security, which can improve human development and trigger economic growth. Many countries, which are economically not so rich, have achieved higher levels of human development through improved water security, with right infrastructure, institutions, and governance systems in place. More importantly, in the case of hot and arid tropics, increasing per capita water storage through large reservoirs is crucial for improving water security, which in turn helps all-round development. Especially, the microlevel evidence from Gujarat shows that allocating large volume of water in semiarid regions for surface irrigation helps in all-round growth, through enhanced agricultural production, increased farm income of gravity and well irrigators and rural wage laborers, improved sustainability of drinking water wells, reduced revenue losses to the state power utilities, and higher growth in agricultural GDP.

For countries like India that are characterized by mismatch between water availability and water demand both spatially and temporally (GOI, 1999; Kumar et al., 2012), projects to transfer water from water-rich basins to water-scarce basins are essential for achieving future food and energy security and sustainable water supplies, as population growth, rising incomes, urbanization, and industrialization rapidly increase water demand for competitive uses. This was sufficiently illustrated in Chapter 3. While the social and environmental costs of such large-scale water transfer projects are likely to be significant and should not be overlooked (Shah and Kumar, 2008), what needs to be kept in mind is the fact that the social cost of not implementing such projects is going to be even greater.

In the case of interlinking of rivers/interbasin water transfer projects, the issues are also highly political in nature (Iyer, 2012). The vested interest of political parties from the donor basins is in stopping the implementation of any water transfer links as they use it to flare up emotions of people for creating vote banks by infusing fear in their minds of permanently losing the rights over the water in their rivers. In spite of the fact that a large part of northern Bihar suffers from devastating floods in Kosi River, a tributary of the Ganges, a former chief minister of the state opened declared that not a single drop of water from the rivers of Bihar would be allowed to be diverted to the south under any water transfer projects.

While concerns about ecological health of rivers and social and environmental impact of large water projects are serious and cannot be wished away, the existing institutional regime in India at the state and central level are capable enough to address these concerns. Unfortunately, there is exaggerated fear being created by some civil society groups and academics about the negative ecological and environmental impacts of such projects (Verghese, 2003). Notably, the opponents have not yet offered any viable alternatives to interbasin water transfer to solve the growing water management problems in the country. That said, we need to significantly improve the method of working out the project costs so as to include the real cost of displacement and submergence and evolve appropriate institutional mechanisms for compensating the project-affected people. Robust methods and tools for evaluating the positive externalities (indirect benefit) of projects are needed (Cestti and Malik, 2012), as the absence of it encourage the concerned agencies to overlook some of the project costs. This in turn would encourage the official agencies to consider all costs that often get bushed under the carpet while evaluating the projects.

In the case of droughts like the one in Maharashtra, an impression is created in the minds of the general public by both civil society groups and the governments through constant media campaigning that the problems of severe water shortage during such natural calamities are because of the lack of sufficient rainwater harvesting efforts by the communities, and urban areas wasting a lot of water. This is far from the truth. Occasionally, blame is also put on the irrigation bureaucracy for building dams and allocating water for irrigating the water-intensive sugarcane in the designated command areas. But no mention is made about the fact that farmers tend to grow sugarcane because they do not incur any substantial cost because the use of large amount of water for growing this crop on the one hand and the crop gives high returns on the other. This was discussed in Chapter 4.

The "academics-bureaucrat-politician" nexus also helps propagate the myth about the ways to solve water crisis. The academics advise the politicians and bureaucrats that providing free power, fuel subsidies, free power connections, and free water would help eradicate poverty and bring more votes and market-based solutions would harm the interests of the poor. The politicians and bureaucrats are also being advised that "small check dams" are viable alternatives to building large water storage infrastructure. Such an approach is being resorted to in a haste to find a place for their ideas in political and policy circles and to show quick policy impacts of their research. This "politics-bureaucracy-academics" nexus is costing the economies hugely.

Such myths had helped the successive governments of drought-prone states in covering up their failure in building adequate infrastructure for transferring water from water-rich regions to chronically drought-prone regions that are naturally water-scarce. On the other hand, they help the civil society organizations in pushing their agenda of promoting small-scale water harvesting/recharge and watershed development projects in rural areas.

The much needed reforms in the water sector are compromised for serving the limited agenda of civil society groups. A good example is the Mihir Shah

Committee report for restructuring CWC and CGWB. The committee used the opportunity to put the blame for the current poor state of affairs of the water sector mainly on Central Water Commission, which works only in an advisory capacity, while all the water-related projects are planned, designed, executed, and managed by the state water agencies (water resources departments dealing with irrigation, drainage, and flood control; water supply department; minor irrigation department dealing with tanks; and groundwater department). This was discussed in Chapter 5.

The committee missed a great opportunity to reinforce the need for bringing about the much needed legal, institutional, and policy reforms in the water sector at the state level and went ahead to suggest cosmetic changes in the working of the two agencies by proposing to bring them under an umbrella organization by the name "National Water Commission." Such changes do not serve any purpose of raising the level of transparency, accountability, and incentive structures within the water bureaucracy. The report amply demonstrates that the members of the committee had not understood the working of water administration in the country, wherein water for all practical purposes is a state subject, and are carried away by some of the buzzwords in the water sector such as paradigm shift, community management, aquifer management, and Water User Associations.

The government also finds it convenient to put the blame for the poor performance of the water sector on working of technical agencies rather than the lack of serious reforms in the sector that requires political will.

In Chapter 6, we showed how many of the widely talked about projects being implemented by the state and central government in the fields of water, agriculture, forestry, energy, and climate lacked any hard science and instead were mere examples of the amount of misconceptions that exist in these sectors. One of them is the impact of forest cover on hydrology. The government of Telangana's program for afforestation to cover 10% of the state's geographic area doesn't take cognizance of the basic scientific fact that these trees (2.3 billion in number) once matured would consume a lot of water. No consideration is given to the fact that such massive tree planting would significantly affect the flow into thousands of tanks that are being rehabilitated by the same government, in whose catchment plantation is taken up.

The poor knowledge of groundwater-surface water interactions in river basins and the lack of knowledge about the implications of climate variability for water management interventions are some of the other problems hindering good water management decisions. Another growing concern is the peddling of falsehood by some researchers, who have influence in the media. They claim that their ideas have solved many serious water problems facing the country. In order to prevent this "travesty of science," the institutional regimes that evaluate the research and ideas have to be more robust and should respond to the growing needs of the society.

In the wake of extreme variability in climate, the studies for predicting future climate change need to be grounded to enhance their utility, by developing models that simulate the changes in streamflows and groundwater recharge in river basins in response to historical fluctuations in hydrologic variables (such as rainfall, temperature, and relative humidity) occurring in river basins. Such models can then predict the likely future changes in hydrologic variables due to the predicted changes in rainfall, temperature, and relative humidity. Likewise, the models that analyze the impact of climate change on crop yields also need to be robust enough to capture the effect of interannual variability in rainfall, temperature, and other weather parameters, an issue also raised by many scholars internationally (Hlavinka et al., 2009; Rowhani et al., 2011; Thornton et al., 2014).

Chapter 7 reviewed the Mission Kakatiya for the rejuvenation of tanks in Telangana. The analysis has shown that the program in the current format of taking up every irrigation tank in the state for rehabilitation will not help improve the performance of tanks, as it completely ignores the hydrologic reality of the tank catchments. We argued that the tanks that have high "irrigated area-wetland area ratio" are less likely to get degraded fast. Further, those tanks that are heavily degraded, as indicated by high intensity of cultivation and high density of wells in the catchment, should not be considered for rehabilitation as their hydrologic feasibility will be very poor. For those tank catchments where the feasibility is likely to be high, detailed microlevel hydrologic studies need to be carried out to ascertain the amount of runoff that would be generated from the catchment and demand for water in the command area, in order to determine the extent of desilting work involved.

But there is little acknowledgment of the fact that not much can be achieved by engaging in the rehabilitation of tanks that are degraded due to intensive catchment cultivation, through engineering interventions of bank stabilization, waste-weir construction and clearing of supply channels. The governments do not want to deal with the complex socioecological issues associated with managing tanks and want to approach the problem of deteriorating tank hydrology using the kid gloves of civil works.

In Chapter 8, we showed that the widening gap between the irrigation potential created and actual irrigation cannot be attributed to the poor governance and management of large irrigation projects by the agency, as many other factors on which the irrigation bureaucracy has no control influence how much area can actually be irrigated. It was further argued that the age-old criterion of evaluating the performance of surface irrigation systems on the basis of "irrigation potential utilized" against the "irrigation potential created" is flawed and should be done away with. In the context of Madhya Pradesh, the analysis showed that unlike what was argued by some researchers, the remarkable increase in irrigated area in the recent past was not because of any irrigation sector reform, but the several large reservoir projects completed during that time and the difference in the rainfall between the years used for comparison. By citing the recent cases from Gujarat and Madhya Pradesh, we questioned the practice of attributing the increase in agricultural growth rate witnessed over "short time durations" to any policy interventions of the government, as year-to-year variation in rainwater still plays a significant role in deciding the agricultural outputs over short time horizons, in India, particularly in states that have large area under crops that require irrigation.

In Chapter 9, by discussing the issue of floods in Chennai, a city that is prone to droughts and floods, we showed how ill-informed the government and the civil society groups were, while advocating solutions for solving water problems in large cities, including water scarcity and floods. Without developing a comprehensive understanding of the magnitude of the problems and the technical feasibility and economic viability of various viable options to tackle them, the civil society groups campaigned for rainwater harvesting for achieving drinking water security. With the mounting scientific and empirical evidence showing its ineffectiveness as an alternative water supply source for cities notwithstanding, the idea was picked up by the political leadership to make it mandatory for the city dwellers to build rooftop rainwater harvesting systems through legislation. Rainwater harvesting systems score very low on hydrologic feasibility (water quantity and quality) and economic viability in the urban situations of India. While medium-term solution for water problem was found by the setting up of a huge desalination plant by Chennai metro waters, there was no discussion on the fate of the hundreds of thousands of rainwater harvesting tanks built in the city by the residents. Also, there was no pressure built on the government to address the problems of inadequate drainage resulting from improper design of the storm water drains, which resulted in devastating city floods twice in 2015.

The Chennai example clearly exposed the institutional weakness of urban local governments arising out of the lack of adequate scientific and engineering research on water management in cities on the one hand and the extremely limited perspective of civil society organizations with regard to the nature of water problems in cities and the solutions therein.

Chapter 10 discusses some of the fundamental flaws in climate research in the Indian context, particularly in lieu of the fact that there is significant variability in weather parameters such as rainfall, humidity, temperature, and wind speed between years. This means that hydrologic predictions that do not factor in these variability aspects will not have much value. So is the case with crop modeling studies, if we consider the fact that the crop yield is determined by a complex function, with solar radiation, micro climate, and ecology, and most crop models do not consider the complex interactions between climate parameters at the "local scale." We argued that the much touted solar-energy-run irrigation pumps as climate mitigation intervention have poor physical efficacy and economic viability, whereas interbasin water transfer projects (for transferring water from water-rich basins to water-scarce basins) with gravity irrigation can contribute significantly to climate mitigation by reducing carbon footprint in agriculture. We argued that the metering and pricing of electricity in the farm sector can reduce carbon emissions to the tune of 6.24 m ton per annum while saving the energy economy and conserving the scarce water resources.

Here, again, the governments shy away from bringing about radical reforms in the energy sector, fearing political backlash. On the other hand, the push for solar energy helps them get the environmental groups that vociferously oppose fossil-fuel-based growth on board. For any elected government, this is easier than indulging in water transfer project and large hydel projects that are largely controversial. Clearly, no environmental group has come forward to critique the government's renewable energy policy.

In Chapter 11, we first showed how the comparison between private well irrigation and large-scale, public surface irrigation using simple statistics is flawed and misleading. In the chapter, we highlighted some of the unique virtues of large surface irrigation systems as a multiple use systems, supporting various water uses including fisheries. These are largely ignored by researchers who have a bias against gravity irrigation. The unique advantage of large surface irrigation systems is their ability to address demand-supply imbalances in water spatially and temporally. This is not possible in groundwater irrigation due to poor technical feasibility and economic viability. Only large surface water systems can meet the highly concentrated water demands of large cities. The chapter also highlighted that there are serious equity concerns posed by intensive groundwater irrigation, which is supported by heavy public subsidies for electricity with large farmers appropriating all the benefits, and that water markets become highly monopolistic under the current institutional and policy regimes governing the use of groundwater, characterized by the flat-rate pricing of electricity and the lack of well-defined water rights, whereas introduction of surface irrigation in well-irrigated areas is found to be reducing the monopoly power of water sellers.

Agricultural stagnation in eastern India, a region that has the largest concentration of the world's poor, has caught the attention of agricultural scientists and development economists alike for the past three decades. Intensive groundwater use is suggested by many in the recent past as a "silver bullet" for intensifying irrigation, raising agricultural productivity, and reducing rural poverty in the region. In Chapter 12, we argued that there are macrolevel, physical, and socioeconomic constraints to agricultural growth in the region such as limited arable land availability, an excessively large rural population depending on agriculture, and very small operational holdings of millions of farmers, and climatic and ecological constraints, which cannot be removed by merely tinkering with certain norms and procedures.

For agriculture in eastern India to break the current stagnation, the region needs farming systems that suit its agroecosystems. The region is characterized by numerous wetlands covering large areas, including low-lying areas that are under paddy grown in submerged conditions, areas that are inundated due to floods and tides, and numerous tanks and ponds that have water year round. The areas that are inundated due to floods and tides are suitable for extensive shrimp farming. While the areas that get water from tidal exchange would be suitable for salt-water shrimp, the flood-prone areas would be suitable for freshwater shrimp and numerous varieties of native fish, along with paddy.

In the inland paddy-growing areas, which are in the very-high-rainfall region with shallow groundwater and alluvial plains, water from ponds and lakes could be used to irrigate the paddy fields, and shrimp can be farmed in the fields. In the canal-irrigated areas of coastal Orissa and coastal plains of WB and Bihar, the ponds and tanks can be fed by canals.

In the hard-rock areas of the eastern India, comprising the western parts of WB, western Orissa, and Jharkhand, having tropical, semiarid climates with high rainfall, farmers grow many crops other than paddy especially during winter season. In plots that are not traditionally used for raising paddy, farmers can take up vegetable production. Given the high rainfall, small water harvesting systems such as ponds can also be built, and used for supplementary irrigation of vegetables and fruits, along with fish rearing.

However, ideas such as pump subsidy, power subsidy, and free power connections for agricultural pumping (Mukherji et al., 2012) and solar pump irrigation (Shah et al., 2016) quickly earn political currency given their ability to get popular support for the party in power.

In Chapter 13, we showed that the criterion used by the government of India to classify agricultural areas into "rainfed" and "irrigated," which merely banks on percentage area under irrigation, in spite of the vast differences in the biophysical and socioeconomic characteristics within those areas, is flawed. This criterion fails to consider the agroclimate and hydrometeorology of the area that decide whether crops can be grown under rainfed conditions or require irrigation. Watershed development interventions for agricultural development in rainfed areas are bound to fail when the rainfall is low and aridity is high, strategically.

We argued that the ideal classification of areas for our agricultural development planners should be (1) cropped areas requiring irrigation and (2) cropped areas not requiring irrigation. Within the first category, there can be subcategories depending on the number of seasons of the year during which crops require irrigation to mature.

In the regions where crops require irrigation to mature, there are some areas with medium to high rainfall, where sufficient water in the rivers exist and can be tapped for irrigation. But the rest of the places with low to medium rainfall and high aridity, major expansion in irrigated area will be possible only through import of water from distant water-rich catchments. This can be complemented by interventions to improve water-use efficiency in agriculture. In the areas that do not require irrigation by virtue of the high to excessively high rainfall and subhumid to humid climate, depending on the agroecology, watershed development projects can be taken up.

However, overemphasis on watershed development for all "rainfed areas" as a panacea for poor agricultural productivity and highlighting its virtues help the government defer the major investments needed for irrigation development in such regions.

Chapter 14 was about the ongoing polemics about the impact of drip irrigation on water use in agriculture. The first issue concerns whether there is real water saving per unit of land, and the second concerns what farmers do with the saved water to know whether this is water saving at the farm level. In this chapter, we argued that the water-saving impact of drip irrigation is highly context-specific, with the first type of impact depending on crop type, type of technology, soil, climate, and geohydrology. The second type of impact depends on the relative scarcity of land and water and the overall situation with respect to power supply and water supply availability. We need to do high-quality empirical research involving field measurements of water consumption and compare the same with modeling studies to see whether such modeling results pertaining to the first type of impact are reliable or where there is a need for making these models robust. In any case, the governments need to exercise caution while launching large-scale projects to promote micro irrigation, involving big subsidy support to farmers. Particularly, the local-specific conditions and their implications for water saving and "rebound effect" need to be fully understood. However, a simplistic view of the impacts of drip irrigation also helps the governments run various subsidy schemes for promoting adoption of the system that can be used as vote banks.

India's dairy sector is poised for high growth, driven by a rising demand for dairy products and the greater access to production technologies and market infrastructure. But its long-term impact on water resources has been least studied. In Chapter 15, we argued that dairy production has major impact on India's water resources, with an estimated total water consumption of nearly 250–270 billion cubic meters occurring annually. While reducing water intensity of milk production is very important for making dairy production more sustainable, there is a dichotomy between productivity and production. The regions that give high productivity in relation to water do not offer high production potential. On the other hand, the regions that have high production potential have low milk productivity in relation to water. Since the prospects of engaging in large-scale import of dairy products are unlikely due to its implications for the livelihood security of the tens of millions of small-scale milk producers, future focus should be on improving the productivity of water use, in regions where the production potential is high. This can be through the use of fiscal instruments, as water-use efficiency improvement will not happen merely with the adoption of water-saving irrigation devices.

In Chapter 16, we have drawn certain important lessons from the Indian experience with the processes in water policy making for sub-Saharan Africa in terms of practical agenda for water management for the countries of that region. Also, a few suggestions were offered on how the governments of these young democracies should manage the political process of policy making to make sure that the agenda is not hijacked by external players such as the international aid agencies, NGOs, and civil society groups.

Overall, efficiently and sustainably managing our water and energy resources for agricultural growth, poverty reduction, water security, and energy security is about managing the democratic polity well (Kumar, 2014). Governments are generally enthusiastic about implementing programs that quickly catch popular imagination, and not those that are based on scientific rationale. They do not show the patience to generate sufficient scientific evidence that support decisions in favor of such programs. Planting billions of trees in water-scarce regions to increase the forest cover and bring rains and to increase the base flow in rivers, building hundreds of thousands of small rainwater harvesting systems in basins that are already "closed" or on the verge of closure to capture runoff for rejuvenating the rivers and to adapt to climate change, and promoting drip irrigation through heavy capital subsidies without charging for water and electricity assuming that it would reduce water use in agriculture are just some of the examples. We therefore conclude that the recent government policies in the water, agriculture, and energy sector are not based on hard evidence, but "voodoo science."

The political leaders need to be educated about the long-term benefits of adopting solutions that are based on good science and hard evidence rather than those that are appealing to the masses, promoted by vested interests. There is also a need for strengthening the institutional regime that regulates the quality of academic research in developing countries like India so that peddling of falsehood by the influential writers can be prevented to a great extent. Simultaneously, there is a critical need for academic institutions to develop good theoretical understandings of the science-policy nexus and strategies for managing the nexus that are capable of enhancing the accountability and policy relevance of scientific research while preserving its core of independent inquiry (Graffy, 2008).

Parallelly, academic research institutions also need to be autonomous so as to have independent views on issues so that they can critically analyze government programs and schemes (Li et al., 2016). Also, work in the water, agriculture, energy, and climate sectors needs to be interdisciplinary and multidisciplinary in nature. The increasing role of civil society organizations in government policy making is welcome. However, the credibility of these organizations should be subject to regular scrutiny so as to make sure that they represent the interests and concerns of the society at large and that the larger developmental agenda is not subverted for serving the agenda of a few activist groups.

There are many important learning that can be drawn from Indian experience with policy making in the fields of water, energy, agriculture, and climate and applied to countries that are still at much lower level as compared to India in terms of water and energy security and agricultural development such as those in sub-Saharan Africa. This concerns the ways to go about planning of water resource development projects including that for groundwater; improving rainfed agriculture; planning for decentralized energy systems, especially solar PV systems; planning for agricultural development including dairy development; and drought proofing, especially mitigating the impacts of climate variability on water supplies and crop production.

Water resources and agricultural planners in sub-Saharan Africa must draw lessons out of India's experience with watershed development and rainwater harvesting to use these approaches only in areas where rainfall is dependable, aridity is low, and topography is really conducive. They need to guard themselves against the euphoria being created by some international development agencies and research institutes about what rainwater harvesting can achieve in naturally water-scarce and drought-prone regions. These countries must also tread carefully when it comes to exploitation of groundwater resources in their territories (Foster et al., 2006; Kumar, 2012). While taking up schemes that promote intensive use of groundwater, due consideration should be given to sustainable yield of aquifers with data on the same generated through proper geohydrologic studies. More importantly, technologies and institutions for the management of this complex resource should be developed prior to going for large-scale development so that the resource use by the communities does not pose a governance challenge.

REFERENCES

Cestti, R., Malik, R.P.S., 2012. Indirect economic impact of dams. In: Tortajada, C., Altinbilek, D., Biswas, A.K. (Eds.), Impacts of Large Dams: A Global Assessment. Water Resources Development and Management, Springer-Verlag, Berlin, Heidelberg. https://doi.org/10.1007/978-3-642-23571-9_2.

Foster, S., Tuinhof, A., Garduño, H., 2006. Groundwater Development in Sub-Saharan Africa: A Strategic Overview of Key Issues and Major Needs. Case Profile Collection # 15, The World Bank, Washington DC.

Government of India, 1999. Integrated Water Resources Development: A Plan for Action. Report of the National Commission on Integrated Water Development, Ministry of Water Resources, Government of India, New Delhi.

Graffy, E., 2008. Meeting the Challenges of Policy-Relevant Science: Bridging Theory and Practice. Thinking About Public Administration in New Ways. US Geological Survey.

Hlavinka, P., Trnka, M., Semeradova, D., Dubrovsky, M., Zalud, Z., Mozny, M., 2009. Effect of drought on yield variability of key crops in *Czech Republic*. Agric. For. Meteorol. 149, 431–442.

Iyer, R.R., 2011. National water policy: an alternative draft for consideration. Econ. Polit. Wkly. XLVI (26–27).

Iyer, R.R., 2012. River interlinking project: a disquieting judgement. Econ. Polit. Wkly. 47 (14), 33–40.

Kumar, M.D., 2012. Water management for food security and sustainable agriculture: strategic lessons for developing economies. In: Kumar, M.D., Sivamohan, M.V.K., Bassi, N. (Eds.), Water Management, Food Security and Sustainable Agriculture in Developing Economies. Routledge/Earthscan, London, pp. 211–220.

Kumar, M.D., 2014. Of statecraft: water, energy and food security in developing countries. In: Kumar, M.D., Bassi, N., Narayanamoorthy, A., Sivamohan, M.V.K. (Eds.), Water, Energy and

Food Security Nexus: Lessons From India for Development. Routledge/Earthscan, London, pp. 208–228.
Kumar, M.D., Sivamohan, M.V.K., Narayanamoorthy, A., 2012. The food security challenge of the food-land-water nexus in India. Food Security 4 (4), 539–556.
Li, X., Yang, K., Xiao, X., 2016. Scientific advice in China: the changing role of the Chinese Academy of Sciences. Palgrave Commun. 2, 16045. https://doi.org/10.1057/palcomms.2016.45.
Mukherji, A., Shah, T., Banerjee, P., 2012. Kick-starting a second green revolution in Bengal. Commentary. Econ. Polit. Wkly. May 05, 2012.
Rowhani, P., Lobell, D.B., Linderman, M., Ramankutty, N., 2011. Climate variability and crop production in Tanzania. Agric. For. Meteorol. 151, 449–460.
Shah, M., 2016. A 21st Century Institutional Architecture for Water Reforms in India. Final Report Submitted to the Ministry of Water Resources, River Development & Ganga Rejuvenation, Government of India, New Delhi.
Shah, T., Pradhan, P., Rasul, G., 2016. Water challenges of the Ganga basin: an agenda for accelerated reform. In: Bharati, L., Sharma, B.R., Smakhtin, V. (Eds.), The Ganges Basin: Status and Challenges in Water, Environment and Livelihoods. Earthscan/Routledge, London.
Shah, Z., Kumar, M.D., 2008. In the midst of the large dam controversy: objectives, criteria for assessing large water storages in the developing world. Water Resour. Manag. 22 (12), 1799–1824.
Thornton, P.K., Ericksen, P.J., Herrero, M., Challinor, A.J., 2014. Climate variability and vulnerability to climate change: a review. Glob. Chang. Biol. 20 (11), 3313–3328.
Verghese, B.G., 2003. Exaggerated Fears on Linking Rivers. http://www.himalmag.com/2003/.

Index

Note: Page numbers followed by *f* indicate figures, and *t* indicate tables.

G

H

I

J

K

L

M

N

Printed in the United States
By Bookmasters